Clocking in Modern VLSI Systems

Series on Integrated Circuits and Systems

Series Editor: Anantha Chandrakasan
Massachusetts Institute of Technology
Cambridge, Massachusetts

For other titles published in this series, go to
http://www.springer.com/Series/7236

Thucydides Xanthopoulos
Editor

Clocking in Modern VLSI Systems

Springer

Editor
Thucydides Xanthopoulos
Cavium Networks
Marlboro, MA 01752
USA

ISSN 1558-9412
ISBN 978-1-4419-0260-3 e-ISBN 978-1-4419-0261-0
DOI 10.1007/978-1-4419-0261-0
Springer Dordrecht Heidelberg London New York

Library of Congress Control Number: 2009928618

Printed on acid-free paper

Springer is part of Springer Science+Business Media (www.springer.com)

Στη Μαργαρίτα, το Νικόλα και τη Μαρία-Ελένη που μας έπιασε στα μισά

:)

To Margarita, Nicholas and Maria-Helen who joined us half-way through this book

:)

Preface

... εἰωθότες οἱ ἄνθρωποι οὗ μὲν ἐπιθυμοῦσιν ἐλπίδι ἀπερισκέπτῳ διδόναι, ὃ δὲ μὴ προσίενται λογισμῷ αὐτοκράτορι διωθεῖσθαι.

THUCYDIDIS HISTORIAE IV:108
C. Hude ed., Teubner, Lipsiae MCMXIII

῾Οι ἄνθρωποι, ἄλλωστε, συνειθίζουν νὰ ἐμπιστεύωνται εἰς τὴν ἀπερίσκεπτον ἐλπίδα ἐκεῖνο ποὺ ἐπιθυμοῦν καὶ ν' ἀποκρούουν δι' αὐθαιρέτου συλλογισμοῦ ἐκεῖνο ποὺ ἀποστέργουν.

ΘΟΥΚΥΔΙΔΟΥ ΙΣΤΟΡΙΑΙ Δ:108
Κατά Μετάφρασιν Ἐλευθερίου Βενιζέλου
Δ. Κακλαμάνος Εκδ.
Σμυρνιωτάκης, Αθήνα

It being the fashion of men, what they wish to be true to admit even upon an ungrounded hope, and what they wish not, with a magistral kind of arguing to reject.

Thucydides (the Peloponnesian War Part I), IV:108
Thomas Hobbes Trans., Sir W. Molesworth ed.
In *The English Works of Thomas Hobbes of Malmesbury, Vol. VIII*

I have been introduced to clock design very early in my professional career when I was tapped right out of school to design and implement the clock generation and distribution of the Alpha 21364 microprocessor. Traditionally, Alpha processors exhibited highly innovative clocking systems, always worthy of ISSCC/JSSC publications and for a while Alpha processors were leading the industry in terms of clock performance. I had huge shoes to fill. Obviously, I was overwhelmed, confused and highly confident that I would drag the entire project down. When a few years later

Carl Harris asked me to do a book on clocking for the Springer Integrated Circuits and Systems Series, I readily agreed with the hope that I could save young and aspiring clock designers substantial time and frustration by providing leads and maybe answers to the questions that I had when I was embarking on the Alpha clock design quest. As my choice of opening quotation would suggest, clock design can be a minefield of misconceptions based on little more than a reluctance to apply Kirchhoff's laws, basic constituent relationships, and a little bit of common sense.

In addition to my personal design experience, the choice of material for this book has been heavily informed by my long tenure in the International Solid-State Circuits Conference (ISSCC) program committee. The subjects covered reflect to a large extent the collective interests and foci of both industry and academia with respect to clocking based on ISSCC submissions. The only exception is that there is no coverage of phase locked loop design since there are a number of recent texts available on this subject matter.

It is my hope that this book will help engineers and students interested in clock design obtain the appropriate mental models and design viewpoints, capture design trends that have appeared over the last few years, and provide a comprehensive list of references for further study. I am indebted to my co-authors for providing precise, structured and complete coverage in their respective chapters in addition to maintaining a viewpoint that is very up to date and highly reflective of current trends in the industry. I hope that the reader will not find "ungrounded hopes" and "magistral arguings" in this book.

Carl Harris and Katelyn Stanne of Springer deserve special thanks for helping me throughout the preparation of the manuscript. I wish to acknowledge a number of colleagues at Cavium Networks for their helpful and stimulating discussions and excellent feedback: Scott Meninger, Ethan Crain, David Lin, and Suresh Balasubramanian. I would like to thank my bosses at Cavium Networks Anil Jain and Syed Ali for building a great semiconductor company from the ground up and an excellent working environment that fosters creativity and innovation in addition to maintaining a sharp focus on product development and company value. I would especially like to thank Anil Jain for entrusting me with the Alpha clocking project while being my boss at Compaq Computer which helped me acquire the background and skills necessary to produce this book. Above all, I would like to thank my wife Margarita not only for putting up with my constant working on this book but also for typesetting the entire manuscript in LaTeX and retouching figures as needed. I could not have done this without her.

Boston, Massachusetts *T.X.*
December 2008

Contents

List of Figures

List of Tables

List of Contributors

Georgios Konstadinidis
Sun Microsystems

Scott Meninger
Cavium Networks

Simon Tam
Intel Corporation (SC12-408)

James Tschanz
Intel Corporation

Nestoras Tzartzanis
Fujitsu Laboratories of America

James Warnock
IBM T.J. Watson Research Center

Thucydides Xanthopoulos
Cavium Networks

1

Introduction and Overview

Thucydides Xanthopoulos

Cavium Networks

Clock frequency is a major attribute of any microprocessor design. Early on, during product definition, it constitutes a major business or marketing decision and it is usually the result of a trade-off among customer needs, competitive landscape, and time-to-market. As soon as the frequency target is handed down the food chain to silicon implementation, it will affect all project design aspects from the day that the project is kicked off until it tapes out (and in most cases well beyond this point too). It is not surprising therefore that the job of generating, distributing, and analyzing the clocks in complex chips is considered to be an important and visible assignment. Clock design has traditionally been an area of innovation and has been in the spotlight in technical conferences and journals.

Why is clock frequency such an important microprocessor aspect? For a number of applications it is only loosely correlated with performance with other design aspects such as memory system, parallelism, and hardware acceleration being equally or even more effective. Nevertheless, it is a single number that is widely understood by both technical and nontechnical audiences and in certain situations has strong correlation with single-thread performance.

Clock frequency, although very important, is only one aspect of clock design. Other aspects include power dissipation, efficient clock signal distribution in large and complex chips, coping with variation and uncertainty, managing multiple clock domains in the context of highly integrated system-on-a-chip (SoC) designs, and multicore integration, providing good voltage/frequency scalability to support a wide product roadmap, tuning capabilities for yield enhancement and postsilicon optimization, and sophisticated active power management features.

The purpose of this book is to introduce a designer to important aspects of state-of-the-art clock design by exposing methodology steps and analytical modelling techniques, providing design examples and case studies and enumerating a long list of references for further study.

T. Xanthopoulos (ed.), *Clocking in Modern VLSI Systems*, Integrated Circuits and Systems, DOI 10.1007/978-1-4419-0261-0_1,

1.1 The Clock Design Problem

The plots of Figs. 1.1–1.3 provide good insight into the problem faced by clock designers today. Moore's plot (Fig. 1.1) states that integration keeps on increasing at historical rates. Transistor density is spearheaded by large multicore server processor chips with large caches, integrated memory controllers and multiple high-speed *I/O* links. Although the datapoints in Fig. 1.1 only extend to 2008, in ISSCC 2009 a server processor [1] has been introduced that contains 2.3B transistors and exceeds all previously reported chips in transistor count. For the clock designer, increasing integration means more far reaching and complex clock distribution, more clock domains, and more testability and yield enhancement features.

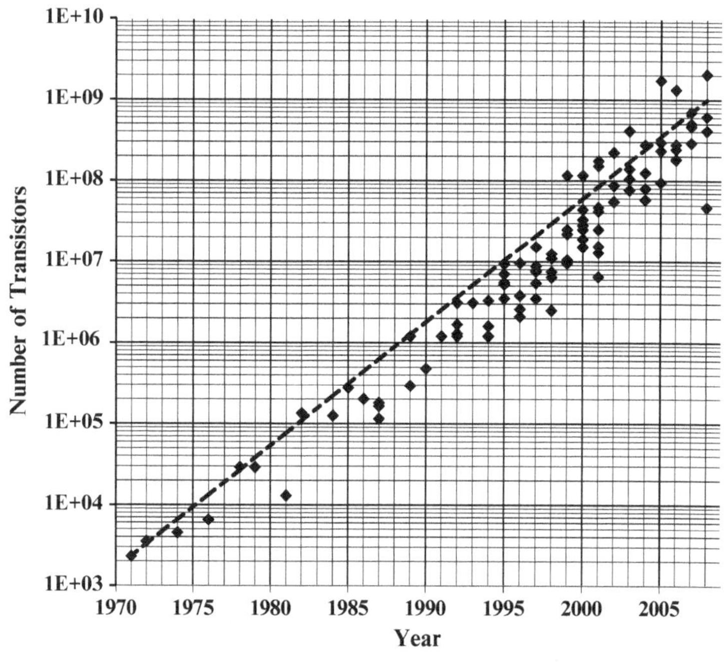

Fig. 1.1. Microprocessor transistor number trend over time [Data compiled by the author from ISSCC proceedings 1973–2008 and publicly available vendor information. Trendline is "visually" fitted and signifies a rate of doubling every 2 years]

On the other hand, frequency and power seem to be levelling off (Figs. 1.2 and 1.3). Microprocessors in excess of 3–4GHz are rarely reported these days. Frequency stopped scaling due to excessive power dissipation penalties. Instead, the industry chose to keep on scaling performance through multicore integration. On the power side, the industry seems to converge to a fixed power envelope for both server (130W) and desktop (60–70W) processors. As far as clock design is concerned, this

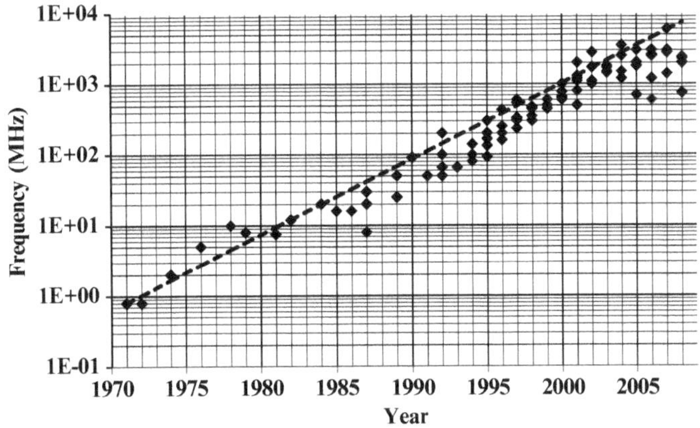

Fig. 1.2. Microprocessor frequency trend over time [Data compiled by the author from ISSCC proceedings 1973–2008 and publicly available vendor information. Trendline is "visually" fitted and signifies a rate of doubling every 3 years]

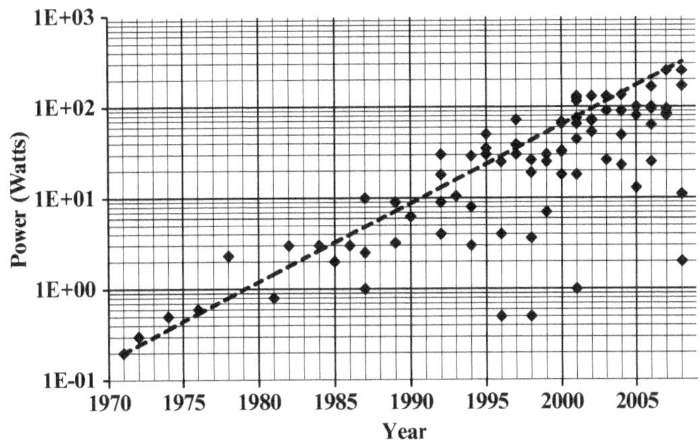

Fig. 1.3. Microprocessor power trend over time [Data compiled by the author from ISSCC proceedings 1973–2008 and publicly available vendor information. Trendline is "visually" fitted and signifies a rate of doubling every 3.5 years]

translates to increased number of features and complexity to support fine-grain clock and power gating, variable voltages, and frequencies in addition to other sophisticated active power management related attributes.

In addition to the above evolving specifications, one must add increased variation in device and interconnect characteristics present in advanced process nodes and increased voltage and temperature variation due to higher integration and more complex interactions. We end up with a multidimensional design problem requiring substantial resources for each product generation.

1.2 Some Subjective Milestones in the History of Microprocessor Clocking

In the last 15–20 years there have been major innovations in microprocessor clocking that have resulted in large increases in frequency as well as substantial clock design methodology changes. This section lists some of those important milestones so that we can observe the industry trends and speculate on potential future directions.

1.2.1 Integrating the PLL

An integrated PLL for a microprocessor application was first reported in 1992 [2]. The motivation for an integrated PLL at the time was the desire to keep the processor and the external bus in phase so as to minimize timing constraints and reach the maximum possible system frequency given the synchronous nature of the system. Additional goals were to clock the processor at even higher frequencies ($2x$) than the bus in addition to running the VCO at twice the processor frequency with a factor of two postscaling for duty cycle fidelity.

Some of the original problems that the designers faced were the digital noise which is highly prevalent on a microprocessor die, in addition to low quality passive devices and overall sensitivity to process, voltage, and temperature variations. The supply noise problem was addressed by the adoption of a differential CML-based VCO with high supply noise rejection capability.

Integrating the PLL was a major step in general purpose microprocessor clocking because it paved the way for the increased frequencies and complex clocking schemes that were to follow.

1.2.2 Clock Distribution Moves to the Forefront: The Dawn of the GHz Race

The original DEC Alpha microprocessor [3] moved clock design to the forefront. It was operating at 200MHz, approximately a factor of 2 faster than other processors at the time (Fig. 1.2). The DEC Alpha design introduced the low-skew and high-power grid clock distribution coupled with detailed RC-based skew simulation and construction of skew contour maps that are taken into account during timing closure: The Alpha pipeline was based on level-sensitive latches. Race-through in this context is a major functional concern. The radial profile of clock skew from the center of the chip to the periphery was taken into account while floorplanning the pipelines in order to guarantee that the skew would improve the functional race margins.

The design also featured important contributions from a process and physical design perspective. The process technology featured a thick low resistance metal 3 layer used exclusively for power, clock, and a handful of critical signals. On-chip decoupling capacitors built with thin oxide devices were also used in close proximity to the clock drivers in order to address Ldi/dt concerns arising from driving a highly capacitive final clock node with a very fast edge rate. A 10:1 ratio of decoupling capacitance to switching capacitance was maintained throughout the design. Many of these contributions are still in use today in current processor designs.

Arguably, the DEC Alpha designs started the GHz race among microprocessor vendors that culminated in the deep pipelines and multiGHz designs of present high performance chips.

1.2.3 Delay Lock Techniques

Simple first-order mechanisms to achieve phase lock [4] have been used extensively in processor designs in the last 10–15 years in order to simplify the clock distribution problem: A large distribution throughout a big die is broken into pieces tailored for each chip partition, and each partial distribution is phase locked using delay locked loops. There are multiple such examples discussed in Chap. 2. Such a partitioning helps with design time, power dissipation, testability, and manufacturing yield. It is definitely possible to achieve the same goal with higher order systems (i.e. distributed PLLs) and this has been demonstrated in the literature [5, 6]. Yet, nothing beats the simplicity of a digital delay line controlled by a basic finite state machine. More details on DLL design will be presented in Chap. 6.

1.2.4 Exploiting Inductance for Oscillation and Distribution

The notion of being able to return energy back to the clock generator has been rather intriguing and holds a lot of promise. Resonant clock drivers have been originally introduced as an off-chip solution in the context of powering adiabatic circuits [7, 8]. In the past several years, there has been renewed interest in this technique due to the fact that digital frequencies have become consistent with resonant frequencies of fully integrated passive devices. The strong motivation is the potential of saving substantial clock power by using LC resonance for clock pulse generation. LC techniques have been recently augmented with transmission-line-based techniques (traveling waves [9], standing waves [10] and salphasic [11]) that address both clock generation and low (or highly predictable) skew distribution. Commercial applications of these techniques are already prevalent. Chapter 4 explores this subject in more detail.

1.2.5 Variable Frequency (and Voltage)

A server processor design [12] introduced the idea of a constant power envelope and variable voltage/frequency. This active power management scheme uses an integrated ammeter that monitors incoming current, a clocking scheme that can generate variable frequencies with fast adjustment time and an on-chip micro-controller that monitors power/ temperature and controls frequency and core supply voltage through an external regulator. This technique is discussed in more detail in Chap. 2.

Variable frequency clocking methods are rapidly becoming mainstream [13, 14] as part of sophisticated power management methodologies designed to control thermal design power in large multicore chips. All digital methods ensure repeatability in a production environment.

1.2.6 Frequency Increase (or Supply Lowering) Through Resiliency

Commercial designs have substantial frequency margins to address issues such as lack of total coverage in production tests, unanticipated corner cases and noise patterns, noncompliant system specifications and device aging. This margin can be substantially reduced or even eliminated if the underlying hardware has error detection and correction capabilities. The margin can then be used as performance benefit by increasing frequency, power benefit by dropping voltage below the specified V_{MIN}, or even yield enhancement by populating existing frequency bins with parts that under different scenarios would not make it.

The Razor technique [15] addresses the frequency/voltage margin issue by instituting the capability of timing error detection and correction in the processor pipeline. Performance can be maximized (or power minimized) by increasing the frequency (or lowering the supply) up to the point where the overhead of error correction will start exceeding the performance or power benefit. The Razor technique is addressed in more detail in Chaps. 3 and 7.

It is not easy to predict the future but given the current industry trends one can conclude that the clocking system will be part of an increasingly sophisticated active power management scheme: Highly sophisticated firmware threads implementing complex control algorithms will be running in parallel with the application. They will be receiving input such as on-chip and system temperature, current measurements, error rates, moving averages of architectural events, and cues from the application and they will control supply voltage, clock frequencies, and higher level architectural events such as clock and power gating, instruction issue rate, and pipeline stalls. To some extent, this is already happening.

1.3 Overview of this Book

Chapter 2 introduces the fundamental setup and hold constraints, and defines basic clock attributes such as skew, jitter, latency, and duty cycle distortion. It introduces basic clock distribution methods such as balanced tree, central spines, and grids and examines them from a performance and power perspective. Numerous case studies from commercial microprocessors are presented and a number of advanced topics such as global and local skew compensation, on-die attribute measurements, various techniques for locating critical paths and synchronization methods are discussed.

Chapter 3 constitutes a detailed discussion of clocked elements (level-sensitive latches and flip-flops) from the viewpoint of latency, hold time requirements, power dissipation, and testability. The focus is primarily on state-of-the-art designs with advanced topics such as process variation and reliability addressed in detail.

As mentioned in Sect. 1.2.4, exploiting inductance for clock generation and distribution makes perfect sense: Inductance can produce oscillations with lower energy since an LC-based system inherently recycles energy between the capacitive and inductive elements. Moreover, oscillator phase noise is less because all active-device-related noise sources do not exist. Chapter 4 presents detailed background information on integrated inductors and transmission lines. Furthermore, it discusses examples of LC-based oscillators and transmission-line-based clock generation and distribution schemes.

Jitter analysis is very important in clock system design. If not properly managed, jitter can be the limiting factor in both core and *I/O* clocking. Chapter 5.1 defines all jitter types relevant to clock design and establishes their relationship to phase noise using Parceval's theorem. Furthermore, it enumerates all noise sources inside a clock generator and clearly shows with numerical examples how a PLL transfers jitter from input to output. Based on this analysis, it establishes the importance of reference clock phase noise and jitter regarding the quality of the multiplied output clock. The domain seamlessly moves from frequency to time using mathematical "filter" functions to transform phase jitter to period jitter which is more relevant for core clock generation. Since jitter has a random component, which is theoretically unbounded, the chapter establishes an MTBF-based statistical analysis for estimating the effect of jitter on critical paths. A serial link discussion is also presented, which shows how reference clock jitter can be removed from the total link budget under certain conditions.

Chapter 6 is an attempt at textbook-like coverage of digital delay locked loops that are used extensively in clocking systems. The chapter is design-oriented and contains detailed discussions and analyses on all DLL components and presents a number of relevant applications. It also contains a detailed analysis of metastability in the context of phase detection and a simple analytical model for supply-induced jitter on long delay lines and/or clock buffers.

Advanced process nodes exhibit large variation and uncertainty in device and interconnect parameters. Chapter 7 presents methods of addressing this issue on the design front, manufacturing, and postsilicon tuning front and also by using resilient methods involving hardware timing error detection and correction.

Finally, Chap. 8 addresses process, voltage, and temperature variation issues from a physical design perspective. Clock skew components in the context of setup and hold constraints are redefined with a statistical approach and all variation sources are taken into account. Sources of transistor and interconnect variation are enumerated, quantified, and explained. Methods of accounting for voltage and temperature variations are discussed. In the end, important physical design guidelines are presented to minimize uncertainty and variation in clock-related circuits.

References

[1] S. Rusu, S. Tam, H. Muljono, J. Stinson, D. Ayers, J. Chang, R. Varada, M. Ratta, and S. Kottapalli, "A 45 nm 8-core enterprise Xeon® processor," in *Digest of Technical Papers IEEE International Solid-State Circuits Conference (ISSCC 2009)*, 2009, pp. 56–57.

[2] I. A. Young, J. K. Greason, and K. L. Wong, "A PLL clock generator with 5 to 10 MHz of lock range for microprocessors," *IEEE Journal of Solid-State Circuits*, vol. 27, no. 11, pp. 1599–1607, Nov. 1992.

[3] D. Dobberpuhl, R. Witek, R. Allmon, R. Anglin, D. Bertucci, S. Britton, L. Chao, R. Conrad, D. Dever, B. Gieseke, S. Hassoun, G. Hoeppner, K. Kuchler, M. Ladd, B. Leary, L. Madden, E. McLellan, D. Meyer, J. Montanaro, D. Priore, V. Rajagopalan, S. Samudrala, and S. Santhanam, "A 200-MHz 64-b dual-issue CMOS microprocessor," *IEEE Journal of Solid-State Circuits*, vol. 27, no. 11, pp. 1555–1567, Nov. 1992.

[4] M. Johnson and E. Hudson, "A variable delay line PLL for CPU-coprocessor synchronization," *IEEE Journal of Solid-State Circuits*, vol. 23, no. 5, pp. 1218–1223, 1988.

[5] G. A. Pratt and J. Nguyen, "Distributed synchronous clocking," in *Proceedings of Sixteenth Conference on Advanced Research in VLSI*, 27–29 March 1995, pp. 316–330.

[6] V. Gutnik and A. Chandrakasan, "Active GHz clock network using distributed PLLs," *IEEE Journal of Solid-State Circuits*, vol. 35, no. 11, pp. 1553–1560, Nov. 2000.

[7] C. L. Seitz, A. H. Frey, S. Mattison, S. D. Rabin, D. A. Speck, and J. L. A. van de Snepscheut, "Hot-clock NMOS," in *Proceedings of Chapel Hill Conference VLSI*, 1985, pp. 1–17.

[8] R. Feynman, T. Hey, and R. Allen, *Feynman Lectures on Computation*. Westview Press, Boulder, CO, 2000.

[9] J. Wood, S. Lipa, P. Franzon, and M. Steer, "Multi-gigahertz low-power low-skew rotary clock scheme," in *Digest of Technical Papers IEEE International Solid-State Circuits Conference (ISSCC 2001)*, 2001, pp. 400–401, 470.

[10] F. O'Mahony, C. Yue, M. Horowitz, and S. Wong, "A 10-GHz global clock distribution using coupled standing-wave oscillators," *IEEE Journal of Solid-State Circuits*, vol. 38, no. 11, pp. 1813–1820, Nov. 2003.

[11] V. L. Chi, "Salphasic distribution of clock signals for synchronous systems," *IEEE Transactions on Computers*, vol. 43, no. 5, pp. 597–602, May 1994.

[12] S. Naffziger, B. Stackhouse, and T. Grutkowski, "The implementation of a 2-core multi-threaded Itanium family processor," in *Digest of Technical Papers IEEE International Solid-State Circuits Conference (ISSCC 2005)*, 2005, pp. 182–183, 592.

[13] R. Kumar and G. Hinton, "A family of 45nm IA processors," in *Digest of Technical Papers IEEE International Solid-State Circuits Conference (ISSCC 2009)*, 2009, pp. 58–59.

[14] A. Allen, J. Desai, F. Verdico, F. Anderson, D. Mulvihill, and D. Krueger, "Dynamic frequency-switching clock system on a quad-core Itanium® processor," in *Digest of Technical Papers IEEE International Solid-State Circuits Conference (ISSCC 2009)*, 2009, pp. 62–63.

[15] D. Ernst, N. S. Kim, S. Das, S. Pant, R. Rao, T. Pham, C. Ziesler, D. Blaauw, T. Austin, K. Flautner, and T. Mudge, "Razor: a low-power pipeline based on circuit-level timing speculation," in *Proceedings of 36th Annual IEEE/ACM International Symposium on MICRO-36 Microarchitecture*, 2003, pp. 7–18.

2

Modern Clock Distribution Systems

Simon Tam

Intel Corporation (SC12-408)

2.1 Introduction

Modern clock distribution design continues to face challenges in spite of significant advances in the last decade. We can distinguish three primary challenges. The first is the need to support higher clock frequencies based on the strong correlation between frequency and chip performance. Figure 2.1 shows processor clock frequency trend suggesting a continuous exponential increase in clock frequency with variable rates. Second, process technology scaling allows higher level of integration and larger die size leading to higher clock loading and larger distances the clock network needs to traverse. The final challenge is that technology scaling leads to an increase in on-die variations that may degrade clock performance if not properly addressed.

In order to address these design challenges successfully, it is necessary to understand the fundamental clocking requirements, key design parameters that affect clock performance, different clock distribution topologies and their trade-offs, and design techniques needed to overcome certain limitations. In this chapter, the following topics are presented:

- Definitions and Design Requirements
- Clock Distribution Topologies
- Microprocessor Clock Distributions
- Clock Design for Test and Manufacturing
- Elements of Clock Distribution Circuits
- Clock DFX (Design-for-Test and Design-for-Manufacturing) Techniques
- Multiclock Domain Distributions
- Future Directions

T. Xanthopoulos (ed.), *Clocking in Modern VLSI Systems*, Integrated Circuits and Systems,
DOI 10.1007/978-1-4419-0261-0_2, © Springer Science+Business Media, LLC 2009

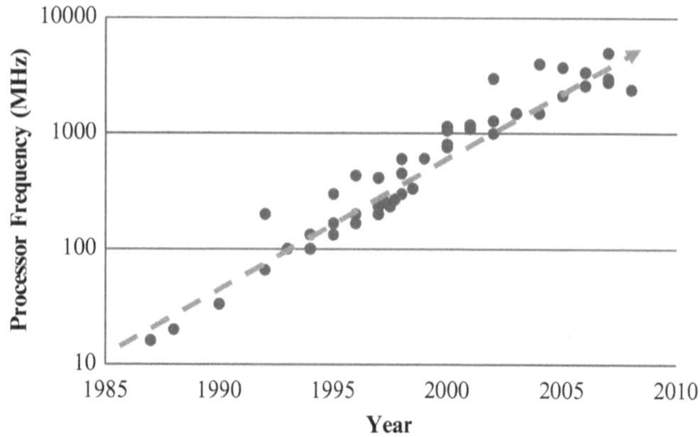

Fig. 2.1. Processor clock frequency trend [1–30]

2.2 Definitions and Design Requirements

Synchronous circuits may be simplified to have two timing limitations: setup (MAX delay) and hold (MIN delay). Setup specifies whether the digital signal from one stage of the sequential structure has sufficient time to travel to and "set-up" before being captured by the next stage of the sequential structure. Hold specifies whether the digital signal from the current state within a sequential structure is immune from contamination by a signal from a future state due to a fast path. Figure 2.2 shows a typical synchronous sequential structure bounded by two flip-flops with a logic circuit that exhibits a circuit delay of value T_d. The sequential elements are clocked by a source clock Ck1 and a destination clock Ck2.

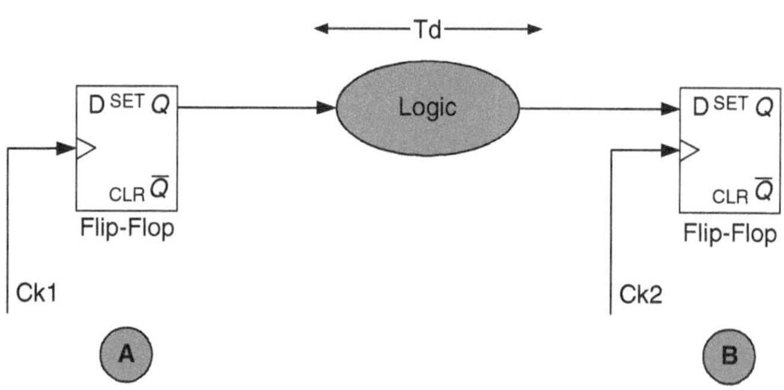

Fig. 2.2. Sequential structure bounded by flip-flops

Clocks Ck1 and Ck2 can be spatially far apart on die as shown in Fig.2.3. In this illustration, clocks Ck1 and Ck2 have their root at a common point (Clock Gen.) and are routed through the on-die clock distribution before arriving at their respective destinations. Locations A and B constitute the source and destination of the sequential path. The transit time (clock latency[1]) of CK1 and Ck2, their latency difference, their variations, and the design structure to minimize the above are the main topics of discussion in subsequent sections. As will be shown later, the timing uncertainty and the timing differences of Ck1 and Ck2 will play a fundamental role in determining whether the setup and the hold constraints can be robustly met.

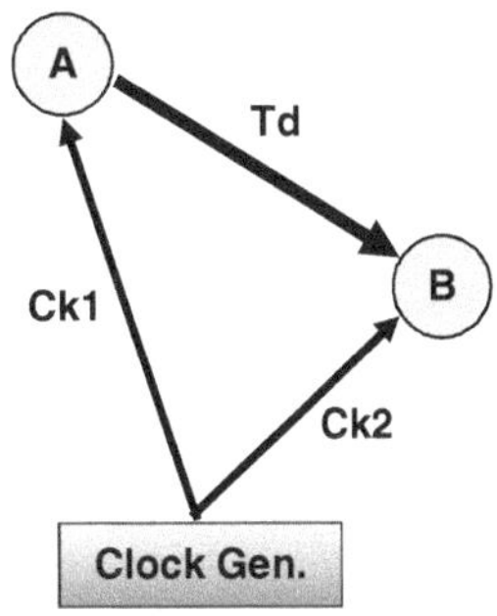

Fig. 2.3. Sequential path showing explicit clock distribution

2.2.1 Setup and Hold Timing Constraints

This section will present a brief outline of the formulation and the key parameters affecting the setup and the hold constraints.

The setup constraint specifies how data from the source sequential stage at cycle N can be captured reliably at the destination sequential stage at cycle $N+1$. This situation is illustrated in Fig. 2.4 in which the source clock Ck1 is shown to lag behind the receive clock Ck2 due to clock uncertainty. The constraint for the source data to be received reliably by the receiver is defined in inequality 2.1, where $T_{\text{d-slow}}$ is the slowest (maximum) data path delay, T_{su} is the setup time for the receiver flip-flop, T_{per} is the clock period, T_{Ck1} and T_{Ck2} are the arrival times for clocks Ck1 and Ck2 (at cycle N) respectively.

$$T_{\text{per}} \geq T_{\text{d-slow}} + T_{\text{su}} + |T_{\text{Ck1}} - T_{\text{Ck2}}| \,. \tag{2.1}$$

In the setup constraint situation, the available time for data propagation is reduced by the clock uncertainty defined as the absolute difference of the clock arrival times. This uncertainty $|T_{\text{Ck1}} - T_{\text{Ck2}}|$ can originate from various sources and their classification will be discussed in subsequent sections. In order to accommodate the clock uncertainty and meet the inequality in (2.1), either clock period must be extended or path delay must be reduced. In either case, power and operating frequency may be affected.

The hold constraint is shown in Fig. 2.5. This case specifies the situation where the data propagation delay is fast, and clock uncertainty makes the problem even

[1] The latency is referenced to the root of the distribution.

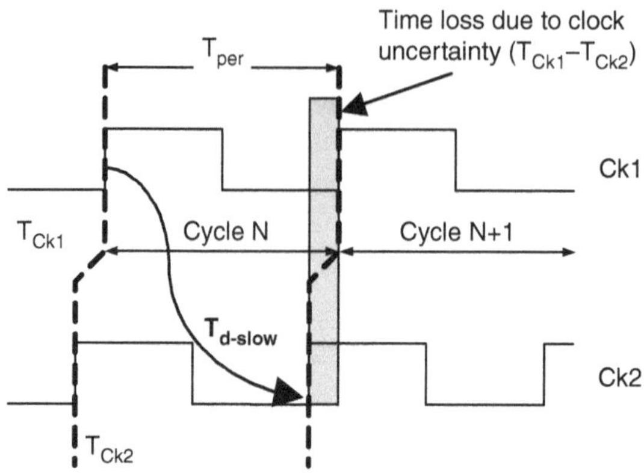

Fig. 2.4. Timing diagram for the setup constraint

worse and the data intended to be captured at cycle $N+1$ is erroneously captured at cycle N, corrupting the receiver state. In order to ensure that the hold constraint is not violated, the design has to guarantee that the minimum data propagation delay is sufficiently long to satisfy inequality (2.2):

$$T_{\mathrm{d-fast}} \geq T_{\mathrm{hold}} + |T_{\mathrm{Ck1}} - T_{\mathrm{Ck2}}|, \tag{2.2}$$

where T_{hold} is the hold time requirement for the receive flip-flop.

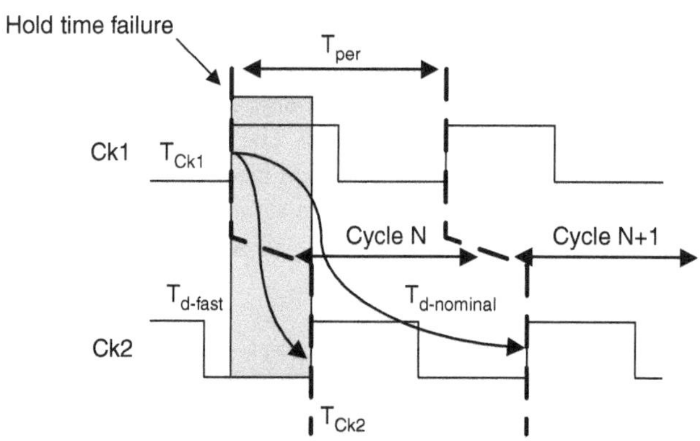

Fig. 2.5. Timing diagram for the hold constraint

In the discussion above, the following relationship is expected to hold (inequality 2.3):

$$T_{\mathrm{d-fast}} < T_{\mathrm{d-nominal}} < T_{\mathrm{d-slow}}. \tag{2.3}$$

Meeting the hold constraint in (2.2) with large clock uncertainty could result in setup violation due to (2.3) since the slowest manifestation of the same path could violate the delay requirement in (2.1). Such two-sided constraints are not uncommon in modern design if the clock uncertainty is high.

Central to the discussion above is the clock uncertainty defined by the absolute difference of delays T_{Ck1} and T_{Ck2}. A typical clock distribution structure (Fig.2.6) relies on buffer[2] stages to amplify the clock from the clock generator to the respective receivers (shown as the sequential elements FF in Fig.2.6). In general, when measuring the clock arrival time at the end points of a clock distribution, the clock latencies with respect to the distribution common point (T_{DELAY}) will exhibit a statistical distribution as shown in Fig.2.6. This statistical distribution is attributed to various static or dynamic sources. For example, design mismatches and on-die process variations will result in static delay mismatches. Clock generator (e.g. PLL) jitter or dynamic voltage variations can introduce dynamic clock uncertainties. Minimizing T_{DELAY} will also minimize clock uncertainty and improve setup and hold margins.

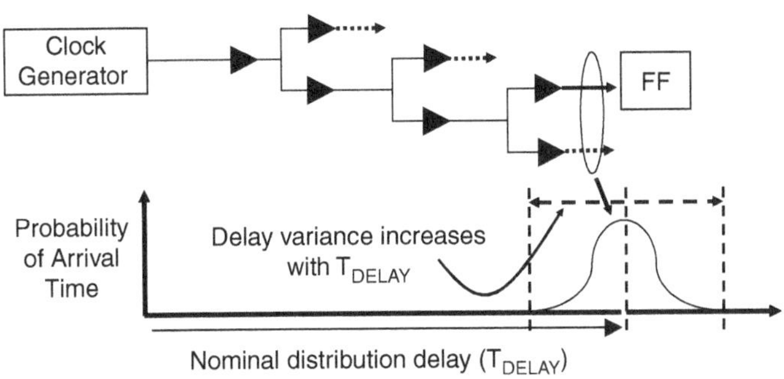

Fig. 2.6. Statistical nature of clock arrival times

2.2.2 Clock Attributes

We use this term to denote clock characteristics that affect the timing constraints described in Sect.2.2.1. The key attributes are:

1. Clock uncertainties (skew and jitter)
2. Clock distribution latency
3. Clock duty cycle

The first and the last play an explicit role in the timing constraints of a synchronous design. The clock latency by itself does not affect the sequential timing constraints but plays a critical role in determining the other two.

[2] In this context, a buffer stage could be any gate that exhibits gain.

Static and Dynamic Clock Uncertainties

Clock uncertainties can be classified as static or dynamic. Static uncertainty does not vary or varies very slowly with time. Process variation induced clock uncertainty is such an example. On the other hand, dynamic uncertainty varies with time. Dynamic power supply induced delay variation is an example of a dynamic uncertainty.

In Fig. 2.7, the clock attributes T_{skew} and T_{jitter} are defined on clock waves Ck1 and Ck2. Taking the wave Ck1 as an example, when one of the clock edges is repeatedly sampled with an ideal reference, a timing histogram will result. A timing histogram exists for every clock edge and is characterized by a mean value and a peak-to-peak range (Fig. 2.7). The difference between the mean of two corresponding clock edges (example: between edge A and edge B) is defined as skew (T_{skew}) and is treated as a static uncertainty. The peak-to-peak range of a single edge is specified as the jitter (T_{jitter}) and its character is dynamic.

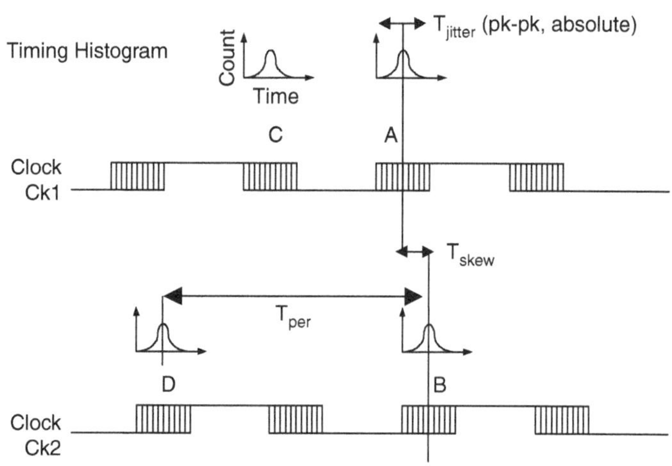

Fig. 2.7. Clock skew and jitter definitions

Table 2.1 highlights the sources of the static and dynamic clock uncertainties. Sources of static clock uncertainties are:

1. Intentional or unintentional design mismatches
2. On-die process variations
3. Loading variations (mismatch) at the intermediate or final stage of the clock distribution

Design mismatches arise because of a number of factors. For example, a nonbalanced clock distribution may be necessary due to floorplan constraints. A poorly chosen distribution topology could lead to structural design mismatches. In other situations, the clock arrival times at certain receivers are intentionally skewed to facilitate time borrowing across sequential boundaries due to nonuniform data path lengths. On-die device mismatch due to on-die process variations is a significant factor. Additionally, nonuniform clock loading is common in highly integrated designs. All skew sources mentioned above remain constant over time (except through the slow process of transistor aging) and are treated as static. Figure 2.8 shows an empirical breakdown of skew contributors.

Table 2.1. Sources of static and dynamic clock uncertainties

Clock uncertainties	Sources
Static (skew)	Intentional or unintentional design mismatches On-die process variations Final or intermediate loading variations
Dynamic (jitter)	Voltage droop and dynamic voltage variations Temperature gradient due to activity variations Clock generator jitter

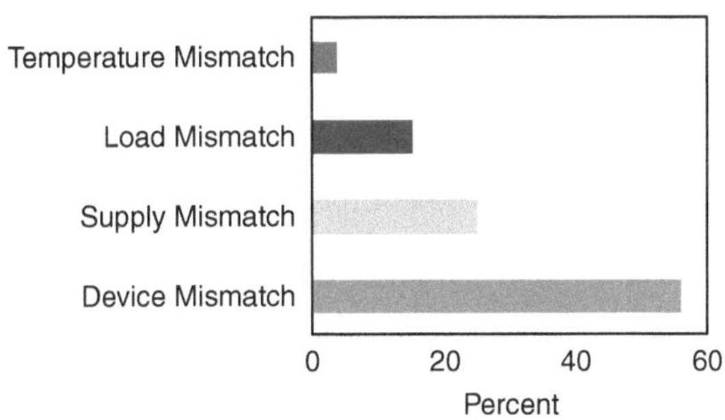

Fig. 2.8. Factors affecting clock skew. Among all the sources, device mismatch is the dominant contributor. Reproduced with permission from [31], ©1998 IEEE

Inherently static clock uncertainties can be corrected either by careful pre-silicon analysis and design or post-silicon adaptive compensation. Accurate pre-silicon analysis can be time consuming and iterative. On the other hand, post-silicon adaptive compensation is flexible and significantly more suited to high volume manufacturing.

Figure 2.9 shows clock skew as a percentage of cycle time vs. processor frequency for a number of recent designs. On average, the trend suggests that the skew as a fraction of the clock cycle time stays at about 4.5–5%. The ability of the trend to continue is attributed to the adoption of clock distribution topologies that are skew tolerant, more robust design flow, and more importantly the incorporation of robust post-silicon compensation techniques.

Clock uncertainties caused by voltage variation, temperature variation, and clock generator jitter are dynamic in nature. We use the term jitter to encompass all

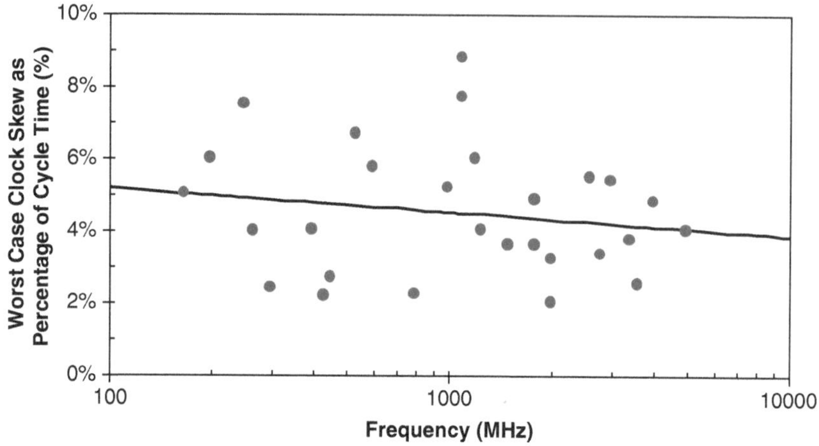

Fig. 2.9. Clock skew as percentage of cycle time vs. processor frequency [8, 15–17, 19–21, 24–26, 28–30, 32–46]

dynamic uncertainties. Voltage variation is the dominant source and it can be due to local switching events affecting specific areas of the clock distribution in a nonuniform fashion. A mathematical model of supply-induced jitter based on additive sinusoidal supply noise is developed in Section 6.8. Global voltage droop and clock generator jitter are common to the entire distribution and contribute to the setup constraint by modulating the cycle time. Clock generator jitter is addressed in detail in Chapter 5. Temperature variation has a long time constant and its impact is usually minor as seen in Fig. 2.8. Figure 2.10 shows the trend of peak-to-peak clock jitter as a fraction of cycle time. The average reduction of effective clock cycle time due to jitter is about 5.5% and minimizing this is critical for performance reasons.

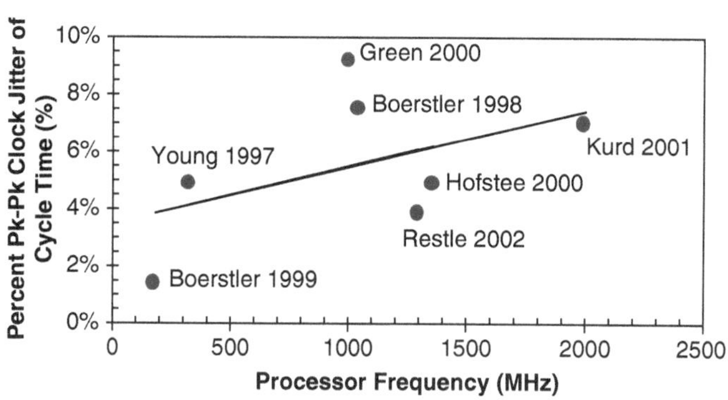

Fig. 2.10. Pk–pk clock jitter as a fraction of clock cycle time vs. processor frequency [6, 17, 18, 36, 47–49]

We now present a simple mathematical skew and jitter model based on clock distribution latency (Fig. 2.11). The figure shows a source clock path with M buffer stages and a receiver clock path with N buffer stages resulting in delays of T_{Ck1} and T_{Ck2} respectively. The point-of-divergence (POD) delay is defined as the sum of the source clock delay and the receiver clock delay measured from a common origin. In Fig. 2.11, the POD delay equals the sum of T_{Ck1} and T_{Ck2}. Assuming T_i is the actual delay at buffer stage i and τ is the average delay per stage, the source clock and the receiver clock delays are:

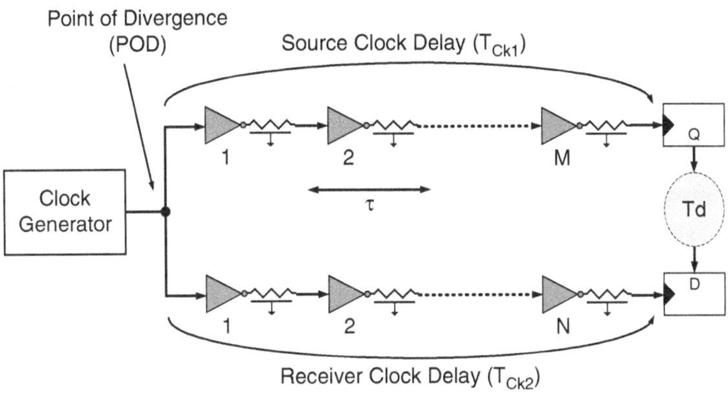

Fig. 2.11. Sample clock distribution for skew and jitter model

$$T_{\mathrm{Ck1}} = \sum_{1}^{M} T_i \cong M\tau, \tag{2.4}$$

$$T_{\mathrm{Ck2}} = \sum_{1}^{N} T_i \cong N\tau. \tag{2.5}$$

The average delay per stage is determined by the drive current of the driver stage (I_{d}), the output loading capacitance (C_{l}), and the output voltage swing (V_{CC}):

$$\tau = \frac{C_{\mathrm{l}} V_{\mathrm{CC}}}{I_{\mathrm{d}}}. \tag{2.6}$$

Using a simple linearized model, the change in delay per stage ($\Delta\tau$) can be formulated as:

$$\Delta\tau = \frac{\partial\tau}{\partial V_{\mathrm{CC}}}\Delta V_{\mathrm{CC}} + \frac{\partial\tau}{\partial C_{\mathrm{l}}}\Delta C_{\mathrm{l}} + \frac{\partial\tau}{\partial I_{\mathrm{d}}}\Delta I_{\mathrm{d}}. \tag{2.7}$$

Evaluating the partial derivatives yields:

$$\Delta\tau = \frac{C_{\mathrm{l}}}{I_{\mathrm{d}}}\Delta V_{\mathrm{CC}} + \frac{V_{\mathrm{cc}}}{I_{\mathrm{d}}}\Delta C_{\mathrm{l}} - \frac{C_l V_{\mathrm{cc}}}{I_{\mathrm{d}}^2}\Delta I_{\mathrm{d}}. \tag{2.8}$$

Finally, substituting the expression for τ (2.6) results in the following expression:

$$\Delta\tau = \tau\left(\frac{\Delta V_{\mathrm{CC}}}{V_{\mathrm{CC}}} + \frac{\Delta C_{\mathrm{l}}}{C_{\mathrm{l}}} - \frac{\Delta I_{\mathrm{d}}}{I_{\mathrm{d}}}\right). \tag{2.9}$$

Equation 2.9 states that the delay variation at each stage can be approximated to be proportional to the stage delay. Additionally, we make the assumption that the delay per stage is a random variable, normally distributed with the following standard deviation:

$$\sigma(\tau) \cong \alpha\tau, \tag{2.10}$$

where α is the proportionality constant predicted by (2.9). We assume that α is on the order of 5%. Under the assumption that each stage delay is independent and identically distributed, the standard deviations of T_{Ck1}, T_{Ck2}, and $|T_{\mathrm{Ck1}} - T_{\mathrm{Ck2}}|$ become:

$$\sigma(T_{\mathrm{Ck1}}) \cong \sqrt{M}\alpha\tau, \tag{2.11}$$

$$\sigma(T_{\mathrm{Ck2}}) \cong \sqrt{N}\alpha\tau, \tag{2.12}$$

$$\sigma(|T_{\mathrm{Ck1}} - T_{\mathrm{Ck2}}|) \cong (\sqrt{M+N})\alpha\tau. \tag{2.13}$$

The standard deviations of skew and jitter are therefore[3]:

$$\sigma[T_{\mathrm{skew}}(\mathrm{Ck1},\mathrm{Ck2})] = \left(\sqrt{M+N}\right)\alpha_{\mathrm{skew}}\tau, \tag{2.14}$$

$$\sigma\left[T_{\mathrm{jitter}}(\mathrm{Ck1})\right] = \left(\sqrt{M}\right)\alpha_{\mathrm{jitter}}\tau, \tag{2.15}$$

$$\sigma\left[T_{\mathrm{jitter}}(\mathrm{Ck2})\right] = \left(\sqrt{N}\right)\alpha_{\mathrm{jitter}}\tau. \tag{2.16}$$

where α_{skew} and α_{jitter} represent the variation coefficients for static and dynamic clock uncertainties, respectively. Typical clock distribution will have $M = N$ and the formulation will be reduced to:

$$\sigma[T_{\mathrm{skew}}(\mathrm{Ck1},\mathrm{Ck2})] = \left(\sqrt{2M}\right)\alpha_{\mathrm{skew}}\tau, \tag{2.17}$$

$$\sigma\left[T_{\mathrm{jitter}}(\mathrm{Ck1})\right] = \left(\sqrt{M}\right)\alpha_{\mathrm{jitter}}\tau, \tag{2.18}$$

$$\sigma\left[T_{\mathrm{jitter}}(\mathrm{Ck2})\right] = \left(\sqrt{M}\right)\alpha_{\mathrm{jitter}}\tau. \tag{2.19}$$

Equations 2.17–2.19 show that the skew and jitter variations will grow as the square-root of the number of distribution buffering stages and linearly with the nominal delay per stage. This formulation can be generalized for any pair of clocks that share a common point of clock divergence. The sum of the variation coefficients for modern process technology and design is between 5 and 10%.

---

[3] T_{skew}(Ck1, Ck2) means the skew between clocks Ck1 and Ck2 and T_{jitter}(Ck#) is the jitter of Ck#.

Distribution Delay

Equations 2.17–2.19 suggest that the clock distribution delay (latency) is a key component in determining the overall clock uncertainties. In order to handle the final clock loading and to traverse the distances needed to reach the loads, a clock network has to rely on a series of clock buffers for gain and signal propagation. In a typical processor, the number of clock buffer[4] stages may exceed 20 resulting in clock latency that can approach 1 ns. Minimizing clock distribution latency is a primary design objective irrespective of the distribution topology.

Duty Cycle

Duty cycle (Fig. 2.12) is the relative percentage of the clock high phase time vs. low phase time. Except for special clocking applications such as pulse generators and clocks for dynamic and memory circuits, a 50% clock duty cycle is considered optimal. This is particularly important for a latch-based designs and memory circuits where any offset between the high phase and low phase can lead to phase paths that are more difficult to meet timing constraints. In a phase path, time lost due to duty cycle distortion will subtract directly from the total available phase time. Cycle-based sequential designs using edge-triggered flip-flops are more immune to clock duty cycle distortion. In a clock distribution, duty cycle distortion is introduced when there is asymmetry between rising and falling edge delays.

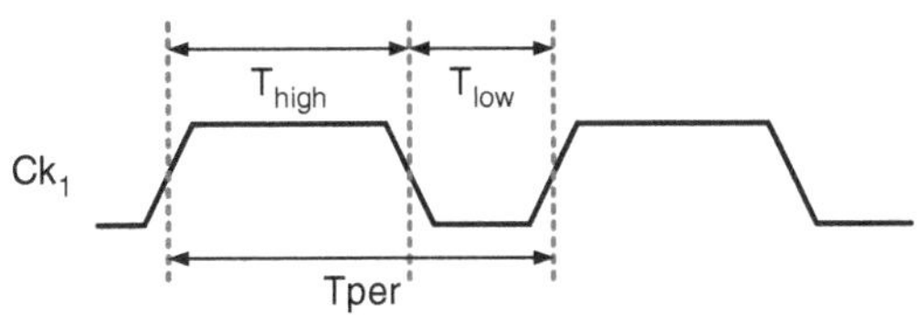

Fig. 2.12. Clock duty cycle

2.2.3 Clock Distribution Power

Power dissipation attributed to the clock distribution has emerged as a critical constraint in multi-GHz multicore processors with large on-die caches. In this section, we will develop a model for clock distribution power. Typically, the total end-of-distribution loading due to sequential elements strongly determines the overall clock network power. Let us consider the switching power of an unconditional clock at the final distribution stage M:

$$Pd_{\mathrm{CK\_}M} = C_{\mathrm{L\_}M} V_{\mathrm{CC}}^2 f, \tag{2.20}$$

[4] In this context, a clock buffer stage represents one unit-inverter stage.

where C_{L_M} is the stage load that encompasses both gates and interconnect, V_{CC} is the power supply voltage, and f is the clock frequency. Total power will include components for short circuit and leakage, but the dynamic component dominates. If the fan out per stage is k, the clock dynamic power consumed at stage $M-1$ is:

$$Pd_{Ck\_M-1} = C_{L\_M-1} V_{CC}^2 f = \frac{C_{L\_M}}{k} V_{CC}^2 f. \tag{2.21}$$

Assuming that the fan out is constant across all M stages, the total clock distribution dynamic power is:

$$Pd_{Ck\_Total} = \sum_{i=1}^{M} Pd_{Ck\_i} = \sum_{i=1}^{M} C_{L\_i} V_{CC}^2 f, \tag{2.22}$$

$$Pd_{Ck\_Total} = V_{CC}^2 f C_{L\_M} \left[\frac{1-\left(\frac{1}{k}\right)^M}{1-\frac{1}{k}} \right]. \tag{2.23}$$

Let us define the clock load multiplier as the ratio of the total clock distribution load capacitance to the end-of-distribution load capacitance:

$$\text{Clock load multiplier} = \frac{\sum_{i=1}^{M} C_{L\_i}}{C_{L\_M}} = \left[\frac{1-\left(\frac{1}{k}\right)^M}{1-\frac{1}{k}} \right]. \tag{2.24}$$

Figure 2.13 shows the clock load multiplier vs. the number of distribution stages. Decreasing the stage fan out will lead to higher total network capacitance that approaches 1.5 at a stage fan out of 3. Note that this parameter is not very sensitive to the number of buffer stages in the clock distribution network. Figure 2.14 shows the

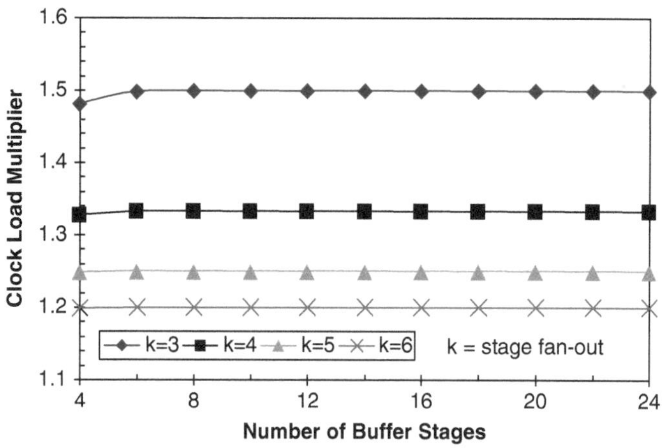

Fig. 2.13. Clock loading multiplier of a clock distribution

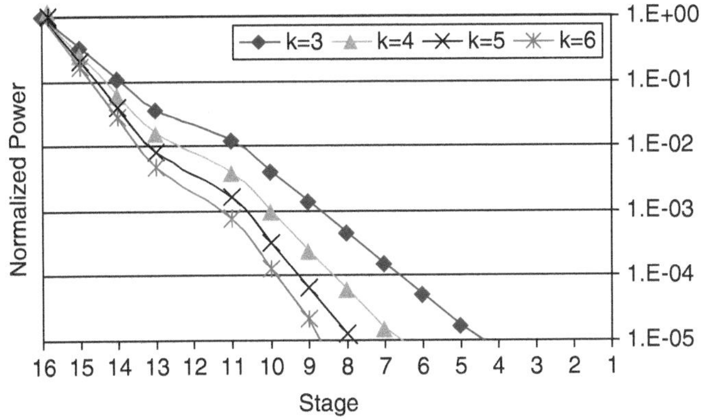

Fig. 2.14. Normalized clock stage power vs. stage number

power dissipation at each of the stage normalized to the power dissipation at the last stage. It can be seen that power dissipation of the clock distribution is dominated by the final end-of-distribution loading and that the last couple of stages in the distribution will account for more than 90% of the total clock power. An implication of this analysis is that the manner in which clock is distributed at the final stages of the distribution will ultimately determine the overall clock power and that the distribution topology upstream will not have a strong impact. In a typical processor, clock distribution (excluding the last stage) should not exceed 10% of total chip power and have a design goal of 5–8%.

2.3 Clock Distribution Topologies

In this section, various clock distribution topologies are described. Table 2.2 lists distribution topologies encountered in modern processors. While discussing these topologies, we will focus on the same attributes described in Sect. 2.2.2. In addition, ease of implementation will be considered. Ease of implementation is subjective and depends heavily on historical design styles and prior art.

2.3.1 Unconstrained Tree

An unconstrained tree style clock distribution is illustrated in Fig. 2.15. It is commonly used in automatic synthesis flows and usually placed with little or no restriction on the number of buffer stages and explicit matching between interconnect delays and the buffer delays. The network design is accomplished with a cost function that minimizes the delay differences across all clock branches. Figure 2.15

Table 2.2. Clock distribution topologies

Style	Description
Unconstrained tree	Automated buffer placements with unconstrained trees
Balanced tree	Multiple levels of balanced tree segments H-tree is most common
Central spine	Central clock driver
Spines with matched branches	Multiple central structures with length (or delay) matched branches
Grid	Interconnected (shorted) clock structure
Hybrid distribution	Combination of multiple techniques Common theme is tree + grid or spine + grid

shows an unconstrained clock distribution tree with K branches. A cost function (ϑ) that minimizes the delay differences can be used in the construction of the network:

$$\vartheta = \sum_{i=1}^{K} \left(T_{\mathrm{Ck}i} - T_{\mathrm{Ck\_Average}}\right)^2, \tag{2.25}$$

$$T_{\mathrm{Ck\_Average}} = \frac{1}{N} \sum_{i=1}^{K} T_{\mathrm{Ck}i}. \tag{2.26}$$

In the primitive form and specifically without explicit structural matching, a clock network with dissimilar buffer and interconnect delay composition may result in radically different branches that will exhibit significant mis-tracking across process, voltage, and temperature variations. More sophisticated optimizing algorithms can be incorporated to improve PVT tracking. Due to this limitation, this style of clock distribution is usually restricted to small functional blocks within a larger design.

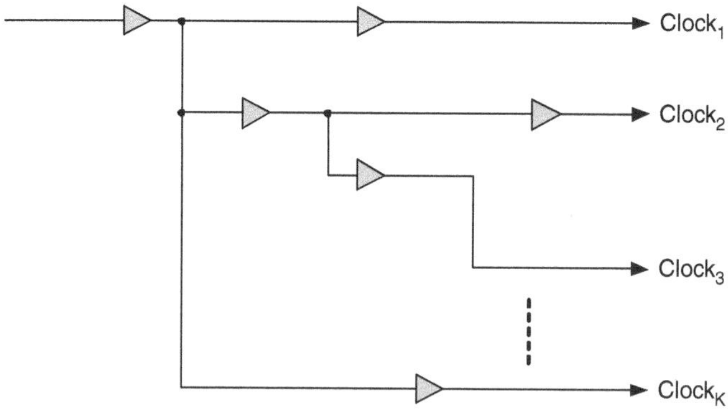

Fig. 2.15. Unconstrained tree clock network

2.3.2 Balanced Tree

Figure 2.16 shows a balanced H-tree clock topology. Due to the structural symmetry, a balanced tree exhibits identical nominal delay and identical buffer and interconnect segments from the root of the distribution to all branches. If the matching is adhered to, structural skew can be zero. With identical buffer and interconnect segments, an idealized balanced tree clock distribution will exhibit good tracking across PVT compared to the unconstrained network described earlier.

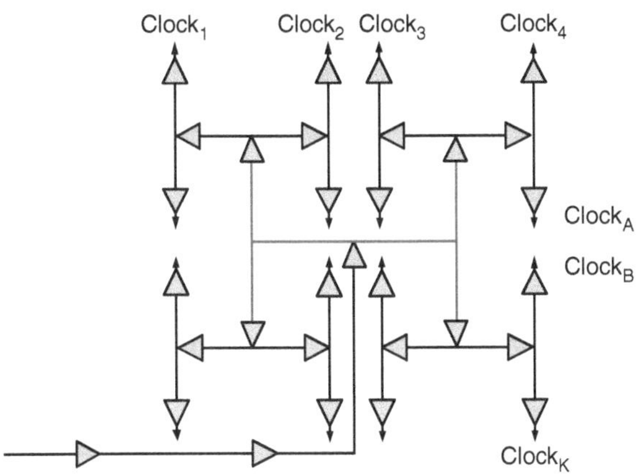

Fig. 2.16. Balanced H-tree clock network

Figure 2.17 shows alternative balanced tree topologies: the X-tree and a tapered H-tree. The X-tree incorporates nonrectilinear clock trunks in the physical implementation but exhibits the same properties as the H-tree. The trunk widths in a tapered H-tree increase geometrically toward the root of the distribution to maintain impedance matching at the T-junctions. One important characteristic of the aforementioned tree structures is that by continuing to expand the buffer hierarchy, balanced trees are capable of delivering the clock to all part of the silicon die. Typically, the clocks at the end-of-distribution branches will serve a small local region. The size and number of the local regions will determine the depth of the tree. A larger number of regions requires a tree with more depth.

Full balanced tree topologies are designed to span the entire die in both the horizontal and vertical dimensions. They are capable of delivering the clock to all regions of the die. A binary tree on the other hand (Fig. 2.18) is intended to deliver the clock in a balanced manner in either the vertical or horizontal dimension.

All branches of a binary tree exhibit identical buffer-interconnect segments, zero structural skew, and similar PVT tracking. In contrast to the H-tree, the buffers in a binary tree can be designed to co-locate in close proximity along a centralized stripe. The closer physical proximity of the buffers in a binary tree can result in

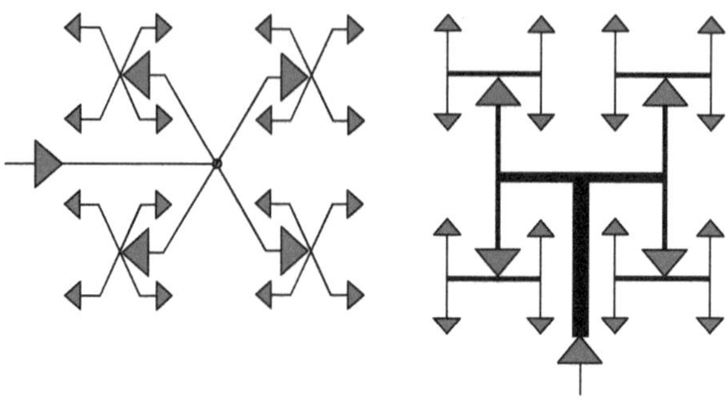

Fig. 2.17. Variations on the balanced tree topology

reduced sensitivity to on-die variation. Moreover, physical placement of the clock buffers in close proximity will minimize floorplan disruptions. On the other hand, the idealized buffer placements associated with an H-tree may be difficult to achieve. Due to these reasons, binary trees are often the preferred structure over an idealized H-tree. Figure 2.19 shows a binary tree distribution with intermediate shorting. The benefits of shorting will be discussed in a later section.

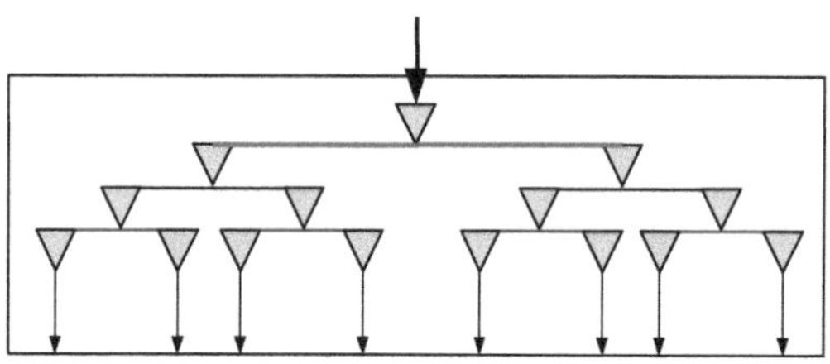

Fig. 2.18. Binary tree clock distribution

A balanced tree will exhibit nonzero clock uncertainty among branches due to nonzero POD delays (2.17–2.19). Let us consider branches $Clock_4$ and $Clock_A$ in Fig.2.16. The point-of-divergence is two buffer-interconnect segments apart. In contrast, branches $Clock_A$ and $Clock_B$ are six buffer-interconnect segments apart resulting in higher POD delay and higher skew uncertainty. Therefore, among pairs of equivalent branches in a balanced tree, nonuniform skew uncertainty will result and will depend on point-of-divergence delay. In summary, a balanced tree is capable of delivering the clock from the root to all regions of the die with good structural

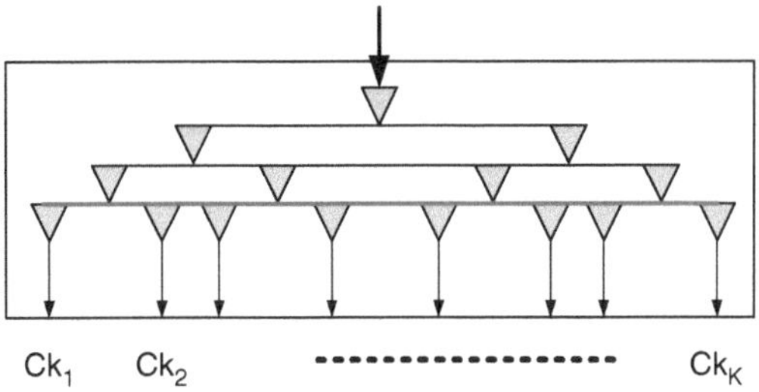

Fig. 2.19. Binary tree clock distribution with intermediate shorting

matching, efficient clock power and low structural latency. On the other hand, it will exhibit nonuniform POD-induced clock skew uncertainty.

Before proceeding it should be noted that a nonsymmetric tree can also be used in this context. A nonsymmetric tree usually maintains the same number of buffer stages but will not have delay matching on a per stage basis. Delay adjustment for overall branch equalization is done in a fashion similar to the unconstrained tree. Intense computational effort usually is needed for this design and its application is less common.

2.3.3 Central Spine

A central spine clock distribution is a specific implementation of a binary tree. Figure 2.20 shows an idealized central spine implementation with the final branches serving all parts of the die. The binary tree is shown to have embedded shorting at all distribution levels and unconstrained routing to the local loads at the final branches. In this configuration, the clock can be transported in a balanced fashion across one dimension of the die with low structural skew. The unconstrained branches are simple to implement although there will be residual skew due to asymmetry (Fig.2.20).

2.3.4 Spines with Matched Branches

An extension of the central spine structure can be realized by replacing the unconstrained end-of-distribution branches with delay matched routes as shown in Fig.2.21. In this implementation, the longest branch determines the delay from the output of the central spine to the end loads. Serpentine routes are added to the shorter branches for delay matching. Figure 2.21 shows a structure with three central spines. Multiple central spines are needed when the routing distance of the local branches is increased. Dividing the chip into several sectors served by multiple spines is a practical topology to ensure small local branch delays.

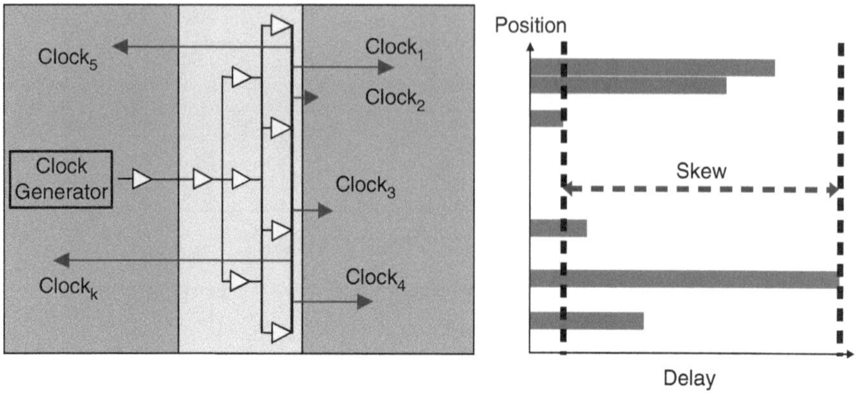

Fig. 2.20. Central clock spine distribution

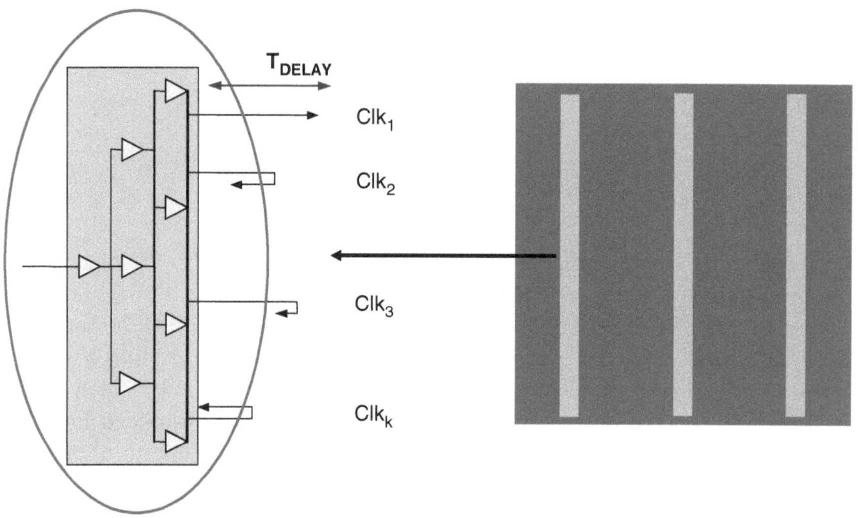

Fig. 2.21. Multiple clock spines with matched branches

2.3.5 Grid

The tree style distributions described in previous sections rely on individual branches to deliver the clock to the local (end-of-distribution) points of clock consumption (e.g. local flip-flops). A processor will have a large number of these local points and will require a large number of branches and therefore a deep distribution tree. A deep distribution tree will exhibit large POD delays and degraded clock performance. Subdividing the die into a smaller number of clock regions and applying a grid to serve each region can be a superior solution.

Figure 2.22 schematically shows a 2-dimensional grid serving one of these clock regions. This clock grid resembles a mesh with fully connected clock tracks in both dimensions and grid drivers located on all four sides. Local loads within a region can be directly connected to the grid. The grid effectively shorts the output of all drivers and helps minimize delay mismatches. Figure 2.22 shows an idealized delay profile of a 2-dimensional grid assuming uniform loading. The shorted grid node helps balance the load nonuniformities and results in a more gradual delay profile across the region. Additionally, since the grid drivers are shorted, the POD delay to all the loads within a region is limited to the interconnect delay of the grid which is typically small and results in lower clock skew uncertainty across the region. Grid drivers may also be placed on two sides leading to a structure and delay profile shown in Fig. 2.23. Critical design parameters for the grid are grid driver locations and pitch in addition to the grid metal pitches and dimensions.

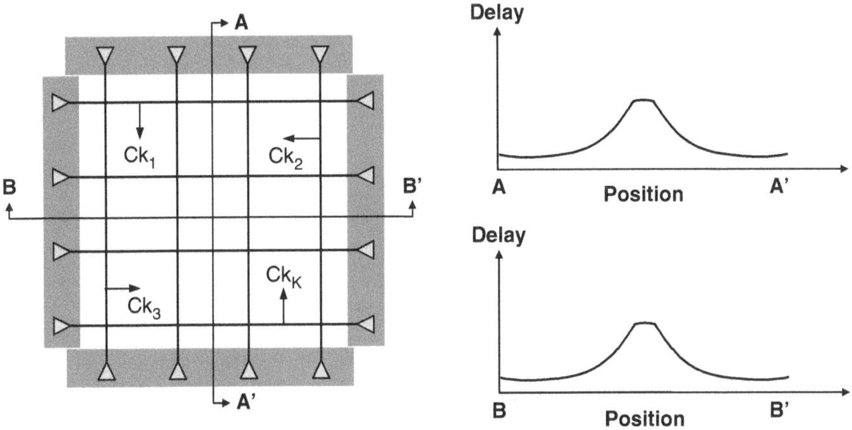

Fig. 2.22. Clock grid with 2-dimensional clock drivers

The recombinant tile structure is an enhancement over the conventional grid structure and incorporates the properties of a balanced tree [29]. Figure 2.24 shows the evolution of the recombinant tile structure from a regular H-tree segment to a tile template and to the final tile assembly. A typical implementation will have uniform interconnect pitches in both the x and y dimensions.[5] The pitches are determined by the intrinsic interconnect segment delays and the edge rate requirements. The uniformity of the segment pitches allows all intermediate buffers to be placed in predetermined locations.

The following analysis highlights the benefits of shorting intermediate stages in reducing clock uncertainties. Figure 2.25 shows two clock branches Ck_A and Ck_B exhibiting input skew $skew_{IN}$. Let us assume that the two branches are identical

[5] Note that the x and y pitches do not need to be equal.

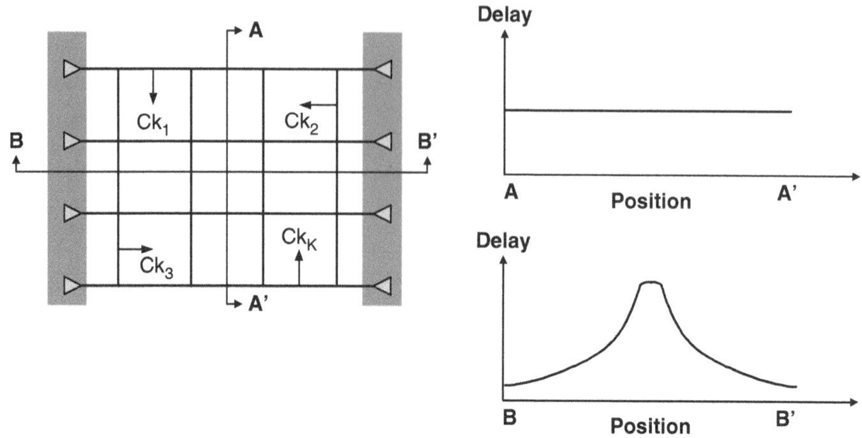

Fig. 2.23. Clock grid with 1-dimensional drivers

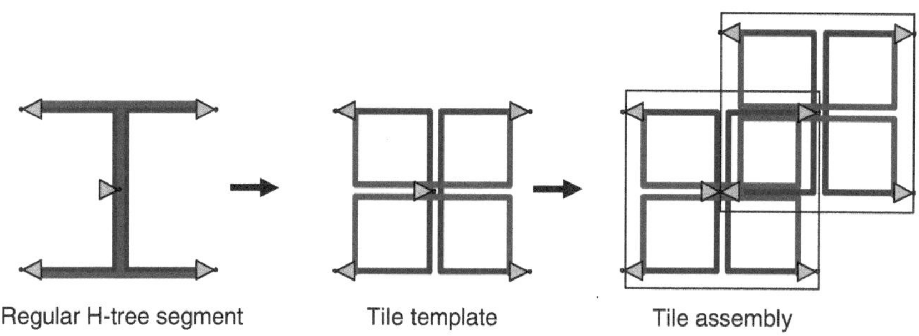

Fig. 2.24. Recombinant tile clock structure. Reproduced with permission from [29], ©2003 IEEE

(A-X-Y and B-U-V) and that R_{SHORT} in Fig.2.25 is an open circuit ($R_{\mathrm{SHORT}} = \infty$). In this case, the skew between $\mathrm{Ck_U}$ and $\mathrm{Ck_Y}$ will be the same as the skew between $\mathrm{Ck_A}$ and $\mathrm{Ck_B}$. With an ideal short circuit ($R_{\mathrm{SHORT}} = 0\Omega$), the skew between $\mathrm{Ck_U}$ and $\mathrm{Ck_Y}$ will be zero. Hence, with a non zero and finite R_{SHORT}, the output skew between $\mathrm{Ck_U}$ and $\mathrm{Ck_Y}$ will be proportional to the incoming skew:

$$\mathrm{skew}(\mathrm{Ck_U}, \mathrm{Ck_Y}) = \gamma \mathrm{skew}(\mathrm{Ck_A}, \mathrm{Ck_B}), \tag{2.27}$$

where γ is a skew averaging coefficient ($\gamma \geq 0$). The averaging coefficient is dependent upon the local POD induced delay mismatch and will scale with the spatial separation between the output nodes. For example, when the shorting resistor R_{SHORT} in Fig.2.25 exhibits near zero resistance, the skew between $\mathrm{Ck_Y}$ and $\mathrm{Ck_U}$ will be close to zero suggesting that γ is near 0. As R_{SHORT} approaches an open, γ will be equal to or larger than 1. For example, in a 90nm technology, a typical R_{SHORT} will

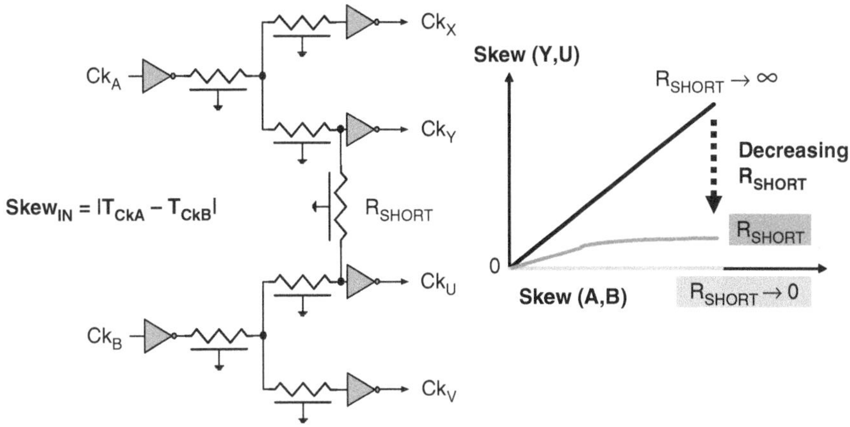

Fig. 2.25. Effect of shorting on clock skew [50]

result in γ in the range of 0.2–0.8 for a few hundred microns of spatial separation [29]. When this structure is cascaded in a clock distribution consisting of S cascaded stages ($S > 20$), the skew at the output of stage S will be (for small γ) [50]:

$$\text{skew}_S \approx \gamma^2(\text{skew}_{IN}) + \gamma(\text{skew}_{IN}). \qquad (2.28)$$

Equation 2.28 states that the skew at the output of the distribution will stay relatively independent of the number of stages and therefore less sensitive to die size. The recombinant tile structure can be easily scaled to accommodate new designs. Comparing the recombinant tile of Fig.2.24 to the grid structures of Figs.2.22 and 2.23, reveals that they share similar skew benefits due to averaging. However, the grid structures require a pregrid clock distribution network to drive the grid or collection of grids. A hybrid clock distribution topology can meet this need.

2.3.6 Hybrid Distribution

A hybrid clock distribution incorporates a combination of earlier described topologies. Common configurations are spines-grid distribution or tree-grid distribution. Figure 2.26 shows the topology of a tree-grid distribution. It employs a multilevel H-tree driving a common grid. Specifically, the multilevel H-tree delivers the clock from the clock generator (PLL in Fig.2.26) to various regions of the die. Regional buffers (labeled as level 4 buffers in Fig.2.26) residing at the end of the multilevel H-tree drive a common grid that includes all local loads. As an alternative, there can be multiple regional grids each served by a branch of the pregrid H-tree. Partitioning the design enables intentional skew rebalancing across the regions.

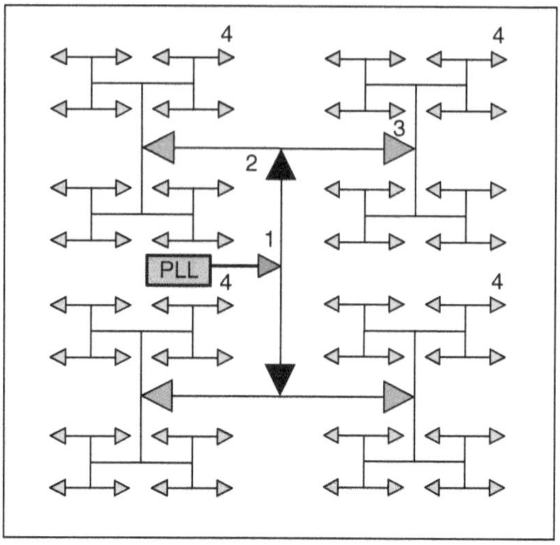

Fig. 2.26. Hybrid clock distribution consisting of balanced H-Tree and Grid

2.4 Microprocessor Clock Distributions

Due to the design complexity and the significant mis-tracking to process, voltage and temperature, a fully unconstrained clock distribution network is rarely (if ever) applied to a processor design.

The closest example is a hybrid combination of symmetric and asymmetric clock trees. Figure 2.27 shows an example of a processor clock distribution with a first level H-tree connected to multiple secondary trees that are asymmetric but delay balanced.

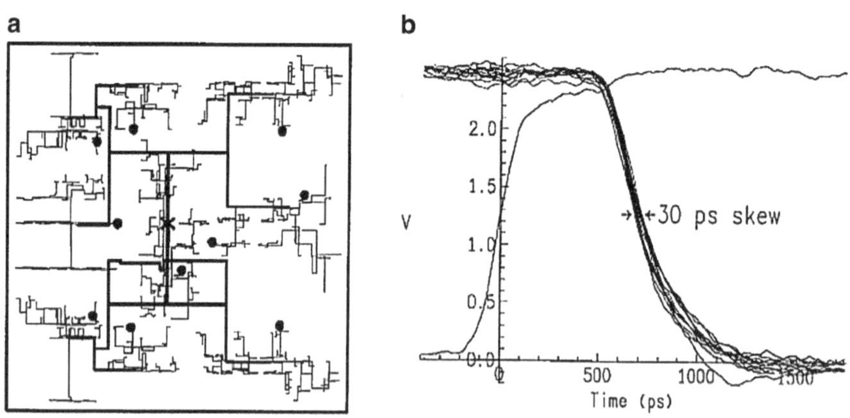

Fig. 2.27. Asymmetric clock tree distribution network based on delay matching. Reproduced with permission from [51], ©1998 IEEE

The construction of the trees was based on a custom methodology that matches the wire delays by tuning the metal widths, spacings, and lengths. Specifically, the first level symmetric H-tree (thick lines in Fig. 2.27a) routes the global clock from the center of the die to 9 sector buffers. The 9 sector buffers rely on multiple delay-matched secondary trees (light lines in Fig. 2.27a) to distribute the clock to 580 global clock receivers. Figure 2.27b shows the measured skew.

Figure 2.28 shows another example of a hybrid multilevel clock tree design [52]. The cache area (un-core) of the processor is partitioned into 13 regional clock zones served by the secondary clock trees. Postlayout extraction-based simulation models were used to perform tree optimization and delay matching. Additional examples of a hybrid multilevel clock tree for processors are found in [10, 12, 13, 43].

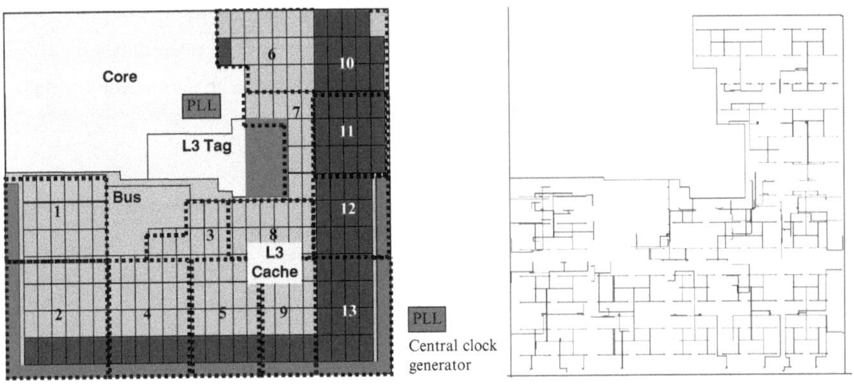

Fig. 2.28. Asymmetric clock tree distribution with multiple regions. Reproduced with permission from [52], ©2005 IEEE

Hybrid clock distributions that consist of multilevel symmetric trees and grids have been applied to a number of processors [17–20]. Figure 2.29 shows an example [18]. In this implementation, the clock from the clock generator is distributed from a central clock buffer through two levels of buffering and three levels of delay tuned H-trees before reaching a main grid covering most of the chip, and two smaller grids covering two units that require delayed clocks.

Figure 2.30 shows another example of a hybrid tree-grid design in [17]. A multi-level tree structure delivers the clock to 64 sector buffers driving a common grid via multiple second level tuned trees.

Figure 2.31 shows the floorplan and the clock skew profile of three generations of Alpha® [6] processors. These designs followed a common strategy of having one or more centralized clock spines to drive a common grid. The first generation design relied on a single spine to support the entire die whereas the third generation design

[6] Other names and brands may be claimed as property of others.

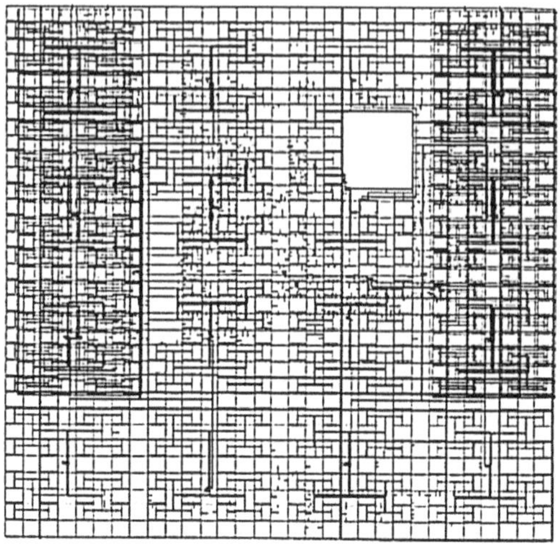

Fig. 2.29. Multilevel symmetric H-Tree distribution. Reproduced with permission from [18], ©2000 IEEE

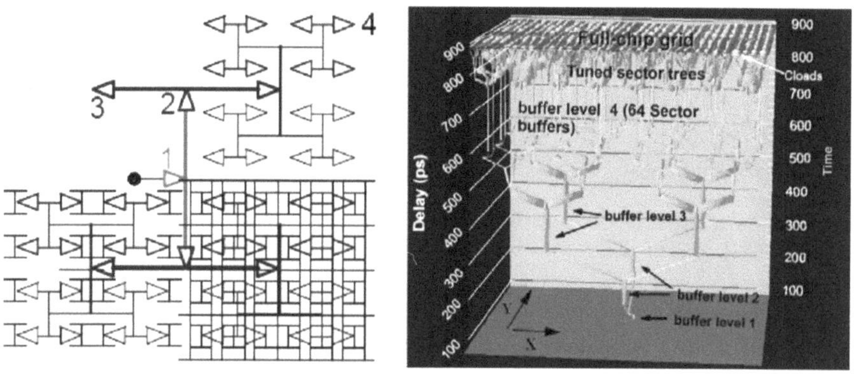

Fig. 2.30. Delay characteristics of a multilevel tree-grid distribution. Reproduced with permission from [17], ©2002 IEEE

utilized 16 central spines organized in a 2-dimensional fashion to drive the common grid. By partitioning the die into smaller regions. the third generation design reduced the clock skew across the grid.

Figure 2.32 shows the application of recombinant tiles to a multi-GHz IA processor fabricated in 90nm [29].[7] The buffers needed for the recombinant tiles are embedded in eight central clock stripes. The recombinant tile distribution consists of

[7] Other names and brands may be claimed as property of others.

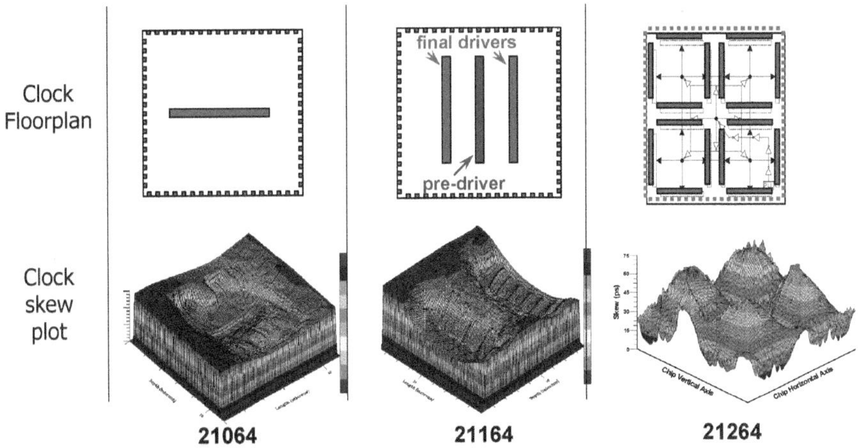

Fig. 2.31. Centralized clock drivers with grids on three generations of the Alpha® microprocessor. Reproduced with permission from [53], ©1998 IEEE

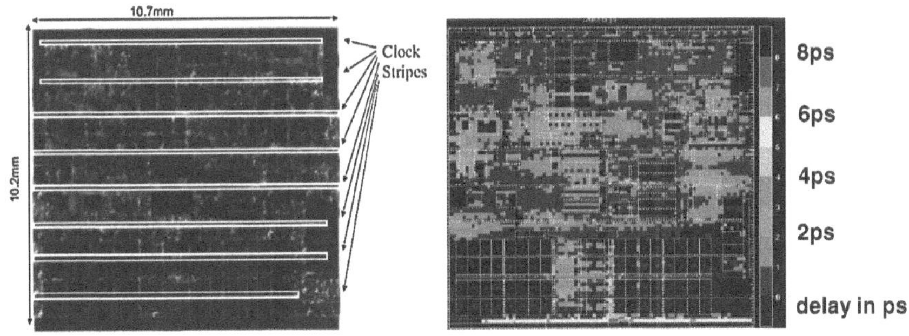

Fig. 2.32. Recombinant clock tiles on a 90nm processor. Reproduced with permission from [50], ©2003 IEEE

27 inversion stages and a total of 1,474 grid drivers (each driver is an inverter). An automated grid driver sizing flow was used to minimize grid driver oversizing for power efficiency. The simulated delay profile is shown in Fig. 2.32. A global skew of less than 10ps was achieved with this design.

An example of centralized spines with delay-matched branches is the clock distribution of the 180nm Pentium® 4 processor [8][8]. Binary distribution trees embedded in three central clock spines drive the local loads with delay-matched branches. The binary trees embedded in middle spine buffer the clock from the central PLL and deliver it in a balanced fashion to the other spines. The final clock drivers use matched

[8] Other names and brands may be claimed as property of others.

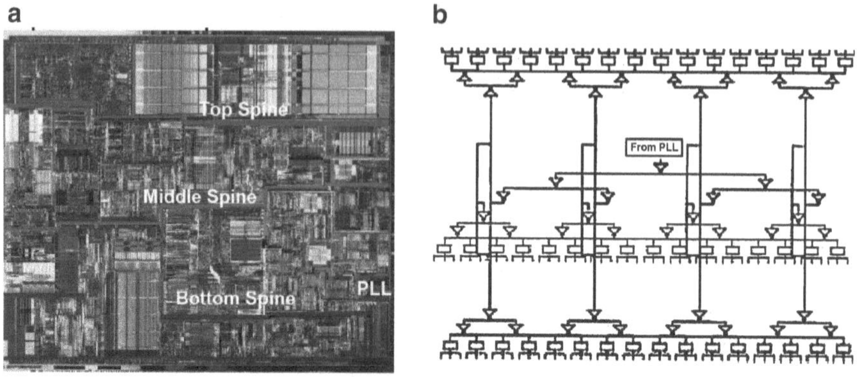

Fig. 2.33. Pentium® 4 processor clock distribution using centralized spines with delay matched final branches. Reproduced with permission from [49], ©2001 IEEE

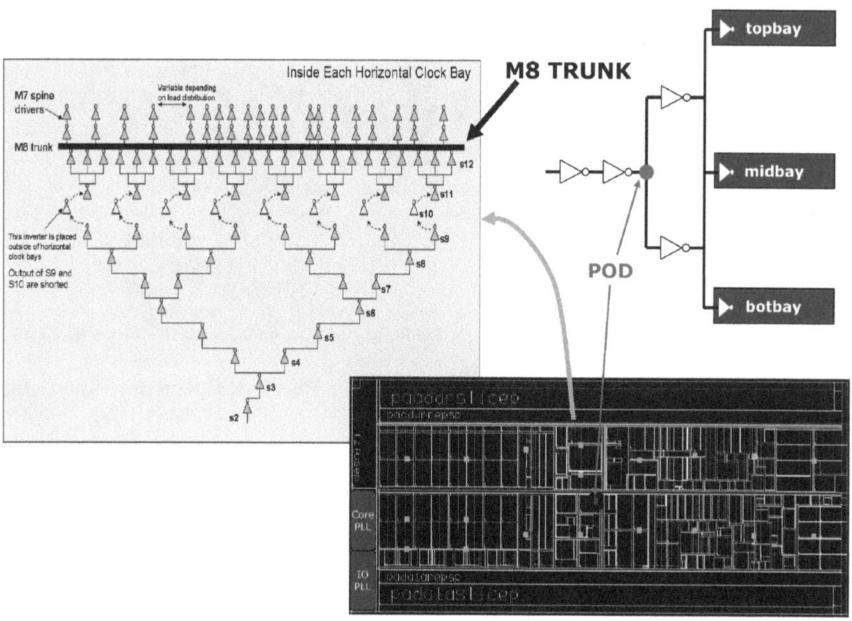

Fig. 2.34. Clock distribution of a low power IA processor consisting of binary trees embedded in the centralized spines. Reproduced with permission from [55], ©2008 IEEE

branches to support the local loads. The final drivers incorporate delay tuning capability to further optimize the skew via post-silicon compensation. Without engaging compensation, the measured skew is ±32ps.

The low power IA processor in [54] is another example of the application of the clock distribution topology consisting of centralized spines and delay-matched branches. The global binary tree with limited clock recombination is embedded in the spines. The spine output drivers are shorted with a high layer metal (M8) to equalize the driver delays. The highly selective application of clock recombination and other power saving schemes enables this design to achieve total clock power dissipation that is less than 10% of the total processor power.

The 65 nm dual-core Xeon® processor employed a hybrid spine-grid clock distribution topology to support the multicore and uncore clock domains [45].[9] Figure 2.35 shows the multiple clock domains and the distribution design of this processor. The processor has two cores operating at the high frequency MCLK. The uncore is supported by the SCLK at half the MCLK frequency and the ZCLK dedicated for the I/O circuits at 4 times the system clock frequency. Binary trees embedded in the horizontal and vertical clock spines in the uncore distribute the clocks to the SCLK and ZCLK grids. The core employs the recombinant tile clock distribution similar to that described in [29]. Operational flexibility is achieved by keeping core and uncore clock regions independent.

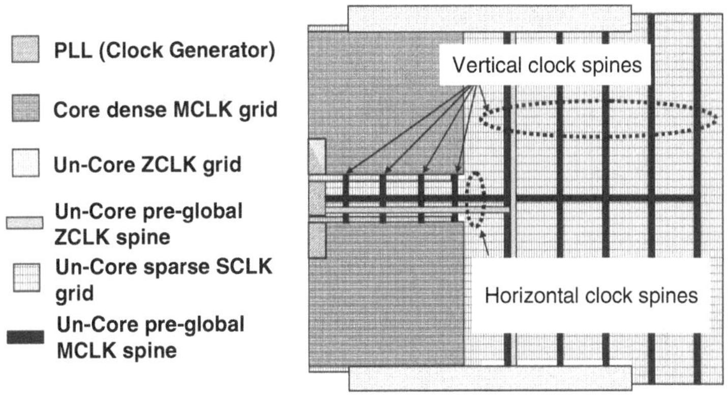

Fig. 2.35. Hybrid spine-grid clock distribution in a dual core processor. Reproduced with permission from [45], ©2006 IEEE

Figure 2.36 shows the details of the uncore grid implementation and the local clocking in [45]. The preglobal clock and the final grid clock driver are embedded in the vertical clock spines. A common SCLK grid covers the entire uncore area to serve the local logic units. Local clocking consists of two buffer stages featuring clock gating and delay tuning. The local clock buffers are placed inside the local logic unit with direct connection to the overlying global grid. Support for multiple local clock flavors is achieved with a family of local clock buffer macros.

[9] Other names and brands may be claimed as property of others.

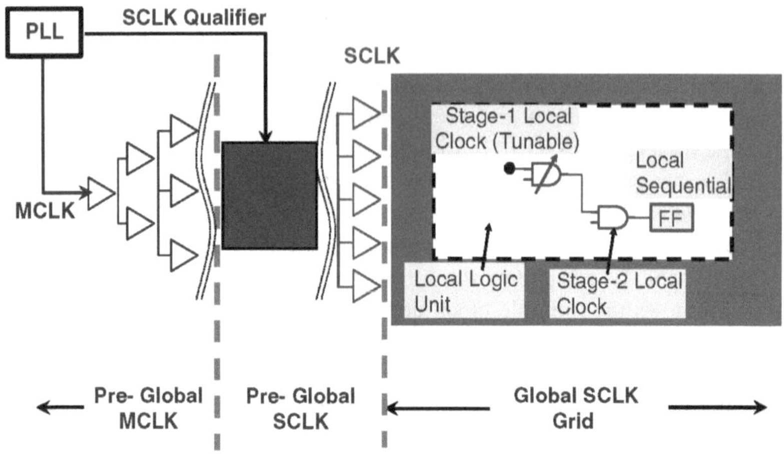

Fig. 2.36. Local clock distribution of the hybrid spine-grid clock distribution. Reproduced with permission from [45], ©2006 IEEE

2.5 Clock Design for Test and Manufacturing

2.5.1 Global and Local Clock Compensations

It should be obvious from the earlier discussions that the primary objective of the aforementioned clock distribution topologies is to deliver the clock to all corners of the die with low skew. For example, the application of averaging in the recombinant tile clock distribution will result in a scalable network that exhibits very low skew. However, implementation of the recombinant tile network will encounter floorplan constraints leading to nonideal driver placements and loss of performance. Floorplan constrains will also affect other topologies such as the H-tree distribution. Moreover, in many situations, intentional clock skew between specific regions of the global clock network is needed to rebalance the path timings. Therefore, a clock distribution network with adaptive delay compensation[10] is superior to the conventional design that has fixed delays, even if the adaptive design may exhibit higher initial skew. Additionally, adaptive delay compensation will adjust to skew caused by process variations and will overcome difficulties related to the construction of a pre-silicon design model and design flow that accurately and exhaustively accounts for all process effects (i.e. SOI dynamic switching effects). As an example, the evolution of the processor clock distribution designs in [17–19] eventually incorporated adaptive clock compensation in the latest implementation [20]. By allowing for slightly high initial skew, the physical design resource needs (e.g. metal tracks, floorplan

[10] The terms "adaptive delay compensation," "active deskew," and "skew rebalancing" are used interchangeably in this discussion.

restrictions, pre-silicon analysis, etc.) for a clock network with adaptive compensation are expected to be lower. In the following sections, we discuss adaptive global and local clock compensation architectures.

2.5.2 Global Clock Compensation Architecture

Figure 2.37 shows a dual-zone adaptive deskew clock distribution architecture implemented in a 450MHz microprocessor [32]. The global clock distribution of this processor is partitioned into two planes supported by the "left" and the "right" clock spines. Binary clock trees embedded in the spines deliver the global clock from the clock generator to the spine drivers. Two digital delay lines with 17 delay adjustment steps and approximately 12ps average delay step size reside near the root of their respective clock spines. A phase detector (PD in Fig.2.37) strategically placed in the microprocessor core compares the phase difference of the clocks between the left and the right planes. A digital control logic unit (Control FSM and Delay SR) interprets the phase detector output and makes adjustments to the delay lines. The construction of the delay line is shown in Fig.2.38. It consists of two cascaded inverters with switchable load capacitors at each stage. The delay shift register (Delay SR) generates a thermometer-coded delay adjustment and sequentially enables the load capacitors between the two stages. A 60ps skew was reported with adaptive deskewing disabled and 15ps with the mechanism engaged.

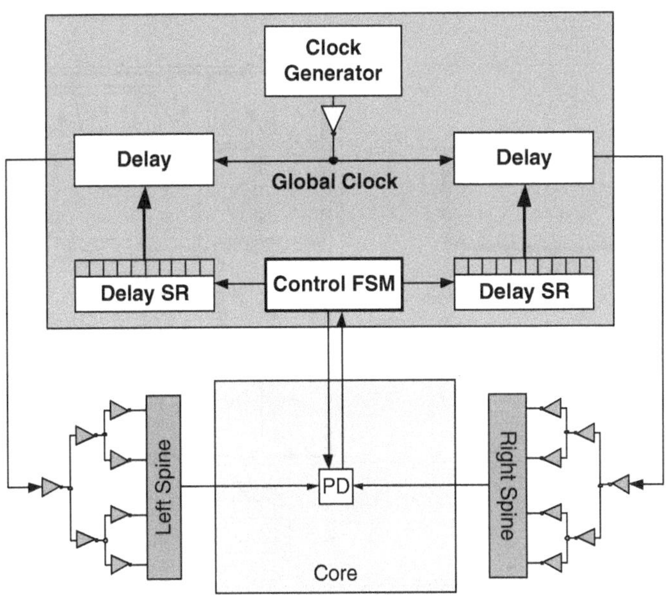

Fig. 2.37. Dual-zone deskew architecture. Reproduced with permission from [32], ©1998 IEEE

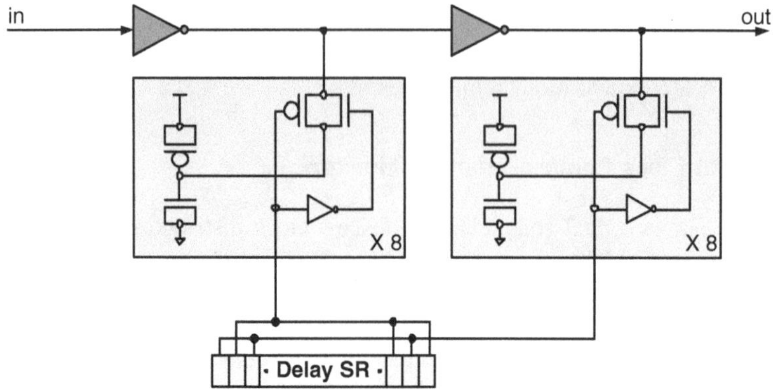

Fig. 2.38. Deskew delay line structure. Reproduced with permission from [32], ©1998 IEEE

The previous implementation [32] embodied only two independent deskew regions. Figure 2.39 shows a deskew architecture in the Itanium® processor[11] that supports 30 independent deskew regions [36, 56]. An H-tree distributes the global clock from the central PLL to eight clusters of deskew buffers (DSK in Fig. 2.39) serving 30 independent deskew regions. Each DSK cluster may contain up to four independent deskew buffers.

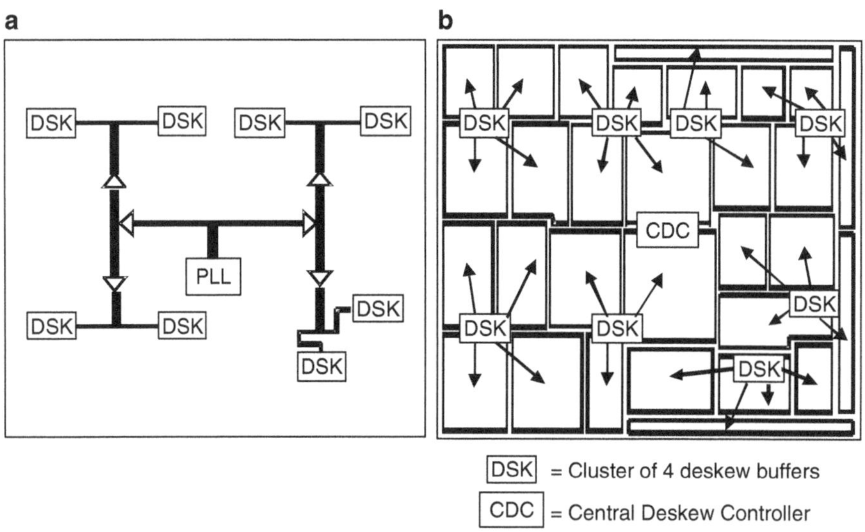

Fig. 2.39. Deskew zones in the itanium® processor. Reproduced with permission from [56], ©2000 IEEE

[11] Other names and brands may be claimed as property of others.

Figure 2.40 shows the detailed clock distribution architecture of [36] that encompasses the H-tree global distribution, the grid structure for the regional distribution, and the local buffers for the local distribution. In this design, the grid drivers (RCD in Fig. 2.40) are located at the top and bottom of grid. In addition to the global clock (main clock in Fig. 2.40), a dedicated and tightly matched reference clock is routed from the central clock generator to the eight DSK clusters (reference clock in Fig. 2.40). The purpose of the reference clock is to act as the reference to deskew the global clock.

Figure 2.41 shows the details of the deskew buffer architecture and the delay circuit design. The delay circuit design is similar to the previous design consisting of two inverter stages with switchable capacitor loads using a 20b thermometer code with a tri-state controllable output stage. In this implementation, the total skew is 28ps with deskew turned on. The skew increases by a factor of 4 with the deskew mechanism disabled. Additional details of this active deskew architecture are discussed in Sect. 7.3.2 in the context of addressing variations in the clock network.

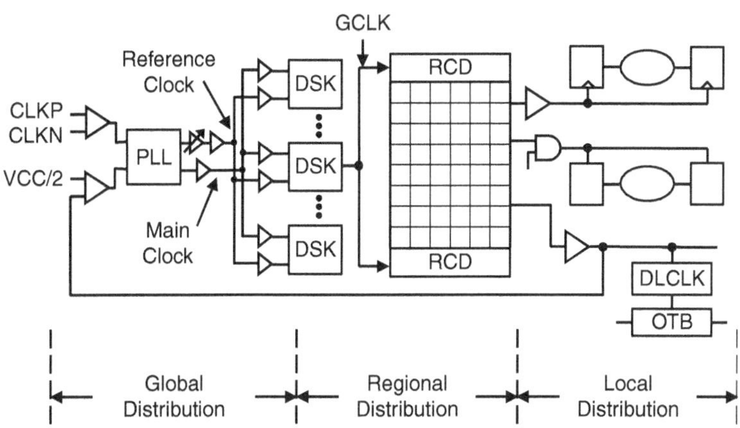

Fig. 2.40. Clocking architecture of the first itanium® processor. Reproduced with permission from [56], ©2000 IEEE

The 180nm Pentium® 4 processor[12] clock distribution described in Sect. 2.4 employs a hierarchical deskew architecture consisting of 47 adjustable clock zones, and 3 levels of deskew hierarchy with 46 phase detectors [49]. The left panel in Fig. 2.42 shows the deskew system architecture whereas the right panel shows the details of the deskew hierarchy. Phase detectors are placed in between the deskew hierarchies. For example, phase detectors between the single primary reference and the secondary references will detect the delay differences between the reference clock zone and the secondary zones. The clock latencies to the secondary zones are adjusted via corresponding deskew buffers (DB2–DB47). Application of hierarchical

[12] Other names and brands may be claimed as property of others.

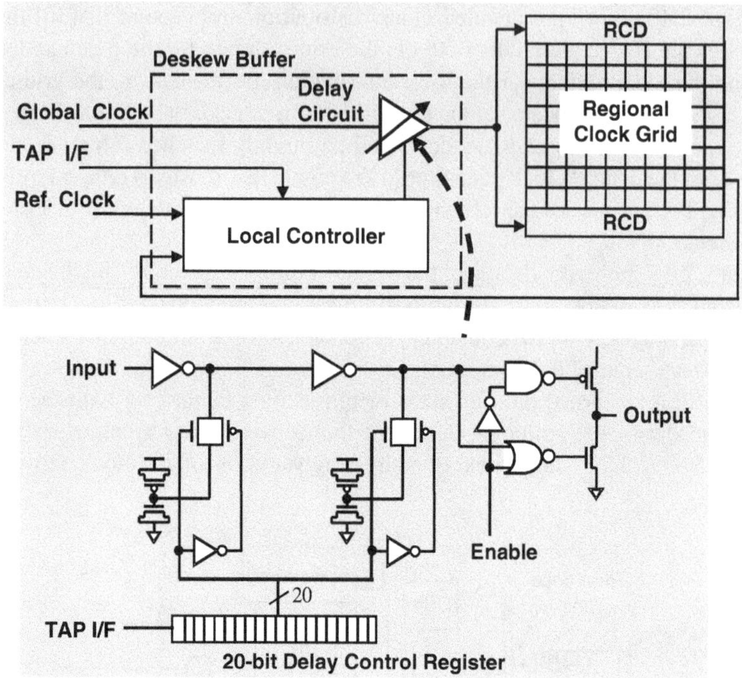

Fig. 2.41. Deskew controller and deskew buffer design. Reproduced with permission from [56], ©2000 IEEE

deskew eliminated the need for a tightly matched reference clock as in [56]. The deskew hierarchy depth and the phase detector quantization error should be kept low to ensure that there is no excessive accumulation of residual mismatches between the primary reference and the final clock zones. In this implementation, the depth of the deskew hierarchy is 3. Figure 2.43 shows the skew profile before and after

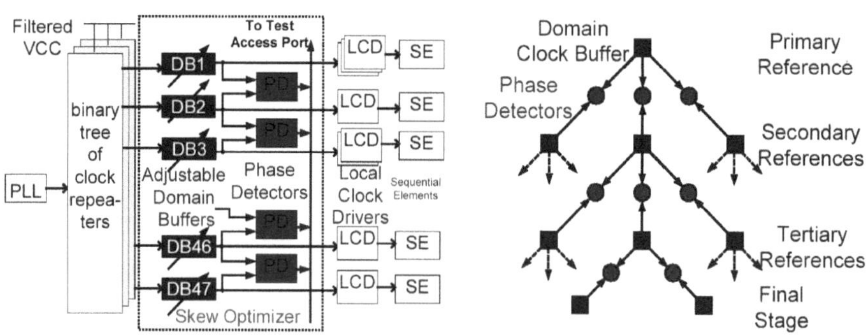

Fig. 2.42. Pentium® 4 processor deskew architecture. Reproduced with permission from [49], ©2001 IEEE

deskew. With deskew enabled, the in-die skew was adjusted to ±8ps and limited by the resolution of the adjustable delay elements. The preadjustment skew was at about 64ps. A hierarchical deskew architecture with deeper hierarchy was used in the 90nm dual-core Itanium® processor [43][13] and is shown in Fig. 2.44.

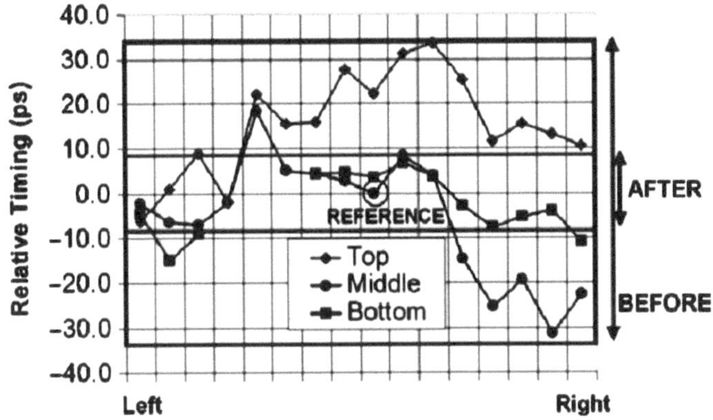

Fig. 2.43. Before and after skew profile of the Pentium® 4 processor. Reproduced with permission from [49], ©2001 IEEE

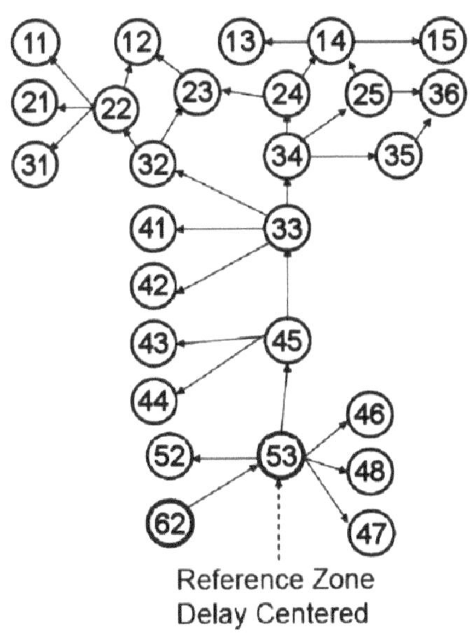

Fig. 2.44. Hierarchical deskew architecture of a dual-core processor. Reproduced with permission from [43], ©2005 IEEE

[13] Other names and brands may be claimed as property of others.

The hierarchical deskew scheme can be generalized to cover any neighboring clock zones in an H-tree style implementation (Fig. 2.45) or a mesh style implementation (Fig. 2.46) [57]. In the H-tree topology shown in Fig. 2.45, deskew is accomplished hierarchically. For example, the level-1 phase detector placed at the boundary between zone D and zone H will determine which of these two clocks is early and the control logic associated with this phase detector will adjust the clock delays to reduce the skew to within some predetermined guard-band. Once the zones associated with the level-1 phase detectors are brought in phase, deskew will continue in zones associated with the level-2 phase detectors. The procedure will continue until it reaches the level-4 phase detector. There are drawbacks associated with the H-tree deskew topology. First, a deskew buffer must be inserted in all the branches associated with a phase detector leading to longer clock distribution delay and higher jitter. Second, there could be large accumulated guard-bands in zones that are physically adjacent but hierarchically far apart (example: zone B and zone C).

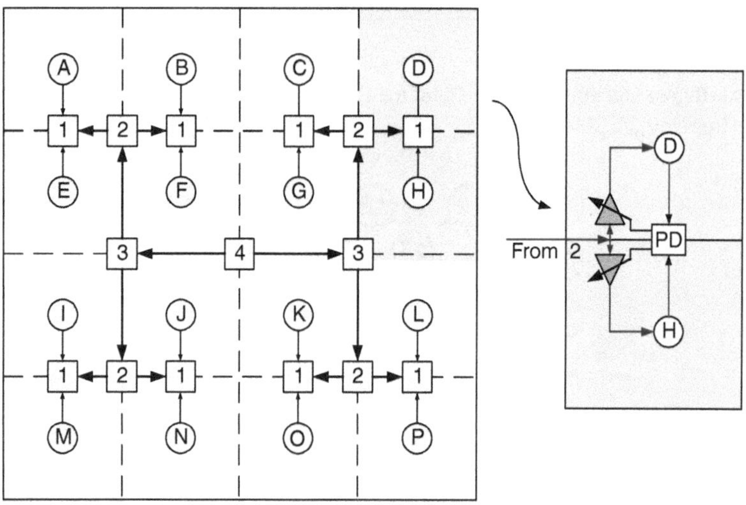

Fig. 2.45. H-tree deskew topology (PD = Phase detector) [57]

The mesh deskew topology addresses the drawbacks of the H-tree deskew topology. In the mesh deskew topology, deskew will be performed on all adjacent zones via averaging. For example, region C in Fig. 2.46 is deskewed with respect to zones B, D, & G simultaneously by averaging their respective phase detector outputs. To ensure stability, a mesh deskew algorithm has been proposed that takes into account potential conflicts and the impact due to guard-band accumulations [57].

Table 2.5.2 summarizes clock distribution characteristics of various commercial processors. The prevalence of adaptive skew compensation techniques is evident.

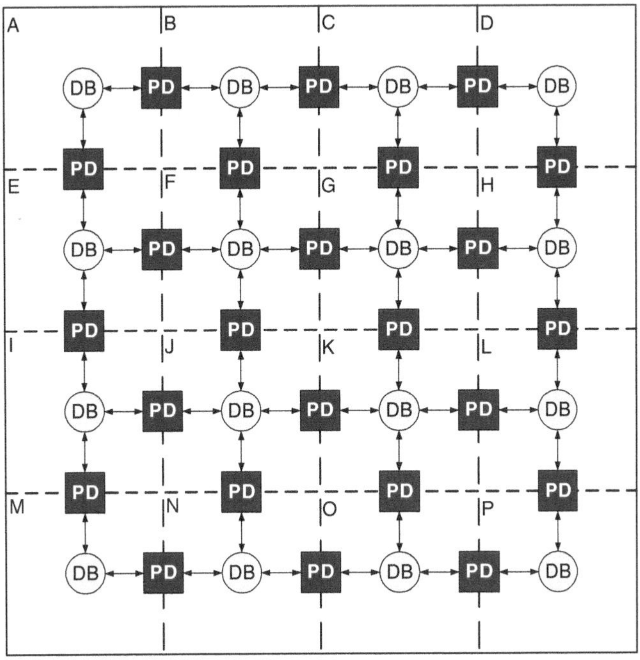

Fig. 2.46. Mesh deskew topology (PD = Phase detector, DB = Deskew buffer) [57]

Table 2.3. Clock distribution characteristics of commercial processors

Name	Ref.	Frequency (MHz)	skew (ps)	Technology (nm)	Clock Dist. style	Deskew
Merom	[30]	3,000	18	65	Tree/Grid	Yes
Power6®	[20]	5,000	8	65	Sym. H-Tree/Grid	Yes
Quad-Core Opteron™	[16]	2,800	12	65	Tree/Grid	
Xeon® processor	[45]	3,400	11	65	Tree/Grid	Yes
Itanium® 2 processor	[43]	>2,000	10	90	Asymmetric tree	Yes
Power5®	[19]	>1,500	27	130	Sym. H-Tree/Grid	No
Pentium® 4 processor	[29]	3,600	7	90	Recombinant tile	Yes
Itanium® 2 processor	[42]	1,500	24	130	Asymmetric tree	Yes
Power4®	[17]	>1,000	25	180	Tree/Grid	No
Itanium® 2 processor	[35]	1,000	52	180	Asymmetric tree	No
Pentium® 4 processor	[8]	>2,000	16	180	Spine/Grid	Yes
Itanium® processor	[36]	800	28	180	H-Tree/Grid	Yes

Note: Other names and brands may be claimed as property of others

2.5.3 Local Clock Compensation Architecture

The concept of global clock compensation outlined in the last section and aimed at optimizing clock skews (intentional or unintentional) among global clock zones can be easily extended to the local level to address path specific clock timings. Specifically, locating critical paths (LCP) buffers with adjustable delays can replace regular local buffers with fixed delays [7, 30, 45]. Figure 2.47 shows an implementation example [30]. The local clock buffers are replaced by LCP buffers to create the LCP domains. They are controlled by the LCP control chain and the chain setting is usually determined post-silicon. Since each LCP buffer is targeted at a limited fan out, the number of LCP domains could be too many to manage. To overcome this difficulty, a processor implementation will cluster the LCP domains resulting in hundreds of LCP zones. The LCP technique is highly effective in resolving clock timing marginalities post-silicon. For example, a MAX timing path between "A" and "B" in Fig. 2.47 can be compensated by intentionally delaying the LCP buffer at the receiver (B). On the contrary, a MIN path marginality between "A" and "B" can be resolved by delaying the LCP buffer at domain A. The highly distributed nature of the LCP methodology permits fine-grain post-silicon clock timing adjustments in contemporary processor implementations.

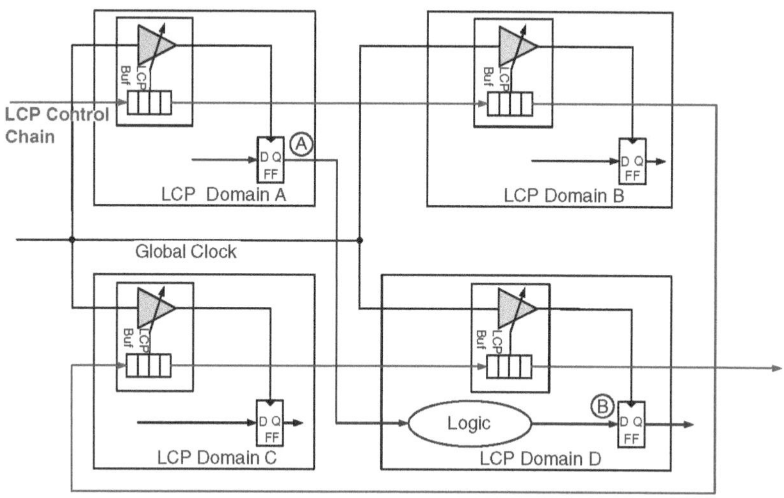

Fig. 2.47. Local Clock Compensation [30]

2.6 Elements of Clock Distribution Circuits

2.6.1 Clock Duty Cycle

Maintaining the clock duty cycle as close to 50% as possible is important and particularly critical for high performance processors. Specifically, high performance processors will have many phase paths in which any duty cycle distortion will unnecessarily penalize the design. Let us assume a clock period of T_{per} with a duty cycle error of Q%. The corresponding high and low phase times become:

$$T_{\text{phH}} = T_{\text{per}} \times \frac{(50+Q)}{100}, \tag{2.29}$$

$$T_{\text{phL}} = T_{\text{per}} \times \frac{(50-Q)}{100}. \tag{2.30}$$

Let $F_{50\%}$ be the maximum operating frequency achievable with a 50% duty cycle. Then, the corresponding high and low phase paths will correspond to effective frequencies F_{maxH} and F_{maxL}:

$$F_{\max\text{H}} = F_{50\%} \times \frac{100}{(100+2Q)}, \tag{2.31}$$

$$F_{\max\text{L}} = F_{50\%} \times \frac{100}{(100-2Q)}. \tag{2.32}$$

Figure 2.48 shows the effective $F_{\max}$ increase necessary for a phase path to the meet timing due to duty cycle distortion. The effective maximum frequency increase is approximately twice that of the percentage duty cycle offset from 50%.

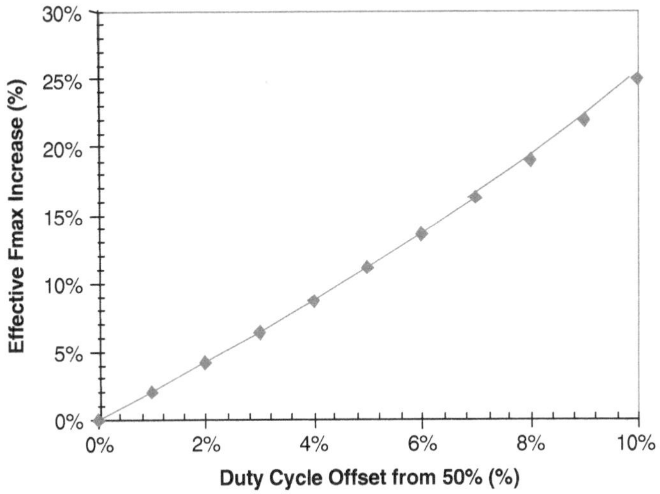

Fig. 2.48. $F_{\max}$ shift caused by duty cycle distortion in a phase-path dominated design

Clock distribution induced duty cycle error is mainly attributed to asymmetry in the clock distribution repeaters. Figure 2.49 compares a buffer-based (two inverters) clock distribution design vs. an inverter-based clock distribution. In a buffer-based design, an incoming clock edge undergoes asymmetric rise and fall edges: Two fall edges experience gate loading only whereas two rise edges experience interconnect loading. In contrast, in the inverter-based implementation, both positive and negative edges experience similar loads. By having loading symmetry between rise and fall edges, an inverter-based clock distribution network is more robust in maintaining duty cycle fidelity.

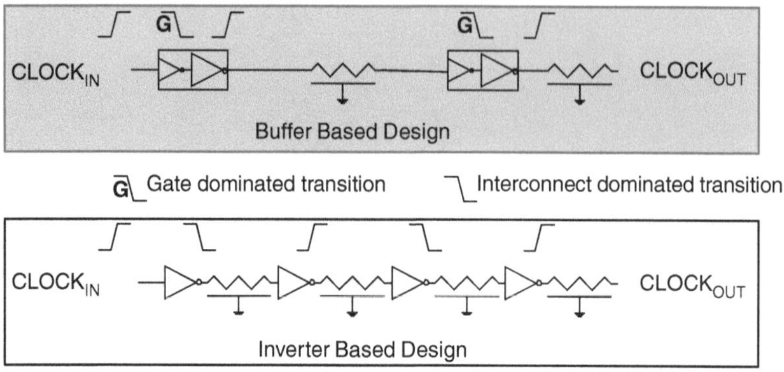

Fig. 2.49. Duty cycle distortion due to asymmetric edge propagation between a buffer-based clock distribution and an inverter-based clock distribution

In applications such as double data rate circuits where attaining a near perfect 50% duty cycle is critical, a differential clock distribution is the preferred solution. Sending the clock differentially improves noise immunity and duty cycle fidelity. Moreover, cross coupled stages can be added to further reduce duty cycle error by introducing constraints between true and complement clock signals. Finally, active duty cycle correction can be added to dynamically adjust the duty cycle across process, voltage, and temperature variations. Figure 2.50 shows an active duty cycle correction system with the corrector and detector circuits shown in Fig.2.51. In a differential clocking environment, the duty cycle can be easily determined by sensing the clock common mode and making the appropriate adjustments.

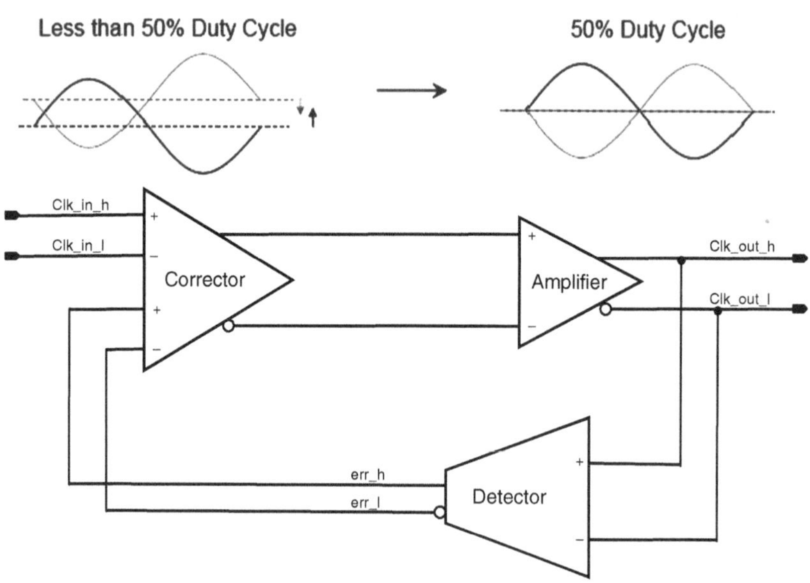

Fig. 2.50. Duty cycle corrector. Reproduced with permission from [13], ©2008 IEEE

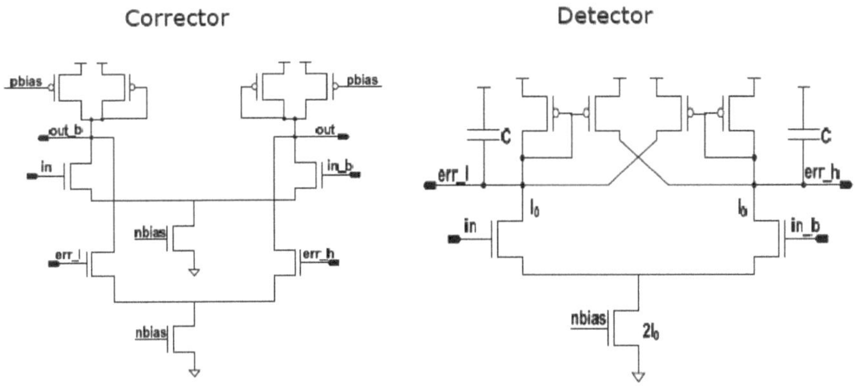

Fig. 2.51. Duty cycle corrector circuits. Reproduced with permission from [13], ©2008 IEEE

2.6.2 Power Supply

In (2.9), it was noted that the clock distribution delay variation is proportional to the distribution latency of which the delay sensitivity to power supply variation is a key factor. Power supply droop due to di/dt and power supply noise due to switching are two significant sources.

Power supply decoupling capacitors are used extensively to reduce the power supply droop and noise effects. Wong et al. [58] studied improving the processor timing margin by enhancing its immunity to power supply noise by compensating the clock delay against the data delay. In addition to on-die decoupling capacitors, on-die power supply filters can reduce the impact of power supply noise. Figure 2.52 shows two possible implementations [16, 49]. The implementation in [49] demonstrated a 5× reduction in power supply noise with filtering (Fig.2.53). It should be noted that the circuits shown in Fig.2.52, if not designed with care, can experience excessive DC power supply drop leading to delay degradations.

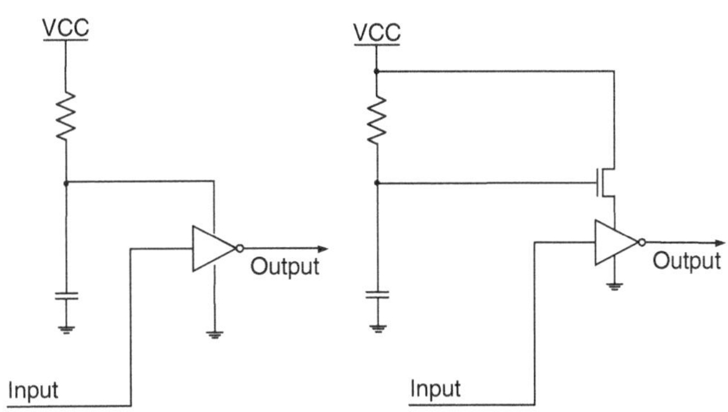

Fig. 2.52. Clock buffer design with power-supply filters. Reproduced with permission in a form similar to that in [49] and [16], ©2001, 2007 IEEE

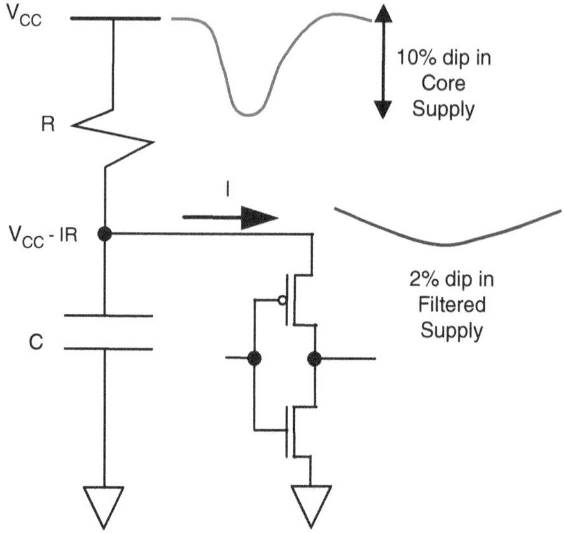

Fig. 2.53. On-die clock tree filter circuit. Reproduced with permission from [49], ©2001 IEEE

The reader is encouraged to review Sect.6.8 for an analytical model of supply noise induced jitter as a function of noise frequency and amplitude and clock network latency.

2.7 Clock DFX Techniques

Clock DFX refers to the design-for-test or design-for-manufacturing features. DFX capabilities are critical for the design of the clock distribution network. For example, methods to measure on-die clock skew and jitter are needed for post-silicon validation of the design. Additionally, the ability to manipulate the clock edges at some specific clock cycle is a critical timing debug feature.

2.7.1 Optical Probing

Due to the increased number of metal layers and flip-chip packaging, optical probing from the back-side of the die is a standard technique to probe die internal nodes [59]. This technique is quite suitable for on-die clock measurement due to the periodic nature of the clock. The back-side optical probing technique monitors the infrared photons that are emitted by switching transistors. Figures 2.54 and 2.55 illustrate the emission mechanism and measurement methodology [60]. The photon emission intensity is proportional to the transistor switching current and is at a maximum during the switching transient. Therefore, the clock transition edge can be determined by correlating the photon emission peak with a time base. By monitoring the clock

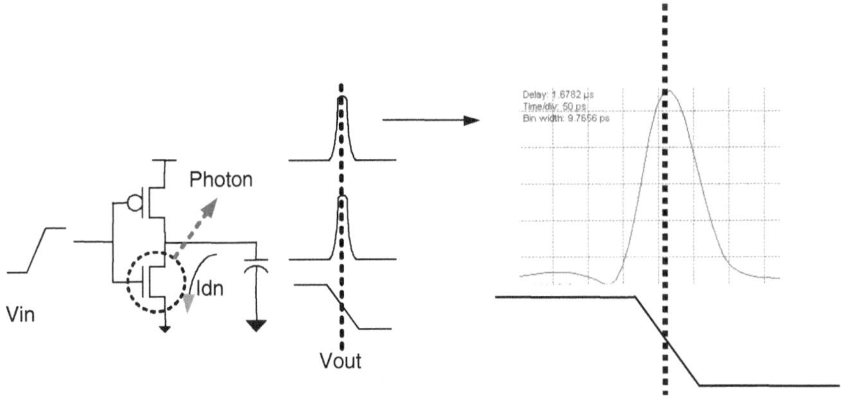

Fig. 2.54. Back-side optical probing technique (a). Reproduced with permission from [60], ©2004 IEEE

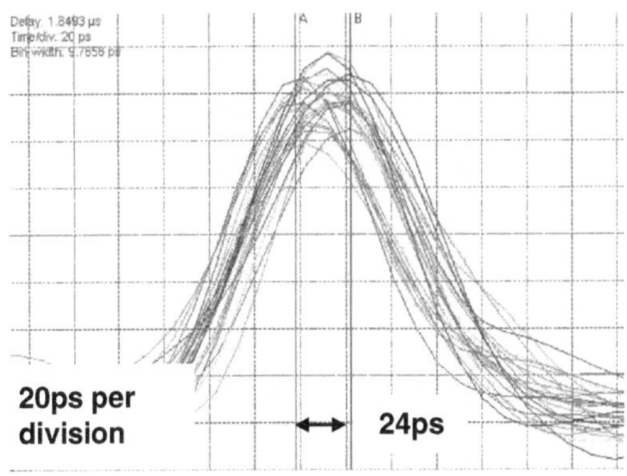

Fig. 2.55. Back-side optical probing technique (b). Reproduced with permission from [60]. ©2004 IEEE

drivers in various die locations, this technique provides static clock skew measurement capability. Limitations of this technique are instrumentation complexity, timing resolution and its inability to measure dynamic clock uncertainties (jitter) due to the intrinsic averaging effect of the measurement.

2.7.2 On-Die Measurement

On-die skew and jitter measurement schemes are more suitable for high volume data collection when compared to the optical back-side probing technique outlined in the last section. Figure 2.56 shows a circuit capable of measuring both skew and jitter [61]. The circuit contains an inverter based delay line (inverters A_1 to A_{K+1}) and a chain of flip-flops configured to sample the taps along the delay line. The inverters are designed to be nominally identical and with low fan out. The length of the delay line and the sampling chain K is designed to be sufficiently long so that the total delay will fully contain at least one clock period at the lowest frequency of interest. If τ is the nominal unit delay for the inverter and $T_{\text{per\_max}}$ is the maximum clock period, then $K\tau > T_{\text{per\_max}}$. In [61], K is set to 128 and the inverter has a nominal delay of 8ps. When Ck_A is selected as the input to the delay line (via the MUX in "$\text{Ck}_{\text{Sample1}}$" in Fig. 2.56), the resulting pattern from the sampling flip-flops is shown in Fig. 2.57 with the clock transition edges denoted by two consecutive 0s or two consecutive 1s. If the number of inverter stages between the clock transition edges is N_x, the phase time T_{ph} is bounded by:

$$N_x\tau \le T_{\text{ph}} \le (N_x+1)\tau. \tag{2.33}$$

If there is no jitter on the incoming clock, the stage count N_x and the pattern will be the same in every cycle. With nonzero clock jitter, the sample pattern will not be identical in every cycle and the sampled locations of the clock edges will vary.

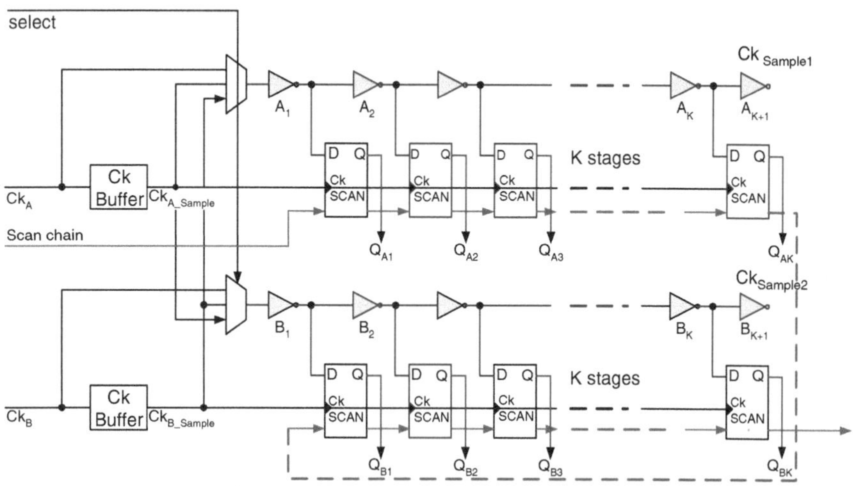

Fig. 2.56. Skew and jitter measurement circuit. Reproduced with permission from [61], ©2004 IEEE

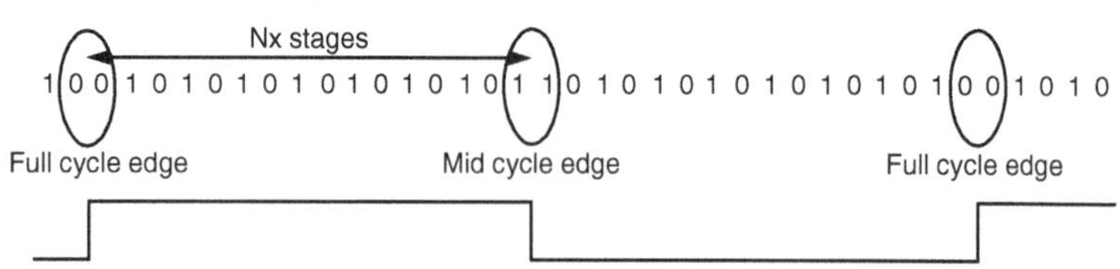

Fig. 2.57. Sampled delay pattern of the skew and jitter measurement circuit

To measure the clock skew between clock Ck_A and Ck_B, the coupled configuration shown in Fig. 2.56 is used. Clock Ck_{B_Sample} is multiplexed to the delay chain in circuit "$Ck_{Sample1}$" and Ck_{A_Sample} is multiplexed to the delay chain in circuit "$Ck_{Sample2}$." If there is no skew between Ck_A and Ck_B, the captured patterns from the two circuits will be identical. With finite skew, the difference in the position of the sampled clock edges is the measured skew.

Shortcomings of the aforementioned technique are the limited measurement resolution due to the discrete unit delay of the inverter and the delay variation among the inverters in the delay line [62, 63]. A vernier delay line (VDL) as shown in Fig. 2.58 can be used to improve the measurement resolution. With $\tau_A \neq \tau_B$ (Fig. 2.58), the timing resolution becomes $\delta = |\tau_A - \tau_B|$. Although this scheme can in principle achieve higher resolution, delay variation among the stages will impose a limitation on the results.

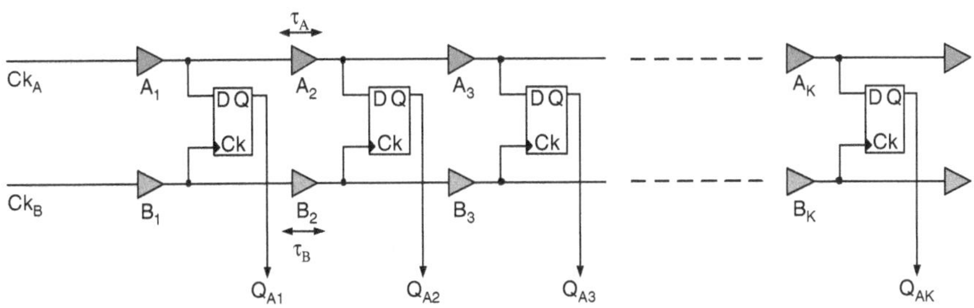

Fig. 2.58. Vernier delay line

Figure 2.59 shows a timing diagram of the Vernier delay line with Ck_A and Ck_B skewed by 3δ where δ is the delay difference between τ_A and τ_B. The sampled

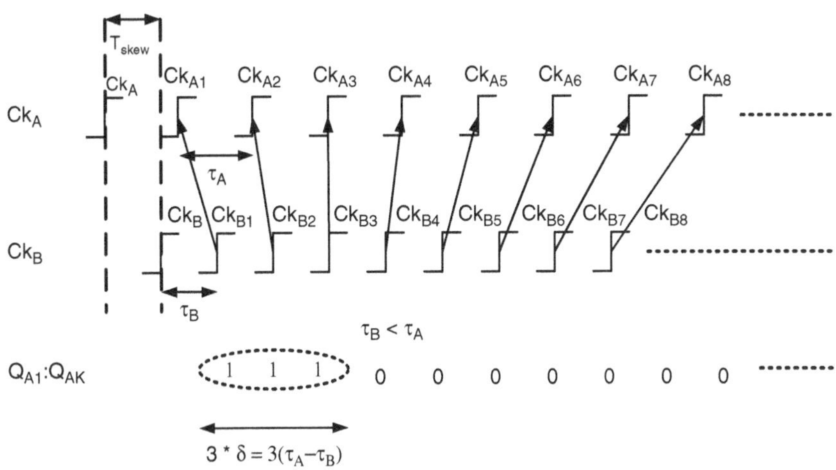

Fig. 2.59. Vernier delay line timing example

pattern Q_{A1} to Q_{AK} will show three consecutive ones representing the skew between the two clocks. In general, if the skew between the clock inputs to the VDL is T_{skew} and the stage delays are τ_A and τ_B respectively, the number of ones (N_y) in the observed pattern will be:

$$N_y = \left\lfloor \frac{T_{skew}}{\delta} \right\rfloor, \tag{2.34}$$

$$\delta = |\tau_A - \tau_B| \tag{2.35}$$

The limitation of the VDL technique is attributed to the stage-to-stage delay variations making measurement resolutions in the sub-picosecond range difficult to achieve without incorporating additional mismatch compensation schemes.

2.7.3 Locating Critical Path

In addition to being an important post-silicon (local) clock compensation technique, the locating critical path (LCP) scheme is also a valuable post-silicon timing and clock debug tool [7, 30]. By intentionally skewing the clock arrival times between local clock zones, physical locations of marginal timing paths can be identified for further analysis. When coupled with the on-die clock shrink method (Sect. 2.7.4), the LCP technique has been shown to be a critical post-silicon clock and timing debug tool.

2.7.4 On-Die-Clock Shrink

In conjunction with the LCP technique, the on-die clock shrink (ODCS) method [56] is an effective clock edge manipulation technique that has been used extensively in post-silicon timing debug. The main concept behind ODCS is the capability to manipulate specific clock edges for phase or cycle expansion or contraction. Figure 2.60

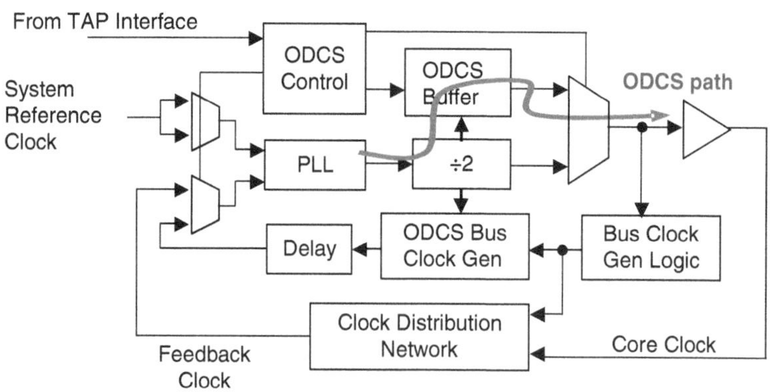

Fig. 2.60. On-die-clock shrink architecture. Reproduced with permission from [56] ©2000 IEEE

shows the clock generation architecture capable of supporting this feature. An ODCS buffer is inserted in series with the core clock generation path. The ODCS buffer is designed to be able to stretch or shrink the phase or cycle time.

Figure 2.61 shows an example of the ODCS operating principle. The input to the ODCS buffer is assumed to have equal high and low phase times. The initial rise and fall default setting are equal (10 arbitrary units in Fig. 2.61). With this setting, the default ODCS buffer output will also have equal high and low phase times. The settings in cycles $N+1, N+2$, and $N+3$ shown in Fig. 2.61 will shrink the low phase at cycle $N+1$ and then restore the clock waveform to the default.[14] Figure 2.62 summarizes the overall capabilities of the ODCS technique. Implementation of ODCS requires the incorporation of ODCS rise and fall registers and a state machine to cycle through these registers in a deterministic manner.

In an actual experiment to locate the source of the timing failure during timing debug, the default operating frequency of the device under test is first relaxed and the ODCS phase or period shrinkage is applied systematically across a specific range of tester cycles to locate the actual cycle of failure and the failure timing margin. In this manner, the failure pattern can be correlated to a specific test cycle to identify root cause.

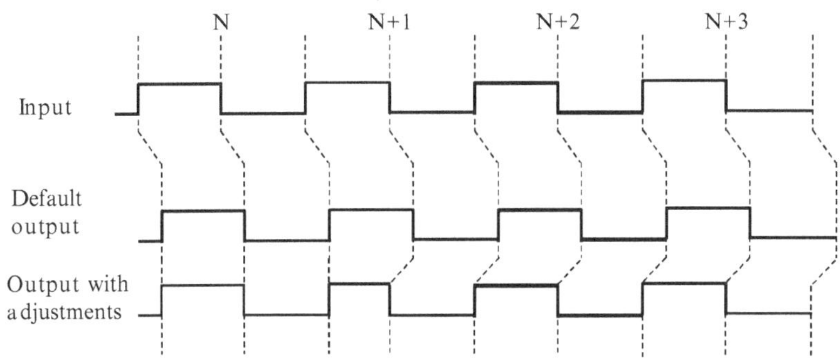

State	Rise Setting (arbitrary time unit)	Fall Setting (arbitrary time unit)
Current	10U	10U
N+1	10U	8U
N+2	8U	8U
N+3	8U	8U

Fig. 2.61. ODCS clock waveform. Reproduced with permission from [56] ©2000 IEEE

[14] In this implementation, the $N+1$ "Rise" setting actually affects the rising edge at cycle $N+2$ since this setting is applied after the fall edge of cycle $N+1$. The $N+1$ "Fall" setting affects the falling edge at cycle $N+1$ because it is also applied after the rising edge of cycle $N+1$.

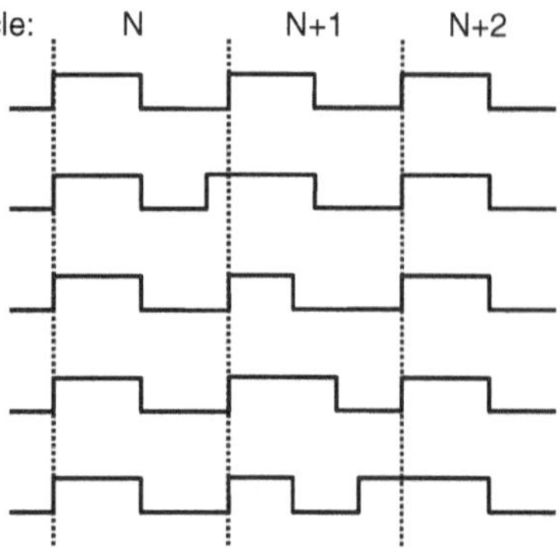

Fig. 2.62. ODCS capabilities. Reproduced with permission from [56] ©2000 IEEE

2.8 Multiclock Domain Distributions

As digital designs move towards multicores and systems-on-chip (SOC) architectures, multiple clock domains will become a prevalent design style and the clock distribution schemes will need to be enhanced to fulfill this need. A multiclock domain design typically embodies multiple islands operating synchronously and served by independent clocks within the domain and with dedicated interfaces to manage the inter-domain communications. In a multicore processor, each of the cores can be a separate clock domain. Similarly, different functional units in an SoC can be on separate clock domains.

In multicore processors and SoC designs, it is advantageous to implement multiple clock domains because this scheme provides functional flexibility for each of the domains to operate at the optimal frequency and to minimize the complexity and power associated with distributing a low skew clock to the entire die. The multidomain clock distribution architectures for multicore processors and SoCs belong to a class of designs called globally asynchronous and locally synchronous (GALS) [64]. Table 2.4 summarizes synchronization categories within the GALS class.

Table 2.4. Clock synchronization categories

Type	Characteristics of distribution
Synchronous	Single distribution point-of-divergence (POD) with known static delay offsets among all the branches and single operating frequency. Encompasses all of the clock distribution styles described in Sects. 2.3 and 2.4
Mesochronous	Single distribution POD but with nonconstant delay offset among the branches
Plesiochronous	Multiple distribution PODs but with nominally identical frequency among all the domains
Heterochronous	Multiple distribution PODs with nominally different operating frequencies among the domains

Figure 2.63 is a generic illustration of the GALS design style. Multiple clock domains are embedded in a single silicon die. The chip may receive multiple copies of the system clock and multiple PLLs are used to generate the clocks for each of the synchronous units. Clocks Ck1,Ck2,...,Ck4 may have different phases or frequencies. The clock distribution within Each synchronous unit will rely on any of the topologies described in Sect. 2.3 to achieve low skew and fully synchronous operation.

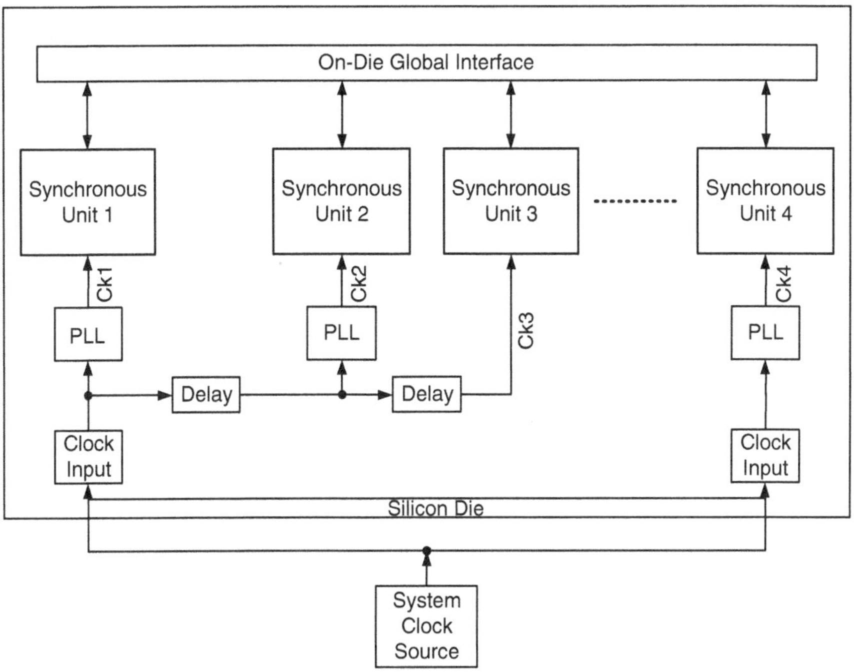

Fig. 2.63. Globally synchronous and locally synchronous architecture. Reproduced with permission from [64], ©2007 IEEE

Since each of the synchronous units can operate at different frequency and phase from the rest, the on-die global interfaces will be responsible for the data transfer among the domains. Many modern multicore processors and SoCs only adopt the loosely synchronous styles of Table 2.4 to avoid the significant complexity associated with truly asynchronous design.

2.8.1 Multicore Processor Clock Distribution

A plesiochronous clock distribution involves multiple distribution PODs but with a nominally identical operating frequency. The 65 nm dual core processor described in [45] is an example. Figure 2.64 illustrates this scheme which consists of two domains for the two cores and the uncore and I/O domain with the interface operating at the same frequency. The system clock input to the three clock generators (PLLs) is routed in the package. There are three independent distribution PODs for the cores and the uncore. A recombinant style distribution for low skew is used in the core whereas the uncore employs a hybrid spine-grid structure for lower power and simpler implementation. Data communication between the cores and the uncore is accomplished via the pipelined deskew logic (PDSL) interfaces operating at the same frequency. Introducing independent clock domains for the cores permitted a modular design.

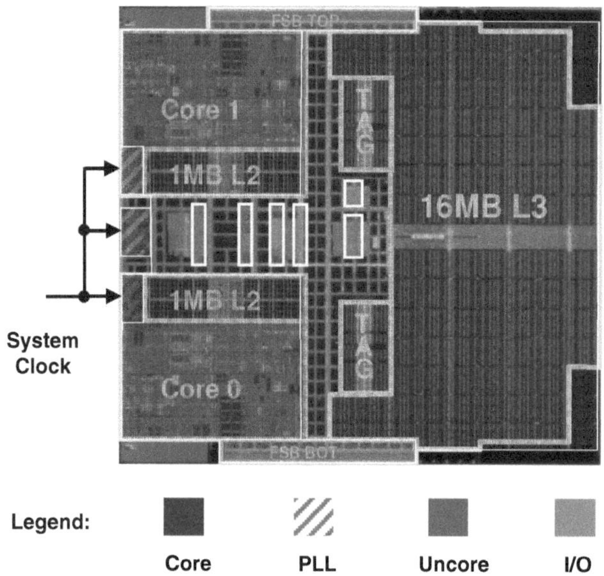

Fig. 2.64. Multidomain clocking in a dual-core processor. Reproduced with permission from [45], ©2006 IEEE

A mesochronous clock distribution has a single distribution POD while exhibiting nonconstant delays to the branches. The 80-tile processor design in [65, 66] is one example of a mesochronous clock distribution. This processor has a single PLL and therefore a single distribution POD (Fig.2.65). Distributing the global clock to the respective tiles is achieved using chains of clock buffers embedded within each tile. The daisy-chaining of the global clock buffers suggests nonequal clock latencies to each processor tile. The systematic clock skews across the processor tiles help spread the clock switching power over an entire cycle. Interprocessor-tile communication is achieved with a skew insensitive FIFO interface [66]. Within each processor tile, a balanced H-tree is responsible for the distribution and all circuits within each processor tile operate synchronously. Figure 2.65 (left) shows details of the distribution scheme. Figure 2.65 (right) shows the relative arrival times of the clocks at the processor tiles. In a processor design having a large number of processor cores operating at the same frequency, this clock distribution style has the potential of reducing the impact of simultaneous switching on the on-chip power delivery network.

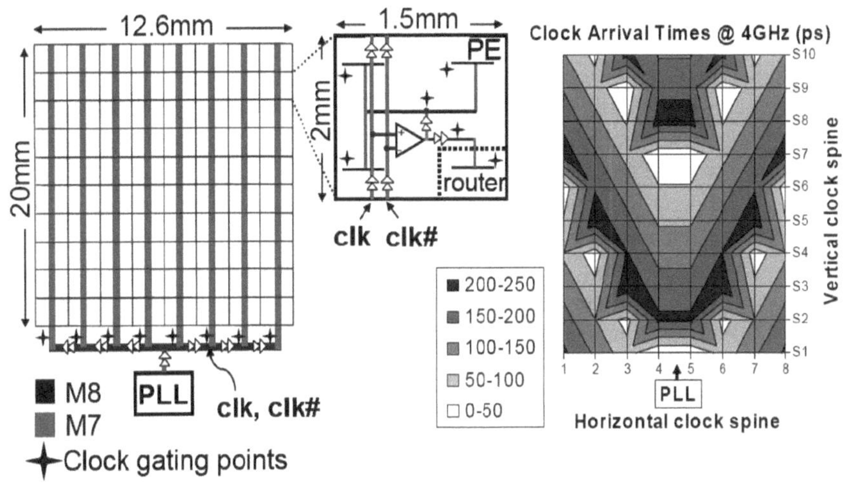

Fig. 2.65. Clock distribution of an 80-tile processor design. Reproduced with permission from [66], ©2008 IEEE

A heterochronous clock distribution has multiple distribution PODs and can support nominally different operating frequencies across the domains [12, 13, 16, 67]. The 90nm 2-core Itanium® processor [12] and the 65nm quad-core Itanium® processor [13] have adaptive core frequency switching implemented with multiple PLLs. Low failure probability FIFOs are placed at the interface of the corresponding clock domains. Similarly, [16] and [67] have independent PLLs for the cores, uncore, and the *I/O* that are capable of operating at different frequencies. Clock domain crossing is accomplished with low-latency FIFO schemes.

Despite the potential design benefits of the loosely synchronous methods described above, the requirement of a deterministic domain crossing will impose higher design complexity at the interfaces. FIFO-like structures are used at the cost of additional transfer latency in order to compensate for the higher crossdomain clock uncertainties. The PDSL interface in [45] incurs two additional cycles of latency when compared to a single-cycle transfer in a fully synchronous design. Similarly, [66] incorporates a 4-deep FIFO to ensure robust crossdomain transfer. Better partitioning that reduces the crossdomain traffic will reduce the performance penalty due to the increased synchronization latency.

2.9 Future Directions

Alternative clock distribution methods have been proposed in the literature. Among them are optical clock distribution [68, 69], package level clock distribution [70], clock network with distributed PLLs [71] and resonant clocking [72, 73].

The main advantage of an optical clock distribution scheme vs. a conventional one is the potential for higher performance and short-term jitter reduction. This is due to the ability of an optical distribution network to have fewer repeaters and be first-order immune to on-chip power supply noise. A significant disadvantage is the design complexity and performance limitations (e.g. power & electrical robustness) of the signal conversion from optical to electrical and vice versa. Another limitation is due to the process complexity involved in the formation of the on-die waveguides.

In a package level clock distribution, high speed package routes replace the on-die clock network. The package interconnect will have comparatively much lower *RC* delay. The main limitation is due to the interface between the package network and the on-die receivers. Additional skew and jitter at the on-die receivers (due to buffer design and ESD requirements) and testing complexities in prepackaged parts are the most important implementation impediments.

Application of on-die resonant clocking is a very promising clock distribution alternative due to lower power dissipation and jitter reduction. A significant hurdle of on-die resonant clocking is in the physical limitation of the network and the naturally narrow frequency operating window. Numerous enhancements to resonant clocking have been proposed and some of these techniques can become practical. A resonant clock distribution network was applied to a 90nm processor by adding a new top layer metal for the *LC* tank inductors and capacitors [74]. This design demonstrated basic functionality and power savings at frequencies above 3GHz. A detailed discussion of resonant clocking techniques is presented in Chap. 4.

All of the alternatives above try to address limitations of electrical clock interconnect such as power and noise. With the strong emergence of multidomain clock distribution in multicore processors and SOCs, constraints due to the very large on-die electrical interconnect are significantly alleviated. Since the industry is moving away from a single large distribution network with long latency and large number of clock buffers, on-die electrical clock distribution will continue to be the predominant clocking technology in the near future.

2.10 Conclusion

In this chapter, the requirements for synchronous clock distributions have been described. Numerous clock distribution topologies have been described with spine-grid and tree-grid topologies emerging as the dominant ones. On-die deskew has became common practice in order to contain skew without excessive design complexity and use of interconnect resources. Various clock debug features have been summarized that encompass circuits for skew and jitter measurement, design for test and critical path location. Additionally, the chapter has also outlined the basics of multidomain clock distribution topologies that are becoming the industry norm. Finally, the chapter concludes by mentioning three alternative clock distribution techniques that may demonstrate practical design benefits in the future.

References

[1] J. Schutz, "A CMOS 100MHz microprocessor," in *Digest of Technical Papers IEEE International Solid-State Circuits Conference (ISSCC 1991)*, 1991, pp. 90–91, 294.

[2] J. Schutz, "A 3.3V 0.6μm BiCMOS superscalar microprocessor," in *Digest of Technical Papers IEEE International Solid-State Circuits Conference (ISSCC 1994)*, 1994, pp. 202–203.

[3] R. Colwell and R. Steck, "A 0.6μm BiCMOS processor with dynamic execution," in *Digest of Technical Papers IEEE International Solid-State Circuits Conference (ISSCC 1995)*, 1995, pp. 176–177, 361.

[4] M. Choudhury and J. Miller, "A 300 MHz CMOS microprocessor with multimedia technology," in *Digest of Technical Papers IEEE International Solid-State Circuits Conference (ISSCC 1997)*, 1997, pp. 170–171, 450.

[5] J. Schutz and R. Wallace, "A 450MHz IA32 P6 family microprocessor," in *Digest of Technical Papers IEEE International Solid-State Circuits Conference (ISSCC 1998)*, 1998, pp. 236–237.

[6] P. Green, "A 1GHz IA-32 architecture microprocessor implemented on 0.18μ technology with aluminum interconnect," in *Digest of Technical Papers IEEE International Solid-State Circuits Conference (ISSCC 2000)*, 2000, pp. 98–99, 449.

[7] J. Schutz and C. Webb, "A scalable x86 CPU design for 90nm process," in *Digest of Technical Papers IEEE International Solid-State Circuits Conference (ISSCC 2004)*, 2004, pp. 62–63, 513.

[8] N. Kurd, J. Barkatullah, R. Dizon, T. Fletcher, and P. Madland, "Multi-GHz clocking scheme for Intel® Pentium® 4 microprocessor," in *Digest of Technical Papers IEEE International Solid-State Circuits Conference (ISSCC 2001)*, 2001, pp. 404–405.

[9] G. Singer and S. Rusu, "The first IA-64 microprocessor: a design for highly-parallel execution," in *Digest of Technical Papers IEEE International Solid-State Circuits Conference (ISSCC 2000)*, 2000, pp. 422–423.

[10] S. Naffziger and G. Hammond, "The implementation of the next-generation 64 b Itanium™ microprocessor," in *Digest of Technical Papers IEEE International Solid-State Circuits Conference (ISSCC 2002)*, vol. 1, 2002, pp. 344–345, 472.

[11] J. Stinson and S. Rusu, "A 1.5GHz third generation Itanium® processor," in *Digest of Technical Papers IEEE International Solid-State Circuits Conference (ISSCC 2003)*, 2003, pp. 252–253, 492.

[12] S. Naffziger, B. Stackhouse, and T. Grutkowski, "The implementation of a 2-core multi-threaded Itanium family processor," in *Digest of Technical*

Papers IEEE International Solid-State Circuits Conference (ISSCC 2005), 2005, pp. 182–183, 592.

[13] B. Stackhouse, B. Cherkauer, M. Gowan, P. Gronowski, and C. Lyles, "A 65nm 2-billion-transistor quad-core Itanium® processor," in *Digest of Technical Papers IEEE International Solid-State Circuits Conference (ISSCC 2008)*, 2008, pp. 92–93, 598.

[14] S. Rusu, S. Tam, H. Muljono, D. Ayers, and J. Chang, "A dual-core multi-threaded Xeon® processor with 16MB L3 cache," in *Digest of Technical Papers IEEE International Solid-State Circuits Conference (ISSCC 2006)*, 2006, pp. 315–324.

[15] M. Golden, S. Arekapudi, G. Dabney, M. Haertel, S. Hale, L. Herlinger, Y. Kim, K. McGrath, V. Palisetti, and M. Singh, "A 2.6GHz dual-core 64bx86 microprocessor with DDR2 memory support," in *Digest of Technical Papers IEEE International Solid-State Circuits Conference (ISSCC 2006)*, 2006, pp. 325–332.

[16] J. Dorsey, S. Searles, M. Ciraula, S. Johnson, N. Bujanos, D. Wu, M. Braganza, S. Meyers, E. Fang, and R. Kumar, "An integrated quad-core Opteron processor," in *Digest of Technical Papers IEEE International Solid-State Circuits Conference (ISSCC 2007)*, 2007, pp. 102–103.

[17] P. Restle, C. Carter, J. Eckhardt, B. Krauter, B. McCredie, K. Jenkins, A. Weger, and A. Mule, "The clock distribution of the Power4 microprocessor," in *Digest of Technical Papers IEEE International Solid-State Circuits Conference (ISSCC 2002)*, vol. 1, 3–7 Feb. 2002, pp. 144–145.

[18] P. Hofstee, N. Aoki, D. Boerstler, P. Coulman, S. Dhong, B. Flachs, N. Kojima, O. Kwon, K. Lee, D. Meltzer, K. Nowka, J. Park, J. Peter, S. Posluszny, M. Shapiro, J. Silberman, O. Takahashi, and B. Weinberger, "A 1GHz single-issue 64b PowerPC processor," in *Digest of Technical Papers IEEE International Solid-State Circuits Conference (ISSCC 2000)*, 2000, pp. 92–93.

[19] J. Clabes, J. Friedrich, M. Sweet, J. Dilullo, S. Chu, D. Plass, J. Dawson, P. Muench, L. Powell, M. Floyd, B. Sinharoy, M. Lee, M. Goulet, J. Wagoner, N. Schwartz, S. Runyon, G. Gorman, P. Restle, R. Kalla, J. McGill, and S. Dodson, "Design and implementation of the POWER5™ microprocessor," in *Digest of Technical Papers IEEE International Solid-State Circuits Conference (ISSCC 2004)*, 2004, pp. 56–57.

[20] J. Friedrich, B. McCredie, N. James, B. Huott, B. Curran, E. Fluhr, G. Mittal, E. Chan, Y. Chan, D. Plass, S. Chu, H. Le, L. Clark, J. Ripley, S. Taylor, J. Dilullo, and M. Lanzerotti, "Design of the POWER6 microprocessor," in *Digest of Technical Papers IEEE International Solid-State Circuits Conference (ISSCC 2007)*, 2007, pp. 96–97.

[21] C. Webb, C. Anderson, L. Sigal, K. Shepard, J. Liptay, J. Warnock, B. Curran, B. Krumm, M. Mayo, P. Camporese, E. Schwarz, M. Farrell, P. Restle, I. Averill, R.M., T. Slegel, W. Houtt, Y. Chan, B. Wile, T. Nguyen, P. Emma, D. Beece, C.-T. Chuang, and C. Price, "A 400-MHz S/390 microprocessor," *IEEE Journal of Solid-State Circuits*, vol. 32, no. 11, pp. 1665–1675, Nov. 1997.

[22] G. Northrop, R. Averill, K. Barkley, S. Carey, Y. Chan, Y. Chan, M. Check, D. Hoffman, W. Huott, B. Krumm, C. Krygowski, J. Liptay, M. Mayo, T. McNamara, T. McPherson, E. Schwarz, L. Siegel, C. Webb, D. Webber, and P. Williams, "600MHz G5 S/390 microprocessor," in *Digest of Technical Papers IEEE International Solid-State Circuits Conference (ISSCC 1999)*, 1999, pp. 88–89.

[23] T. McPherson, R. Averill, D. Balazich, K. Barkley, S. Carey, Y. Chan, Y. Chan, R. Crea, A. Dansky, R. Dwyer, A. Haen, D. Hoffman, A. Jatkowski, M. Mayo, D. Merrill, T. McNamara, G. Northrop, J. Rawlins, L. Sigal, T. Slegel, D. Webber, P. Williams, and F. Yee, "760 MHz G6 S/390 microprocessor exploiting multiple Vt and copper interconnects," in *Digest of Technical Papers IEEE International Solid-State Circuits Conference (ISSCC 2000)*, 2000, pp. 96–97.

[24] D. Dobberpuhl, R. Witek, R. Allmon, R. Anglin, D. Bertucci, S. Britton, L. Chao, R. Conrad, D. Dever, B. Gieseke, S. Hassoun, G. Hoeppner, K. Kuchler, M. Ladd, B. Leary, L. Madden, E. McLellan, D. Meyer, J. Montanaro, D. Priore, V. Rajagopalan, S. Samudrala, and S. Santhanam, "A 200-MHz 64-b dual-issue CMOS microprocessor," *IEEE J. Solid-State Circuits*, vol. 27, no. 11, pp. 1555–1567, Nov. 1992.

[25] B. Benschneider, A. Black, W. Bowhill, S. Britton, D. Dever, D. Donchin, R. Dupcak, R. Fromm, M. Gowan, P. Gronowski, M. Kantrowitz, M. Lamere, S. Mehta, J. Meyer, R. Mueller, A. Olesin, R. Preston, D. Priore, S. Santhanam, M. Smith, and G. Wolrich, "A 300-mhz 64-b quad-issue CMOS RISC microprocessor," *IEEE J. Solid-State Circuits*, vol. 30, no. 11, pp. 1203–1214, Nov. 1995.

[26] P. Gronowski, W. Bowhill, D. Donchin, R. Blake-Campos, D. Carlson, E. Equi, B. Loughlin, S. Mehta, R. Mueller, A. Olesin, D. Noorlag, and R. Preston, "A 433-MHz 64-b quad-issue RISC microprocessor," *IEEE Journal of Solid-State Circuits*, vol. 31, no. 11, pp. 1687–1696, Nov. 1996.

[27] D. Bailey and B. Benschneider, "Clocking design and analysis for a 600-mhz Alpha microprocessor," *IEEE J. Solid-State Circuits*, vol. 33, no. 11, pp. 1627–1633, Nov. 1998.

[28] T. Xanthopoulos, D. Bailey, A. Gangwar, M. Gowan, A. Jain, and B. Prewitt, "The design and analysis of the clock distribution network for a 1.2 GHz Alpha microprocessor," in *Digest of Technical Papers IEEE International Solid-State Circuits Conference (ISSCC 2001)*, 2001, pp. 402–403.

[29] N. Bindal, T. Kelly, N. Velastegui, and K. Wong, "Scalable sub-10ps skew global clock distribution for a 90nm multi-GHz IA microprocessor," in *Digest of Technical Papers IEEE International Solid-State Circuits Conference (ISSCC 2003)*, 2003, pp. 346–347, 498.

[30] N. Sakran, M. Yuffe, M. Mehalel, J. Doweck, E. Knoll, and A. Kovacs, "The implementation of the 65nm dual-core 64b Merom processor," in *Digest of Technical Papers IEEE International Solid-State Circuits Conference (ISSCC 2007)*, 2007, pp. 106–107, 590.

[31] G. Geannopoulos and X. Dai, "An adaptive digital deskewing circuit for clock distribution networks," in *Visuals Supplement IEEE International Solid-State Circuits Conference (ISSCC 1998)*, 1998, pp. 326–327, 468–469.

[32] G. Geannopoulos and X. Dai, "An adaptive digital deskewing circuit for clock distribution networks," in *Digest of Technical Papers IEEE International Solid-State Circuits Conference (ISSCC 1998)*, 1998, pp. 400–401.

[33] A. Chamas, A. Dalal, P. deDood, P. Ferolito, B. Frederick, O. Geva, D. Greenhill, H. Hingarh, J. Kaku, L. Kohn, L. Lev, M. Levitt, R. Melanson, S. Mitra, R. Sundar, M. Tamjidi, P. Wang, D. Wendell, R. Yu, and G. Zyner, "A 64 b microprocessor with multimedia support," in *Digest of Technical Papers IEEE International Solid-State Circuits Conference (ISSCC 1995)*, 1995, pp. 178–179.

[34] G. Konstadinidis, K. Normoyle, S. Wong, S. Bhutani, H. Stuimer, T. Johnson, A. Smith, D. Cheung, F. Romano, S. Yu, S.-H. Oh, V. Melamed, S. Narayanan, D. Bunsey, C. Khieu, K. Wu, R. Schmitt, A. Dumlao, M. Sutera, J. Chau, K. Lin, and W. Coates, "Implementation of a third-generation 1.1-GHz 64-bit microprocessor," *IEEE Journal of Solid-State Circuits*, vol. 37, no. 11, pp. 1461–1469, 2002.

[35] F. Anderson, J. Wells, and E. Berta, "The core clock system on the next generation Itanium^{TM} microprocessor," in *Digest of Technical Papers IEEE International Solid-State Circuits Conference (ISSCC 2002)*, vol. 1, 2002, pp. 146–147, 453.

[36] S. Rusu and S. Tam, "Clock generation and distribution for the first IA-64 microprocessor," in *Digest of Technical Papers IEEE International Solid-State Circuits Conference (ISSCC 2000)*, 2000, pp. 176–177.

[37] M. Golden, S. Hesley, A. Scherer, M. Crowley, S. Johnson, S. Meier, D. Meyer, J. Moench, S. Oberman, H. Partovi, F. Weber, S. White, T. Wood, and J. Yong, "A seventh-generation x86 microprocessor," *IEEE Journal of Solid-State Circuits*, vol. 34, no. 11, pp. 1466–1477, 1999.

[38] D. Draper, M. Crowley, J. Holst, G. Favor, A. Schoy, J. Trull, A. Ben-Meir, R. Khanna, D. Wendell, R. Krishna, J. Nolan, D. Mallick, H. Partovi, M. Roberts, M. Johnson, and T. Lee, "Circuit techniques in a 266-MHz MMX-enabled processor," *IEEE Journal of Solid-State Circuits*, vol. 32, no. 11, pp. 1650–1664, 1997.

[39] G. Gerosa, M. Alexander, J. Alvarez, C. Croxton, M. D'Addeo, A. Kennedy, C. Nicoletta, J. Nissen, R. Philip, P. Reed, H. Sanchez, S. Taylor, and B. Burgess, "A 250-MHz 5-w PowerPC microprocessor with on-chip L2 cache controller," *IEEE Journal of Solid-State Circuits*, vol. 32, no. 11, pp. 1635–1649, 1997.

[40] E. Cohen, J. Ballard, J. Blomgren, C. Brashears, V. Moldenhauer, and J. Pattin, "A 533 MHz BiCMOS superscalar microprocessor," in *Digest of Technical Papers IEEE International Solid-State Circuits Conference (ISSCC 1997)*, 1997, pp. 164–165, 448.

[41] E. Fayneh and E. Knoll, "Clock generation and distribution for intel Banias mobile microprocessor," in *Proceedings of Digest of Technical Papers VLSI Circuits 2003 Symposium*, 2003, pp. 17–20.

[42] S. Tam, U. Desai, and R. Limaye, "Clock generation and distribution for the third generation Itanium® processor," in *Proceedings of Digest of Technical Papers VLSI Circuits 2003 Symposium*, 2003, pp. 9–12.

[43] P. Mahoney, E. Fetzer, B. Doyle, and S. Naffziger, "Clock distribution on a dual-core, multi-threaded Itanium®-family processor," in *Digest of Technical Papers IEEE International Solid-State Circuits Conference (ISSCC 2005)*, 2005, pp. 292–293, 599.

[44] D. Pham, S. Asano, M. Bolliger, M. Day, H. Hofstee, C. Johns, J. Kahle, A. Kameyama, J. Keaty, Y. Masubuchi, M. Riley, D. Shippy, D. Stasiak, M. Suzuoki, M. Wang, J. Warnock, S. Weitzel, D. Wendel, T. Yamazaki, and K. Yazawa, "The design and implementation of a first-generation CELL processor," in *Digest of Technical Papers IEEE International Solid-State Circuits Conference (ISSCC 2005)*, 2005, pp. 184–185,592 Vol. 1.

[45] S. Tam, J. Leung, R. Limaye, S. Choy, S. Vora, and M. Adachi, "Clock generation and distribution of a dual-core Xeon® processor with 16MB L3 cache," in *Digest of Technical Papers IEEE International Solid-State Circuits Conference (ISSCC 2006)*, 2006, pp. 1512–1521.

[46] J. Hart, K. Lee, D. Chen, L. Cheng, C. Chou, A. Dixit, D. Greenley, G. Gruber, K. Ho, J. Hsu, N. Malur, and J. Wu, "Implementation of a fourth-generation 1.8-GHz dual-core SPARC V9 microprocessor," *IEEE Journal of Solid-State Circuits*, vol. 41, no. 1, pp. 210–217, 2006.

[47] D. Boerstler, "A low-jitter PLL clock generator for microprocessors with lock range of 340–612 MHz," *IEEE J. Solid-State Circuits*, vol. 34, no. 4, pp. 513–519, April 1999.

[48] D. Boerstler and K. Jenkins, "A phase-locked loop clock generator for a 1 GHz microprocessor," in *Proceedings of Digest of Technical Papers VLSI Circuits 1998 Symposium*, 11–13 June 1998, pp. 212–213.

[49] N. Kurd, J. Barkarullah, R. Dizon, T. Fletcher, and P. Madland, "A multigigahertz clocking scheme for the Pentium® 4 microprocessor," *IEEE Journal of Solid-State Circuits*, vol. 36, no. 11, pp. 1647–1653, Nov. 2001.

[50] N. Bindal, T. Kelly, N. Velastegui, and K. Wong, "Scalable sub-10ps skew global clock distribution for a 90nm multi-GHz IA microprocessor," in *Visuals Supplement IEEE International Solid-State Circuits Conference (ISSCC 2003)*, 2003, pp. 281–284.

[51] P. Restle, K. Jenkins, A. Deutsch, and P. Cook, "Measurement and modeling of on-chip transmission line effects in a 400MHz microprocessor," *IEEE Journal of Solid-State Circuits*, vol. 33, no. 4, pp. 662–665, April 1998.

[52] J. Chang, S. Rusu, J. Shoemaker, S. Tam, M. Huang, M. Haque, S. Chiu, K. Truong, M. Karim, G. Leong, K. Desai, R. Goe, and S. Kulkarni, "A 130-nm triple-V_t 9-MB third-level on-die cache for the 1.7-GHz Itanium® 2 processor," *IEEE Journal of Solid-State Circuits*, vol. 40, no. 1, pp. 195–203, Jan. 2005.

[53] P. Gronowski, W. Bowhill, R. Preston, M. Gowan, and R. Allmon, "High-performance microprocessor design," *IEEE Journal of Solid-State Circuits*, vol. 33, no. 5, pp. 676–686, May 1998.

[54] G. Gerosa, S. Curtis, M. D'Addeo, B. Jiang, B. Kuttanna, F. Merchant, B. Patel, M. Taufique, and H. Samarchi, "A sub-1w to 2w low-power IA processor for mobile internet devices and ultra-mobile PCs in 45nm Hi-K metal gate CMOS," in *Digest of Technical Papers IEEE International Solid-State Circuits Conference (ISSCC 2008)*, 2008, pp. 256–257, 611.

[55] G. Gerosa, S. Curtis, M. D'Addeo, B. Jiang, B. Kuttanna, F. Merchant, B. Patel, M. Taufique, and H. Samarchi, "A sub-1w to 2w low-power IA processor for mobile internet devices and ultra-mobile PCs in 45nm Hi-K metal gate CMOS," in *Visuals Supplement IEEE International Solid-State Circuits Conference (ISSCC 2008)*, 2008.

[56] S. Tam, S. Rusu, U. Nagarji Desai, R. Kim, J. Zhang, and I. Young, "Clock generation and distribution for the first IA-64 microprocessor," *IEEE Journal of Solid-State Circuits*, vol. 35, no. 11, pp. 1545–1552, Nov. 2000.

[57] C. Dike, N. Kurd, P. Patra, and J. Barkatullah, "A design for digital, dynamic clock deskew," in *Proceedings of Digest of Technical Papers VLSI Circuits 2003 Symposium*, 12–14 June 2003, pp. 21–24.

[58] K. Wong, T. Rahal-Arabi, M. Ma, and G. Taylor, "Enhancing microprocessor immunity to power supply noise with clock-data compensation," *IEEE Journal of Solid-State Circuits*, vol. 41, no. 4, pp. 749–758, April 2006.

[59] S. Rusu, S. Seidel, G. Woods, D. Grannes, H. Muljono, J. Rowlette, and K. Petrosky, "Backside infrared probing for static voltage drop and dynamic timing measurements," in *Digest of Technical Papers IEEE International Solid-State Circuits Conference (ISSCC 2001)*, 2001, pp. 276–277, 454.

[60] S. Tam, R. Limaye, and U. Desai, "Clock generation and distribution for the 130nm Itanium® 2 processor with 6-MB on-die L3 cache," *IEEE Journal of Solid-State Circuits*, vol. 39, no. 4, pp. 636–642, April 2004.

[61] P. Restle, R. Franch, N. James, W. Huott, T. Skergan, S. Wilson, N. Schwartz, and J. Clabes, "Timing uncertainty measurements on the Power5 microprocessor," in *Digest of Technical Papers IEEE International Solid-State Circuits Conference (ISSCC 2004)*, 2004, pp. 354–355.

[62] P. Dudek, S. Szczepanski, and J. Hatfield, "A high-resolution CMOS time-to-digital converter utilizing a Vernier delay line," *IEEE J. Solid-State Circuits*, vol. 35, no. 2, pp. 240–247, Feb. 2000.

[63] T. Hashimoto, H. Yamazaki, A. Muramatsu, T. Sato, and A. Inoue, "Time-to-digital converter with vernier delay mismatch compensation for high resolution on-die clock jitter measurement," in *Proceedings of IEEE Symposium on VLSI Circuits*, 18–20 June 2008, pp. 166–167.

[64] P. Teehan, M. Greenstreet, and G. Lemieux, "A survey and taxonomy of GALS design styles," *IEEE Design & Test of Computers*, vol. 24, no. 5, pp. 418–428, Sept.–Oct. 2007.

[65] S. Vangal, J. Howard, G. Ruhl, S. Dighe, H. Wilson, J. Tschanz, D. Finan, P. Iyer, A. Singh, T. Jacob, S. Jain, S. Venkataraman, Y. Hoskote, and N. Borkar,

"An 80-tile 1.28TFLOPS network-on-chip in 65nm CMOS," in *Digest of Technical Papers IEEE International Solid-State Circuits Conference (ISSCC 2007)*, 2007, pp. 98–99, 589.
[66] S. Vangal, J. Howard, G. Ruhl, S. Dighe, H. Wilson, J. Tschanz, D. Finan, A. Singh, T. Jacob, S. Jain, V. Erraguntla, C. Roberts, Y. Hoskote, N. Borkar, and S. Borkar, "An 80-tile sub-100-w TeraFLOPS processor in 65-nm CMOS," *IEEE Journal of Solid-State Circuits*, vol. 43, no. 1, pp. 29–41, Jan. 2008.
[67] N. Kurd, J. Douglas, P. Mosalikanti, and R. Kumar, "Next generation Intel® micro-architecture (Nehalem) clocking architecture," in *Proceedings of IEEE Symposium on VLSI Circuits*, 18–20 June 2008, pp. 62–63.
[68] A. Mule, S. Schultz, T. Gaylord, and J. Meindl, "An optical clock distribution network for gigascale integration," in *Proceedings of IEEE 2000 International Interconnect Technology Conference*, 5–7 June 2000, pp. 6–8.
[69] A. Mule, E. Glytsis, T. Gaylord, and J. Meindl, "Electrical and optical clock distribution networks for gigascale microprocessors," *IEEE Transactions of VLSI Systems*, vol. 10, no. 5, pp. 582–594, Oct. 2002.
[70] Q. Zhu and S. Tam, "Package clock distribution design optimization for high-speed and low-power VLSIs," *IEEE Transactions on Components, Packaging, and Manufacturing Technology, Part B: Advanced Packaging*, vol. 20, no. 1, pp. 56–63, Feb. 1997.
[71] V. Gutnik and A. Chandrakasan, "Active GHz clock network using distributed PLLs," *IEEE Journal of Solid-State Circuits*, vol. 35, no. 11, pp. 1553–1560, Nov. 2000.
[72] F. O'Mahony, C. Yue, M. Horowitz, and S. Wong, "A 10-GHz global clock distribution using coupled standing-wave oscillators," *IEEE Journal of Solid-State Circuits*, vol. 38, no. 11, pp. 1813–1820, Nov. 2003.
[73] S. Chan, P. Restle, K. Shepard, N. James, and R. Franch, "A 4.6ghz resonant global clock distribution network," in *Digest of Technical Papers IEEE International Solid-State Circuits Conference (ISSCC 2004)*, 2004, pp. 342–343, Vol. 1.
[74] S. Chan, P. Restle, T. Bucelot, S. Weitzel, J. Keaty, J. Liberty, B. Flachs, R. Volant, P. Kapusta, and J. Zimmerman, "A resonant global clock distribution for the cell broadband-engine™processor," in *Digest of Technical Papers IEEE International Solid-State Circuits Conference (ISSCC 2008)*, 2008, pp. 512–513.

3

Clocked Elements

James Warnock

IBM T.J. Watson Research Center

3.1 Introduction

One of the most critical design considerations during the planning of any large VLSI structure is the definition and implementation of the clocked storage elements (CSEs) and the circuitry which drives the local clocks to these elements [1–4]. The nature of the solutions adopted will affect almost every aspect of the design, including its manufacturability, testability, reliability, power consumption, and operating frequency, while the complexity and style of latches and flip-flops employed will affect almost every design automation tool, from high level logic simulation methodology and logic synthesis engines, to circuit tools for detailed device tuning, timing, and testability analyses. A modern microprocessor chip may contain from 0.75 to 1.5M latches and flip-flops [5, 6], and clocked elements may account for 30–40% of the total chip AC power dissipation [7, 8]. Furthermore, the delay overhead or latency of these elements is typically in the range of 2–3 FO4 for modern high-speed designs [5, 9] which may account for 10–25% of the design cycle time for designs spanning the range from 10 FO4 (performance-only optimization) up to about 30 FO4, typically the upper end of the range for power performance optimized designs [10]. Thus the CSE definition is of fundamental importance to any VLSI project; the correct selection, optimization, and implementation will be a basic part of the global design strategy. The goal of this chapter is first to provide a high level overview of the design space of these elements covering the basic design metrics, issues, and trade offs, and second to look at several families of CSEs. This will be followed by a more detailed discussion on aspects of test and testability, design robustness against variability, reliability and soft error rate (SER) considerations.

T. Xanthopoulos (ed.), *Clocking in Modern VLSI Systems*, Integrated Circuits and Systems, DOI 10.1007/978-1-4419-0261-0_3,

3.2 CSE Design Issues

In this section some of the basic CSE design issues are discussed, including latency, hold time, power, and scan testing. This foundation will serve as a useful reference when describing the design space covered by the various CSE families.

3.2.1 Latency

One of the primary considerations of any digital design element is its delay, ideally characterized and compared at a constant, fixed power dissipation. For CSEs, the definition of latency is a little more complicated, due to the interaction of the data with the clock edge. If the data arrives at the CSE well before the activating clock edge, the time delay between when the data arrives at the input (labeled for reference as "*d*") to when it is propagated to the output (labeled for reference as "*q*"), will be very long since the data has to wait for the arrival of the clock before it may propagate to the output of the CSE. The total time from d to q will thus decrease linearly as d arrives later (as the wait time is reduced). However, a point will be reached where, if the data arrives too late, the time from clock arrival to the output transition (q) starts to increase, as the point is approached where the data arrives too late to be captured or propagated to q. Therefore, it can be seen that, when measuring the latency of the CSE there will generally be some optimum specification for the data arrival time relative to the clock edge, which produces the minimum overall delay [11], and this optimum point may be used for design comparisons. Since the clock arrival time typically marks the cycle timing boundary, the total $d-q$ delay is often broken into two segments describing the behavior on either side of the boundary; the setup time (amount of time that the data needs to arrive in advance of the clock), and clock-to-q delay (time measured from arrival of the clock and the appearance of the new data at "*q*").

Having described the conditions under which CSE latencies may be compared, it should be noted that the shape of the latency curves described above is also an important design consideration. This effect is illustrated in Fig. 3.1, for two hypothetical CSEs, plotting the time from d to q vs. the d arrival time relative to the clock. It can be seen from this example that "design A" has a lower latency than "design B," since it offers the minimum latency (d to q time). However, this design (A) exhibits a "hard" cycle boundary behavior. Data cannot arrive much later than at the optimal setup time point before the element will fail, and getting the data to the latch early will provide no benefit to the timing of the downstream logic on the next cycle. Design B exhibits a "soft" cycle boundary. The data arrival time can be targeted to fall in the middle of the "window" where the latency is minimal. Provided that the data arrives somewhere in the window, earlier arriving data will show up at q earlier (since the $d-q$ latency is about constant), to the benefit of the downstream logic on the next cycle. Later arriving data will not cause a failure until the data falls outside this "transparent" window, or until the timing impact on the downstream logic causes a failure.

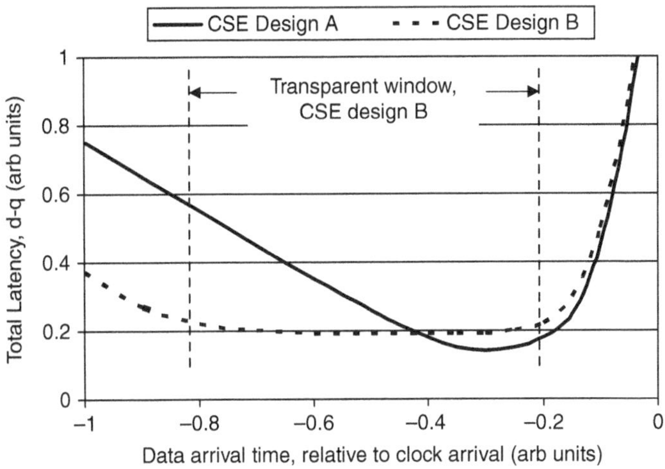

Fig. 3.1. Latency ($d - q$ time) vs. data arrival time for two hypothetical CSE designs

This concept of "soft" vs "hard" cycle boundaries is very important for CSE design, and particularly as technology variability increases, a simple latency metric is insufficient to characterize a particular CSE style. In many situations, with all other things equal, a design like "B" may be preferred over "A," and may actually give better results in hardware, even though the absolute latency is worse.

3.2.2 Hold Time

Just as the data arriving at the CSE must arrive early enough to get properly written, the data must also hold its value for some amount of time (relative to the clock), if the proper value is to be written. Hold time errors may result, for example, from a situation where new data arrives very quickly at the beginning of the cycle, overwriting the data that was supposed to be captured and latched at the end of the previous cycle.

Hold time specifications are not typically considered when comparing power performance characteristics of CSEs. However, long hold times may drive the addition of a substantial number of delay-padding elements to eliminate any possible early transition at the CSE input. Thus indirectly, long hold times may cost both area and power. In addition, the amount of effort devoted to detection, analysis, and the fixing of hold time misses reduces resources available for other forms of design optimization. Finally, design solutions with longer hold times are inherently riskier, and will be more prone to unexpected hardware issues. In many designs there is a natural trade off between the hold time and the size of the transparent window at the cycle boundary. As the window is widened (making the timing on cycle-limiting speed paths less sensitive to clock skew and process variation), the hold time also increases (making the design more prone to clock-skew-induced or variation-induced functional failure through hold time misses).

3.2.3 Power

Power is often considered in conjunction with latency, since the two metrics typically have an inverse dependency on each other. There have been many comparative power performance studies of various latch and flip-flop families [3, 9, 11–14], but the analysis is generally not as straight-forward as it may seem at first. In the same way that the action of the clock signal complicates the latency considerations, it also complicates power comparisons as well. In a given design, the total AC power dissipation is the sum of the power dissipated by clock nets, and other nodes regularly charged and discharged by the clock, and the nodes that switch whenever the input (and output) data change. Thus, when considering a particular latch or flip-flop solution for a specific design application, it is important to characterize the design with the correct weighting between clock switch events and data switch events.

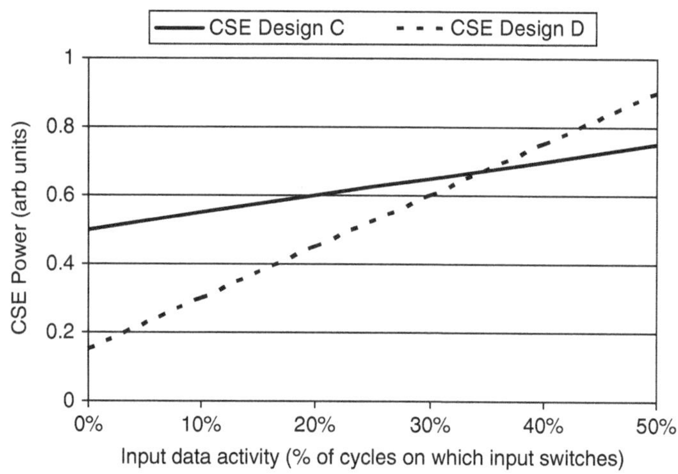

Fig. 3.2. Power vs. input data switching factor for two hypothetical CSE designs

This is illustrated in Fig. 3.2, for two hypothetical CSE designs "C" and "D." Design "C" has a relatively high capacitance on clock nets and/or nodes which charge and discharge every cycle independent of the input data, but uses this dynamic action to limit the amount of capacitance switched by changing data inputs. This sort of behavior is typical of many dynamic CSE designs. Design "D" has much less clock capacitance, but has more capacitance switched when the data changes state. Power comparisons at high data switching factors, for example at 50%, would tend to favor the dynamic design "C," while comparisons at lower switching factors, for example 10–20%, would tend to favor design "D." Complicating the situation further is the fact that in modern microprocessors, the local clocks are gated off whenever it is known enough ahead of time that a group of CSEs does not need to be updated on a given cycle. Thus for power comparisons, it is necessary to understand the average data switching factor on cycles when the clock is active. Depending on the type of CSE, it may also be necessary to consider the power dissipated by data switching activity when the clock is gated off as well.

At this point, it is worth mentioning that the ideal situation, with regard to power consumption, would be to activate the clocks to a given CSE only when the new input data is different from the old data, and there have been several schemes proposed along these lines [12, 15]. However, the power, area, and delay overhead associated with determining whether or not to fire the clocks may be significant, and these techniques have never been widely deployed.

3.2.4 Scan Design for CSEs

Given the complexity of modern VLSI systems, the ability to directly observe the contents of critical system storage elements during design test and debug, as well as the ability to apply arbitrary initialization patterns to these elements, is another important design consideration. Typically this is done by linking the CSEs together in a serial fashion, using a secondary input port as shown schematically in Fig.3.3. Level sensitive scan design (LSSD) [16] is one methodology which accomplishes this goal, although there are other more general ways of providing scan capability in a design.

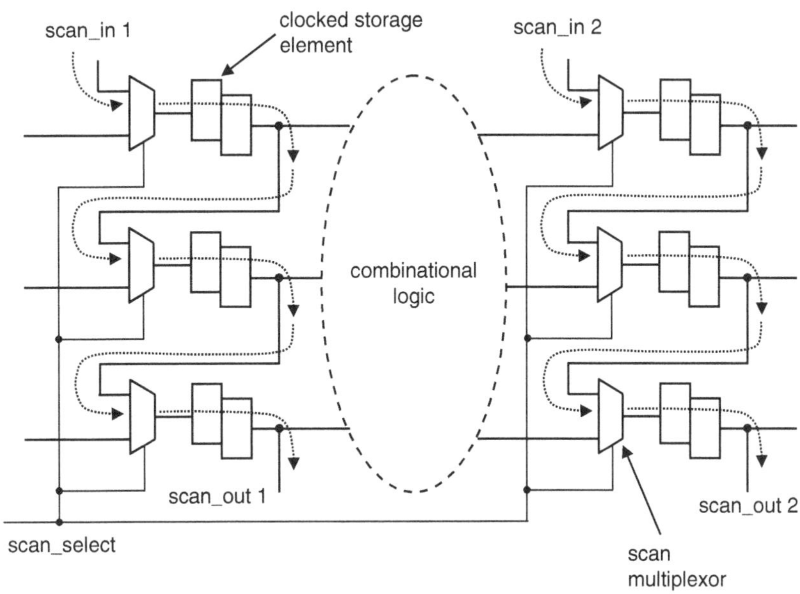

Fig. 3.3. Basic scan design. Scan data flow is indicated by the *dotted lines*

The key features of a scannable design include the scan multiplexor, to allow choosing between scan and normal functional data to load into the CSE, and the scan chain itself which needs to be configured in the style of a shift register to allow movement of the data in and out of the CSEs in an orderly fashion. In practice, the goal is usually to minimize the overhead of the extra logic needed for scan and test functionality, without sacrificing the desired test features; for example the multiplexor in Fig. 3.3 is often integrated directly into the CSE itself. Some of the detailed requirements, and ideal test features are described later in this chapter. In the CSE discussion that follows, for each design style, the implications and overhead associated with scan functionality will be described. In modern VLSI designs, it should be assumed that a significant fraction of CSEs will need to be scannable, and therefore this is a significant design consideration.

3.3 Static Latch Designs

In this section, examples of several basic styles and classes of static latches will be presented and discussed, including a discussion of the local clock generation aspects, comparative merits and general features of each of the families in light of some of the design considerations described earlier.

3.3.1 Master–Slave Latches

For general logic design, the master–slave latch (MSL) is probably the most widely used CSE, with many topological variants and clocking styles described over the years. A typical transmission-gate master–slave latch [17] is shown in Fig. 3.4, using the style and naming conventions of reference [5]. The two clocks, "dclk" (or "data capture clock") and "lclk" (or "launch clock") are roughly 180° out of phase. At the beginning of a cycle (cycle boundary), dclk falls while lclk rises, simultaneously blocking any writes to the master latch while writing the slave latch and propagating

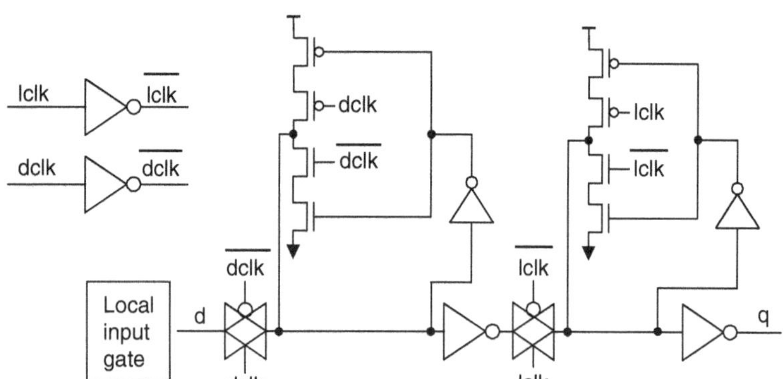

Fig. 3.4. Master–slave latch. Reproduced with permission from [5], ©2006 IEEE

new data to the latch output q. At the middle of the cycle, dclk rises while lclk falls, allowing the master latch to accept new data until the end of the cycle, while blocking any update of the data in the slave latch so that the data is held until the beginning of the next cycle.

In high-speed designs, the dclk and lclk waveforms are designed to overlap (both clocks high) by some amount at the cycle boundary in order to allow late-arriving data to propagate through the MSL with minimal delay by avoiding interference from the clock edges. This technique not only minimizes latency [5], but also provides for soft cycle boundaries; if the data is timed to arrive in the middle of the window when both clocks are high, then later-arriving data can still propagate through, stealing time from the next cycle. Also, the MSL becomes tolerant to some amount of local clock skew depending on the width of this window, and to the extent that the data arrives in the middle of the window, away from the clock edges. The trade off though is that as the falling dclk is delayed to improve latency and widen the window of transparency, the hold time increases, creating race hazards, and requiring delay padding on short latch-to-latch paths. This trade off is a fundamental consideration for MSL operation, and is shown in Fig. 3.5 [5]. Here the delay unit "FO4" refers to a technology independent delay metric, the delay of an inverter with a fan out of 4 [18]. It should be noted here that since the latency of the MSL is highly dependent on the relative timing of the master and slave clocks, any power performance comparisons including MSL elements (often cited as a standard benchmark in comparisons), must provide detailed information on the clocking assumptions made, since otherwise any latency quoted is not very meaningful.

Delaying dclk with respect to lclk also improves the situation at the middle of the cycle. Here, the desire is for the two clocks to be well un-overlapped, with dclk rising well after lclk falls in order to avoid potential race-through of the wrong data at mid-cycle.

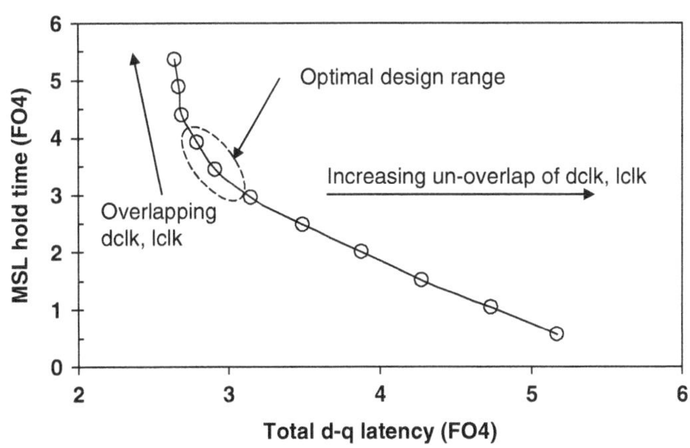

Fig. 3.5. MSL hold time vs. total MSL latency as the cycle boundary overlap of dclk and lclk is varied. Reproduced with permission from [5], ©2006 IEEE

It should be noted here that there are many other MSL circuit topologies, some of which are inherently more robust against race or min-path issues [19]. However, these will generally be higher-latency solutions, and by their very nature will not possess the soft cycle boundary properties which provide protection against delay variability and clock skew. For designs with long cycle times (for example, cycle times larger than 40–50 FO4), it may be more important to have a simple, robust solution, than to get the ultimate latency, or optimal power performance. It is expected though, that modern high-speed microprocessors will continue to need the more difficult design solutions, especially as technology variability (and the benefit of soft cycle boundaries) continues to increase. Increasing sophistication of design analysis tools and techniques will help designers to deal with these more complicated solutions.

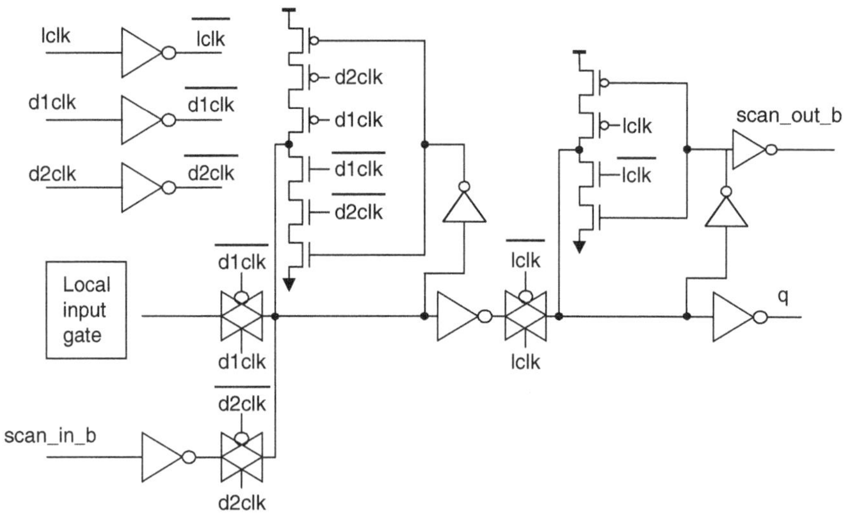

Fig. 3.6. Scannable MSL. Reproduced with permission from [5], ©2006 IEEE

The MSL shown in Fig. 3.4 has no provision for scan testing. One of the most common ways of adding this functionality is to add another port to the master latch, controlled by its own clock. This second port may be controlled by a symmetric twin of the first dclk, as shown in Fig. 3.6. In this case the MSL design can support scan shifts and test sequences at frequencies up to the functional design frequency, although other considerations (such as power, and noise) may often impose practical limits on the scan shift frequency. A set of clock waveforms is shown in Fig. 3.7, illustrating the shifting of data through the scan chains, and functional operation. This type of design solution supports a seamless transition from scan shifting to functional operation, and vice versa.

Other scannable MSL topologies have also been used. One alternate version uses a multiplexor on the slave latch feedback path [20] instead of directly at the input to

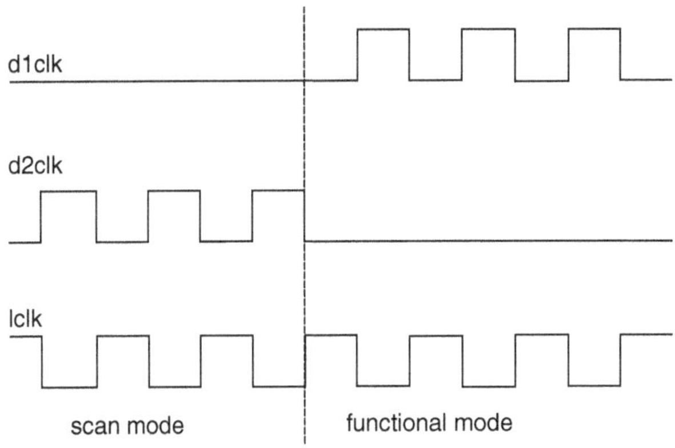

Fig. 3.7. Sample set of clock waveforms for scan shifting and for functional operation, for the MSL of Fig. 3.6

the latch node, as shown in Fig. 3.8. In LSSD methodologies, the second port to the slave latch is generally a low-speed scan clock with its own dedicated distribution network. Even with a low-speed scan clock, it is still possible to do full-frequency AC testing on the part. After scanning the desired data in to all scannable latches and appropriately initializing any nonscan elements, the system can be started synchronously in functional mode at speed, with the first at-speed lclk launching out either the data from the last scan clock, or the functional data waiting at the data port input, depending on the type of test desired. Such LSSD schemes may provide very flexible and robust scan design solutions [21], even if the rate of data shifting through the scan strings is limited by the speed of the secondary LSSD clock distributions.

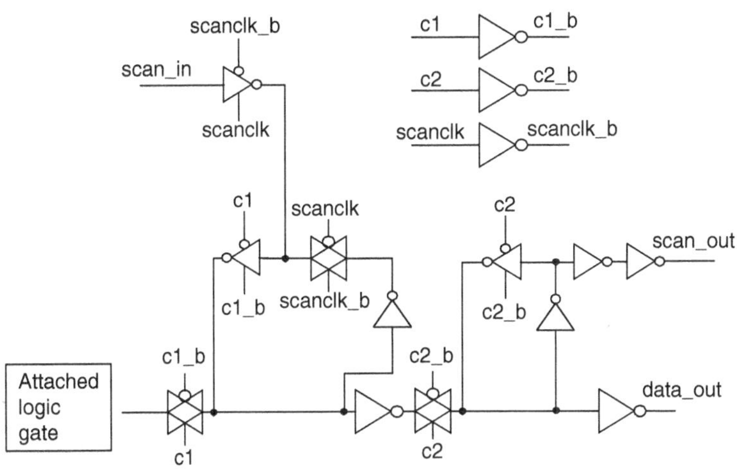

Fig. 3.8. MSL with scan MUX in feedback path. Reproduced with permission from [20], ©2002 IBM

Many VLSI designs have been implemented with MSLs as the main "workhorse" latch, often with sophisticated clock edge controls [22]. However, the variation-tolerant soft cycle boundary concept can only be taken so far because of the hold time increase associated with the transparent overlapped clocks at the cycle boundary. In addition, designers have looked for lower power solutions, or higher speed solutions for specific critical applications. These considerations have led to some of the styles described in the following sections.

3.3.2 Two-Phase Level-Sensitive Latches

It has long been recognized that the use of level sensitive latches, as shown in Fig. 3.9, will improve the clock-skew and delay-variability tolerance of the design, while simultaneously offering more flexibility for balancing logic delays across multiple cycles. This feature, or technique has been referred to as "cycle stealing" [23] since it allows circuits on either side of the latch to "steal" time from the previous or next cycle. In comparison with the MSL style described above, this technique can be conceptualized as the result of splitting apart some, or all of the master and slave latches, with logic interspersed between master and slave, as well as between the slave and master. Any path through the logic will pass alternately through level sensitive latches clocked with opposite phases. It can be shown that, with about one-half of a cycle of logic between each latch, the data arrival can be timed such that data arrives at each latch in the middle of its active clock window. Therefore, this scheme can be made robust to clock skew and local delay variability. It can also be made very robust against hold time issues, by shrinking the active phase of each clock, with no latency penalty as long as the data is timed to propagate through the latches near the middle of each active clock phase.

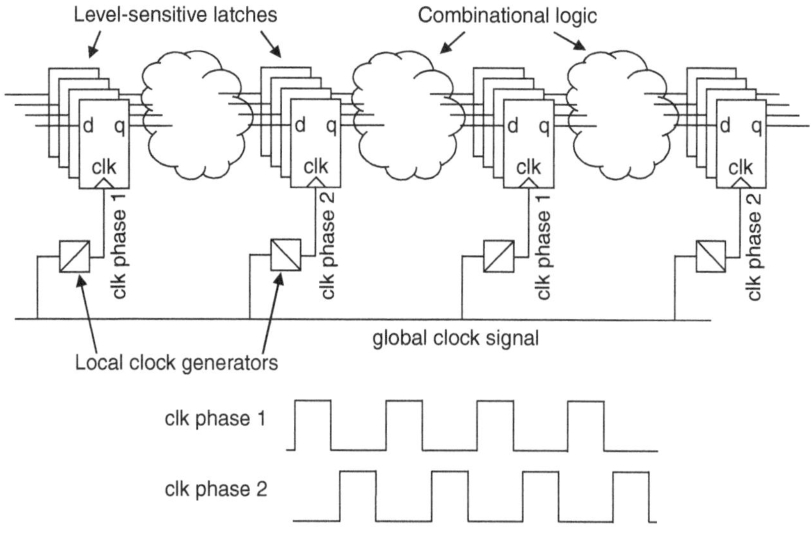

Fig. 3.9. Basic two-phase level-sensitive latch scheme, with sample local clock waveforms

This scheme would seem to offer the most tolerance to process variability and clock skew, and in fact these techniques have been used in some large microprocessor designs [20, 24]. From a power perspective, it would seem that the power dissipated should be comparable to, or somewhat less than that of a design using an MSL methodology, since the extra flexibility offered might help to cut down the overall number of latches needed.

However, there are some negative aspects of this design style which must be weighed in the balance. The timing analysis for split-latch designs can be very complicated, with long multicycle (or multihalf-cycle) critical paths, timing loops, and a general inability to define any clear cycle boundaries [25–27]. In fact, the more optimal the design (in terms of lining up data arrival at the latch with the midpoint of the appropriate clock phase), the more complicated the timing analysis becomes. In addition, the split master and slave latches cannot generally share any of the overhead for test control and clock gating in the generation of the local clocks, and will usually need separate global clock taps. The split latches themselves may impose a certain overhead on the design, especially if small devices are used in the latches in order to save clock power. Each latch may become a "bottleneck," requiring stages of gain afterwards to drive the surrounding logic, or, to avoid this situation, clocked devices may have to be increased in size relative to that of an MSL design.

Finally, there are some test issues that need to be considered. Each scannable split-latch requires an additional scan-latch in order to build a robust scan chain, as shown in the example in Fig. 3.10 [20]. Also, AC test sequences need to consider the

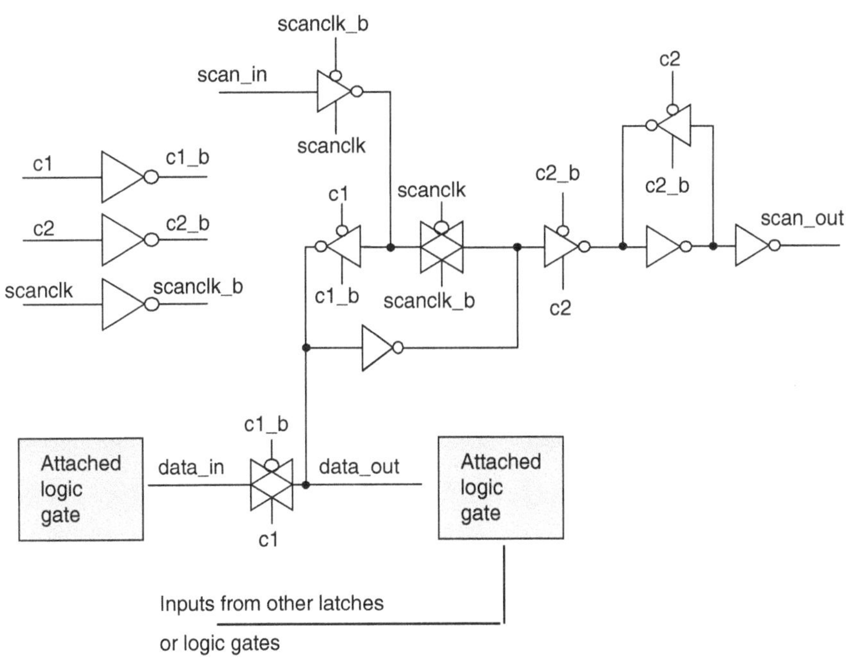

Fig. 3.10. Scannable level-sensitive latch. Reproduced with permission from [20], ©2002 IBM

implications of having the scan data propagating into the downstream logic as soon as the clock fires. This means that the initial timing of signals launched from the level-sensitive latch during test is likely to be different from that during functional operation, when transitions may flush through later in the cycle. Therefore AC test sequences may become more complicated, with several functional clock cycles required.

The above design and test issues probably explain why this design style has never gained widespread acceptance for large, high-speed microprocessor designs, even as design for variability tolerance becomes an increasingly significant consideration.

3.3.3 Pulsed-Clock Static Level-Sensitive Latches

The use of static level-sensitive latches with single-phase pulsed-clocks appears to be gaining favor for use in large microprocessor designs [5, 8, 22], even as device variability and presumably the hazards associated with hold time issues have been increasing. It is worthwhile to study this particular CSE implementation in order to understand the benefits and hazards associated with it.

An example of a very simple static pulsed level-sensitive latch is shown in Fig. 3.11 [5]. This style has minimal overhead, whether measured in terms of latency, power dissipation, or area. The latch itself is very simple, but the clock design is more delicate. A locally-generated, self-timed clock pulse is generated from either a global or local clock at the cycle boundary, and is used to write the latch (or, more generally, a group of latches). The width of the pulse must be wide enough to reliably write the latch across the full process/voltage/temperature (PVT) space, including all margins for local variability, noise, and any modeling uncertainties. However, the wider the pulse, the longer the hold time and the more difficult it becomes to protect against min-path or hold time failures. Therefore, careful design of the pulse-generating circuitry is of utmost importance, and the pulse width is a critical design parameter.

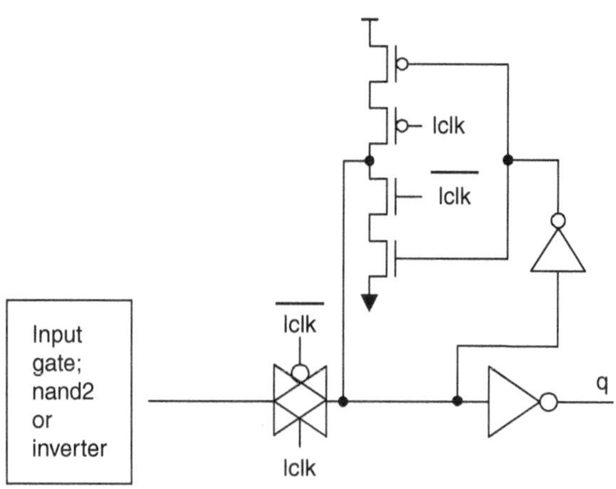

Fig. 3.11. Simple nonscan pulsed-clock latch. Reproduced with permission from [5], ©2006 IEEE

Despite the issues associated with pulsed-clock operation of these latches, there are several significant advantages with this design style. Reduced power consumption is one important benefit. In contrast to the conventional MSL solution, or the 2-phase level-sensitive scheme, only one active clock is needed, cutting clock power by about a factor of 2. In fact, this may be the most important benefit of pulsed clocking. One power-saving strategy along these lines is to use regular MSLs, but force the master clock to stay high all the time, while pulsing the slave clock [5, 22, 28]. This approach delivers significant power savings while preserving a fall-back strategy to conventional two-phase clocking in case problems are seen in the hardware. However, this approach does not provide the benefits of reduced area and latency offered by dedicated pulsed-clock latch solutions.

Another important feature of the pulsed latch is its ability to provide a soft cycle boundary, similar to that of the MSLs with overlapped clocks. Unlike the MSL situation though, the amount of cycle boundary transparency cannot be tuned below a certain value (set by the minimum pulse width needed). Thus, even as the benefit of such a pulsed-clock soft cycle boundary will increase as the technology variability increases, this benefit will be offset by the difficulty of simultaneously ensuring writeability and maintaining margin against data races. The design cycle time figures prominently in this trade off. The tolerable upper limit for pulse width would be expected to scale with the overall logic depth, with a practical maximum pulse width of probably about 1/4 to 1/3 of the design cycle time. With wider pulses, the difficulty of padding all the potential short data paths, while avoiding any impact to the cycle-limiting paths, is likely to become unmanageable. At a constant design cycle time (as measured in FO4) increasing technology variability may push pulse-widths up to their practical limit, necessitating larger devices (with faster write-speed but more power dissipation) or other design restrictions for continued use of pulsed-clock latches.

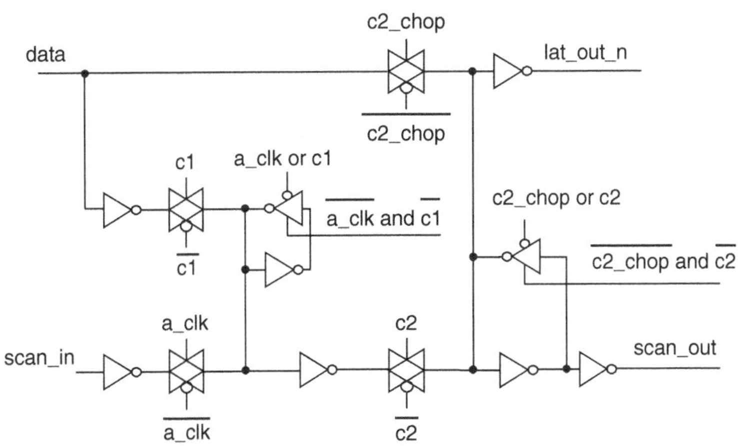

Fig. 3.12. Scannable pulsed-clock (c2_chop) latch with built-in MSL-mode fallback (using c1 and c2 clocks). Reproduced with permission from [28], ©2007 IBM

There are some special test issues for pulsed-latch clocking. In general, it would not be desirable to have to pad the scan path, so usually an additional scan latch is needed. This makes the serial shift operation along the scan chain look like a series of MSLs, whose clocks may be unoverlapped as desired to avoid any possibility of race issues during scan. Thus most of the area advantage of the pulsed-clock solution is likely to be lost in situations where a scannable design is needed. An example of a scannable pulsed latch is shown in Fig. 3.12 [28]. This particular design achieves a low latency typical of dedicated pulsed-clock latches, but still maintains the ability to revert back to an MSL mode of operation if problems develop, although at a reduced performance level. In this sense, the extra scan latch added not only converts the scan chain into a series of MSLs, but also can provide a similar function on the data path as well, if conditions warrant.

3.4 Flip-Flop Designs

Flip-flops are inherently edge-triggered designs, as opposed to the latches described in the previous section, which are inherently level-sensitive structures. There are a large variety of flip-flop designs which have been employed over the years, and it would be impossible to cover all of them here. This section will describe some of the more common styles that have been used, comparing some of the design issues and the merits of each.

3.4.1 Sense-Amp Style Flip-Flop

The proto typical sense-amp flip-flop (SAFF) is shown in Fig. 3.13 [29–31]. This type of design consists of a sense-amplifier coupled with a slave latch to hold the output data when the sense-amp is being reset. This style is more naturally suited for use with dynamic logic on the input side, where data values will hold until they are reset, and where this data reset will occur at about the same time that the flip-flop itself is being reset. Furthermore, with dual-rail dynamic logic, if both true and complement inputs are precharged low, then the activating transition may arrive somewhat later than the clock edge, effectively borrowing time from the next cycle. Since there are internal nodes floating when both inputs are low, and the clock is high, the amount of time borrowing may be limited, depending on the process, noise, and environmental conditions. Also, with dual-rail logic the extra overhead for deriving the other data phase may not be an issue.

Given the generally high power dissipation of dynamic circuit styles, especially dual-rail, sensitivity to noise and process variations, and the general difficulties associated with using such circuits on a large scale in a modern microprocessor, it is unlikely that the industry will see a widespread resurgence in the use of dynamic circuits for general purpose processors. Therefore, this type of design should really be considered in the context of the static circuits which would usually surround it. In a static design, it can be seen that the critical path through the flip-flop traverses 4 stages of logic (the local inverter to generate the complement data, the pull down of one side and subsequent pull-up of the output, and then the pull down of the other side), and so it would not be expected to have a very low latency. Also, power dissipation is likely to be an issue unless very small (and therefore slow, and more variable) devices are used, since one side or the other will need to be charged and discharged in each cycle that the clock is active, independent of the data input pattern. Furthermore, with static input signals, this design does not provide for soft cycle boundaries; whatever data is present when the clock edge arrives is written into the latch. The corollary though is that the hold time for this type of CSE is generally small.

Finally, it can be seen that if the input data were to change before the clock were to fall, internal nodes would be left floating, potentially resulting in a failure to retain the proper state at the output. This particular issue is relatively easy to remedy, by the addition of a single additional small NMOS transistor between nodes A and A' in Fig. 3.13, with gate tied to the supply [30]. This fix will, however, have some impact on the power/performance characteristics. The additional transistor not only adds device and interconnect capacitance to both sides of the sense amplifier (one side of which must be charged and discharged every cycle), but also partially discharges the nonactive side of the sense-amp on every cycle. In addition, there will be a slight degradation of the differential resolution due to the weak coupling between the opposite arms of the sense-amp structure, leading to a corresponding increase in setup time.

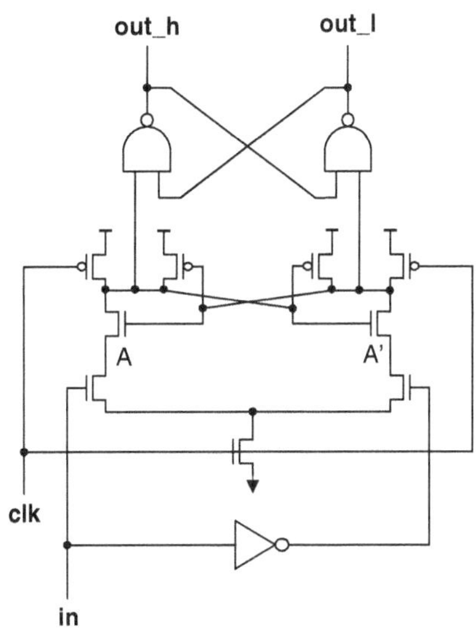

Fig. 3.13. Sense-Amp flip-flop. Reproduced with permission from [31], ©2006 IEEE

For improved testability, modifications have been described to make these designs scannable [32], an example of which is shown in Fig. 3.14. This technique can be optimized for AC test, since the paths from input to output are similar for both the functional data input and the scan input. The scan data input is implemented

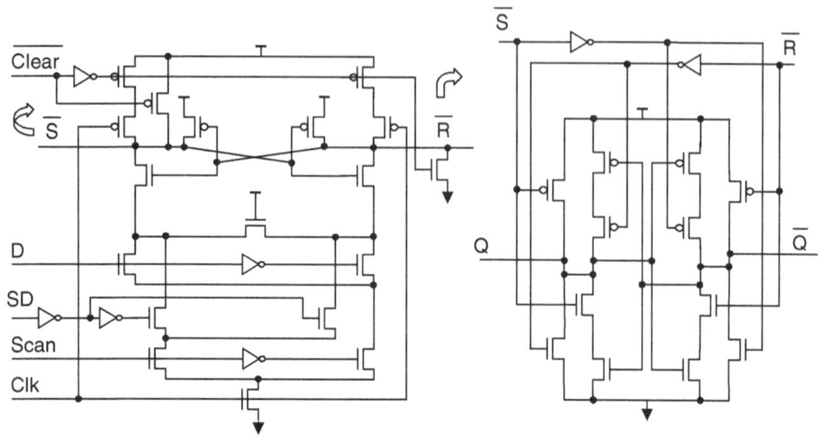

Fig. 3.14. Improved scannable sense-amp flip-flop, with asynchronous reset. Sense-amp "pulse-generator" on the *left* drives the latchings stage on the *right*. Reproduced with permission from [32], ©2000 IEEE

as a second port in the first "pulse-generator" stage of the flip-flop However, this will add capacitance to nodes which need to be precharged every cycle, thereby increasing power dissipation. Other methods [13] may avoid this power cost, but may be less well suited for AC test. In addition, there have been many modifications proposed to this basic design including techniques for "conditional precharge" to try to avoid charging and discharging of internal nodes when input data is not changing from cycle to cycle [33, 34] as well as other variations aimed at improving the power performance characteristics [32, 35, 36]. The resulting designs tend to be more complicated, and may still suffer from one or several of the drawbacks mentioned above. Therefore it seems likely that these SAFF will remain confined to only special-purpose applications in the future.

3.4.2 Hybrid Latch Flip-Flop

The goal of the hybrid latch flip-flop (HLFF) [37] was to try to combine some of the best features of flip-flops, including edge-triggered sensing, low latency, and low input clock load, with some of the best features of latches, including a soft cycle boundary. The result, shown in Fig.3.15 is a flip-flop-style front end, activated by what is effectively a locally generated clock pulse, with a static capture latch back-end.

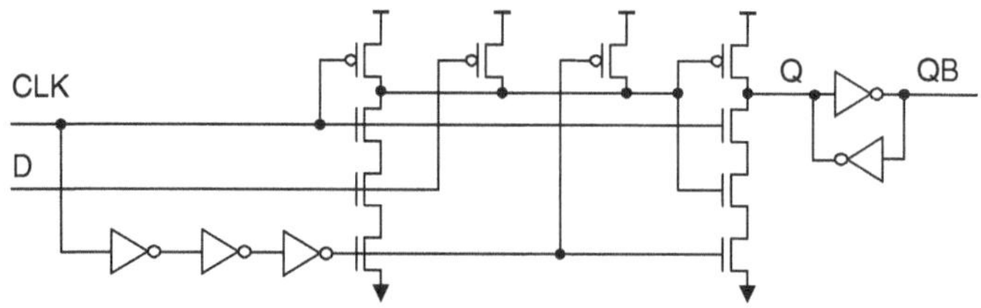

Fig. 3.15. Hybrid latch flip-flop. Reproduced with permission from [37], © 1996 IEEE

With only two gate delays from input to output, low latency can be achieved. Also, the local delay chain sets up a transparent window at the cycle boundary, allowing some amount of clock skew/delay variability tolerance. This particular design suffers from the fact that the static capture latch must be small enough to be quickly overwritten in either direction, meaning that the output will be very sensitive to any noise when only the capture latch holds the state of the output. In realistic applications, an additional local output gate would probably be needed. Also, this type of design will consume significant power, due to the precharge/discharge which occurs every cycle whenever the input data is high, and also the switching in the local delay chain inverters. Finally, the output is subject to glitching when the input data is high, since the output stage will begin to discharge when the clock edge is received, pulling back high only after the first stage output transitions low. Various improvements on this design have been reported [38, 39], but in general microprocessor designers have not found this type of flip-flop to offer any compelling benefits, at least in its original form.

3.4.3 Semi-Dynamic Flip-Flop

The term semi-dynamic flip-flop (SDFF) [40] was coined to refer to a design style which includes a dynamic front end followed by a static latch [5, 28, 41, 42]. In some ways, this is very similar to the HLFF approach discussed above, but now expanded to provide a means of incorporating a stage of high-speed dynamic logic at each cycle boundary. This technique still provides an easy interface to surrounding static logic, accepting normal static inputs, and providing static outputs. A typical design is shown in Fig. 3.16. These designs rely on some form of a pulsed clock to limit the hold time at the input dynamic stage. This may be accomplished by providing an explicit pulsed clock [28, 40, 41], or by ANDing in a delayed complement clock either in the dynamic pull down tree [40], or somewhere in the cone of logic for all the data inputs [5]. The dynamic stage is followed by a static set–reset latch (SRL), which holds the output while the dynamic stage is being precharged for the next cycle. For situations when the dynamic stage pull down path is cut off while the clock input is still high, a full keeper is usually used on the dynamic node in order to ensure that the dynamic node is held solidly low after being discharged. This keeper can be gated with the clock input to avoid drive fights during the precharge operation.

The advantage of this design is that it provides a way to incorporate a stage of dynamic logic into an otherwise fully-static design. It can also be extended to add additional dynamic logic stages after the first stage. In this case, only the first stage, with static inputs, needs a footer device and a pulsed evaluation clock. The SRL is moved downstream to the final dynamic stage, providing dynamic-to-static signal conversion. This technique can be applied at both clock edges [43], so that in such a two-phase system, pulsed clocks are no longer necessary.

The SDFF is a very useful construct for extremely critical paths in a design, where a wide OR, or wide multiplexor is required at the cycle boundary. However, there are some design costs associated with this solution. There is a relatively large capacitance which may be charged and discharged every cycle, even when there is

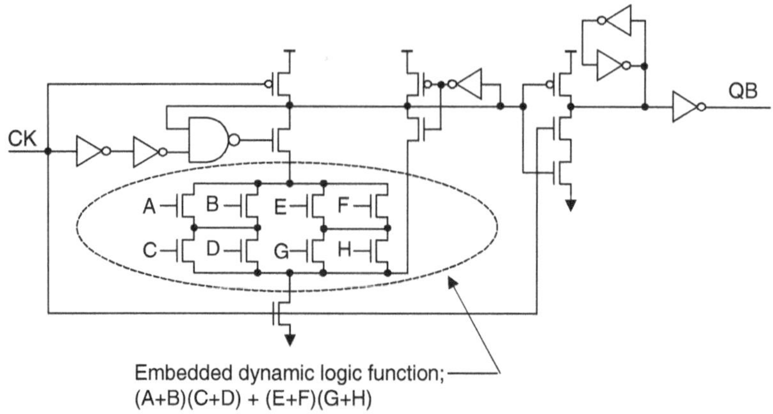

Fig. 3.16. Semi-dynamic flip-flop with embedded dynamic logic. Reproduced with permission from [40], ©1999 IEEE

no switching activity present on the data inputs. Thus, power consumption tends to be high for these designs, especially when used in contexts which do not see particularly high data switching factors, or when efficient clock gating is difficult. Power consumption may also be increased by glitching at the output when holding a 1. This glitch is similar in nature to that observed in the HLFF, occurring since the path to pull the output down (via the reset action of the static latch) is usually faster than the path to force the output high (via the dynamic pull down). The flip side of this glitch though, is that the latch will generally write a 0 faster than a 1, and downstream static logic can be skewed to take advantage of this fact (and also absorb the glitch).

The cycle boundary also is not easily softened to absorb timing variability, in contrast to the HLFF, due to the dynamic input stage. In principle, late rising input transitions may arrive at the dynamic gate after the clock rises, provided that they still have enough time to discharge the dynamic node, but falling transitions must meet a strict setup criterion. Similarly, the hold times for these designs tend to be relatively long, and are also asymmetric due to the action of the dynamic gate. Inputs only need stay high long enough to be able to discharge the dynamic node, but inputs initially low must stay low until near the end of the clock pulse. Just as for the pulsed level-sensitive latches, the clock pulse width must be wide enough to reliably write the latch across the process and application space of interest, but not so wide that the length of the hold time becomes an issue. For the SDFF, it is necessary to consider the range of dynamic input gates used in order to determine the minimum pulse width required.

For test, there are several strategies for making the design scannable [5, 28, 41]. One method involves adding a port into the static latch, and also adding in an additional scan latch, as was required for the static pulsed designs [28, 41]. An example of this approach is shown in Fig. 3.17. With this scheme it should be noted that for AC test, the launch of scan data from the static latch bypasses the input dynamic stage, and so if the test sequence involves a launch from the scan clock, the timing characteristics may differ from that in functional operation. It is possible to more closely match the functional path with the scan path by using an additional input into the front-end dynamic stage [5]. In this case, all the pull down paths through the functional data ports need to be disabled to avoid interference with the scan path.

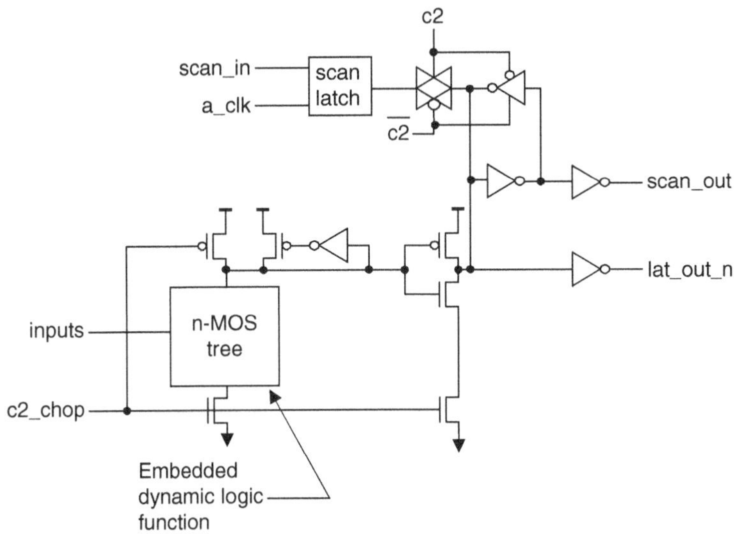

Fig. 3.17. Scannable semi-dynamic flip-flop. "c2_chop" is a pulsed clock input. Reproduced with permission from [28], ©2007 IBM

3.5 Test and Debug Considerations

As microprocessor frequency scaling has run up against severe thermal, physical and electrical constraints, the industry has turned to multicore architectures in order to allow continued performance gains from generation to generation [44]. As a result, microprocessor chips are now being fabricated with over 2 billion transistors [45], and with density scaling still expected to continue for at least several more generations in the future [46], observability and testability, including both DC and AC test coverage, have become of paramount importance. For this reason, large microprocessor design projects have by and large adopted scan design methodologies [5, 28, 41, 47, 48] and it is expected that scan design and test/debug methodologies will continue to need careful consideration in future designs. It will be important to minimize the overhead of these scan methodologies, but it will not be tolerable to make significant sacrifices to the testability of the design.

For DC test coverage, it may not be important exactly how the scan data is written into the latch, since the timing of the launch into the downstream logic is unimportant. Also, in the downstream logic, it is unimportant how each node is switched; the activating transition is unimportant, as long as a pattern can be found which tests the node in each state. However, for high-speed AC test, there are a number of key features which are desirable for a robust and flexible test methodology. The highest test coverage is obtained if each CSE is designed with an extra scan-only storage element, as shown in Fig. 3.18a, used only for test purposes, which can store an independent data value to be used to launch a transition into the downstream logic [49]. In this way, each scannable latch can be independently configured to launch an arbitrary transition into the downstream logic. Referring to the initial input vector as "V1" and the transition vector as "V2," it can be seen in this case that "V1" and "V2" are independent and unconstrained. This enhanced scan design provides the maximum test coverage, but the overhead is generally quite high. Although this technique can be confined to be used only on critical CSEs [50], and selective deployment has been reported in critical areas of some microprocessors [41], the cost is too high for techniques such as these to be used commonly in the industry.

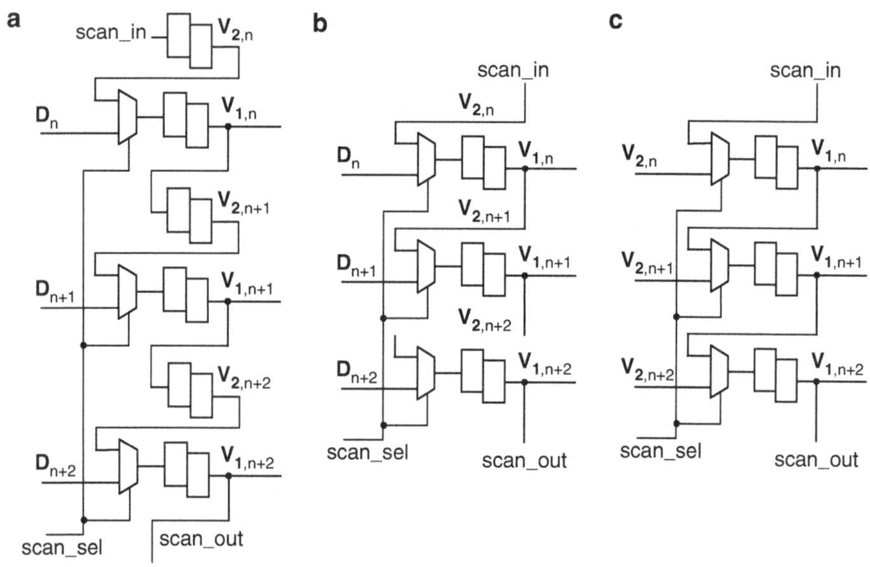

Fig. 3.18. Scan test configurations (**a**) enhanced scan, (**b**) skewed load, (**c**) broadside test

At the other extreme, the simplest AC test sequence relies on loading scan data into all scannable latches, then switching to functional mode to launch and capture data for test. In this case, V1 is unconstrained, but V2 is the next state determined from combinational logic response to the V1 vector, i.e., the functional data at the inputs of the CSEs given the V1 values at their outputs, as shown in Fig. 3.18c. This has been called a "broadside test" [51] sequence, and places minimal constraints on the design of the scannable CSEs. The speed at which the scan data can be written into the CSEs is unimportant, as are the details of the switch from scan mode to functional mode. The downside of this method is that AC test coverage can be low, with few, or no options for improvement through DFT-related design changes. In addition, analysis and debug are more complicated, since V2 depends on the response of all the upstream combinational logic and the V1 state of all CSEs in the cone of logic, increasing simulation time for test analysis and making it harder to change V2 in a systematic way.

A compromise between the two above approaches is the concept of a "skewed load" sequence [52]. In this type of sequence, V1 is again scanned into the CSEs, but now V2 comes from the upstream data in the scan string as shown in Fig.3.18b. During the high-speed test sequence, data must be loaded into each CSE from the scan port, and launched at speed into the downstream logic. On the next clock cycle, data is captured in each CSE through the functional data port. Sample clock waveforms for such a sequence are shown in Fig.3.19a for the scannable MSL shown in Fig.3.6, contrasting with those for the broadside load sequence (Fig.3.19b). To make this sequence work properly, all CSEs must be designed such that the clock-Q delay from the scan port is a close match to that from the data port, otherwise the AC characteristics of the test will not match that of the functional operation. Also, the CSE clock control system has to be capable of switching from "scan mode" to "functional mode" within a single cycle, in order to launch scan data, and then to capture data from the functional port. This will generally require an accurate pipelining and distribution scheme for at least this one high-speed global test control signal.

One advantage of the skewed load test sequence is that it provides enhanced AC test coverage compared to the broadside test, although there are still coverage limitations which may arise from the adjacency of V1 and V2 in the scan string. Furthermore, the test patterns applied are very flexible, easy to modify in a systematic

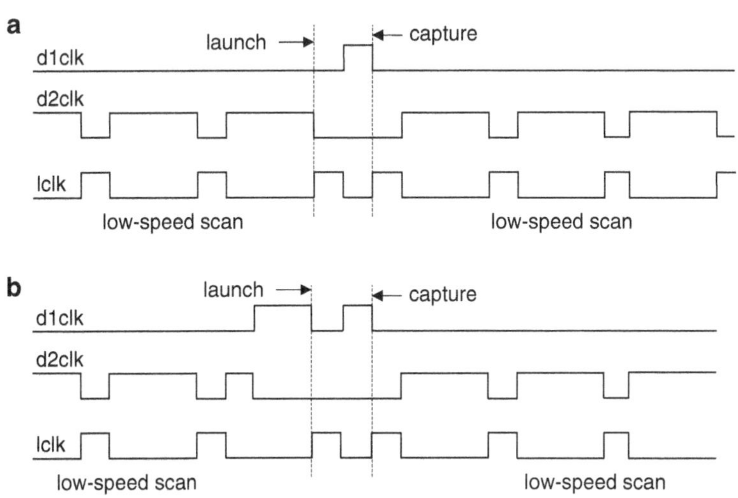

Fig. 3.19. Sample clock waveforms for AC test using the MSL from Fig.3.6. (**a**) skewed load test. Note that data from scan port (d2clk) is launched. (**b**) Broadside test. Note that data from the functional port (d1clk) is launched

fashion, and failure analysis needs only to consider a single cycle's worth of logic instead of two as in the broadside case.

Finally, even though much of the discussion above has focused on single-cycle test sequences, which offer an appealing simplicity from a test point of view, various other CSE design factors may result in the requirement for longer at speed test sequences. Inclusion of nonscan elements (which may help to reduce chip area, latency, and power dissipation) will force the use of longer sequences to test all paths in and out of these components. Also, the use of soft cycle boundaries may also affect AC test operation, since certain critical paths may effectively become more than one cycle long.

3.6 CSE Design for Variability

Design for variability has become an increasingly important consideration as the relative level of parameter and device variability has been increasing recently and is expected to continue to increase rapidly with future technology scaling [53]. The importance of this fact is magnified by the expected continuation of device density scaling; this increased variability is manifested in ever larger collections of devices on a single chip. There are many ways in which variability may degrade the quality of a design, and this topic is addressed in greater detail in Chap. 7 of this book. The discussion here will focus on two particular aspects of variability-induced degradation which are relevant to CSE design, namely operating frequency degradation, which involves the cycle limiting paths in the design, and functional failures, which may result from racing paths with insufficient margin, or other design vulnerabilities.

3.6.1 Variability-Induced Frequency Degradation

The key aspect of cycle limiting paths in any given design is that there is usually a relatively large number of logic gates involved, including not only the logic between the launching and capturing CSEs, but also the CSEs themselves, and the circuitry driving the local clock signals to these CSEs. Thus local, uncorrelated, random fluctuations in device parameters are less likely to have a large impact on the timings of these paths, since these will average over the large numbers of devices involved. Problems are more likely to arise from more global, systematic variations which apply to all the circuits in a given region of the design. Some examples here might include PFET to NFET strength ratios, wire speed to device speed ratios, design pattern-factor-induced effects, chip mean device threshold voltage variability, etc. In principle, given an accurate knowledge of all the parameter distributions, and a complete statistical timing methodology [54], it would be possible to predict more accurately which speed paths are most likely to limit the design, and in fact it is expected that this sort of analysis will become more prevalent in the future. However, the underlying variabilities and uncertainties will not go away, even as the ability to accommodate them improves, and so it is important to understand how specific CSE design techniques might improve the situation.

The concept of soft cycle boundaries has already been discussed at some length, and this is an important method of providing some protection against variability or unknown factors. The 2-phase level-sensitive latch methodology probably provides the most benefit in this regard, but in light of the issues described earlier, it is unlikely that the usage of this style will become widespread as a solution for improved variability tolerance. Rather, it is expected that efforts to soften the cycle boundaries are more likely to be concentrated on MSFF and pulsed level-sensitive designs. Furthermore, given that an MSFF with overlapped clocks and pulsed level-sensitive latches have similar hold time issues for a given transparent window size, but unequal power and latency characteristics [5], it seems that pulsed clocking techniques would be the preferred method of providing a soft cycle boundary to protect against variability-induced frequency degradation. However, the local clock edges for the MSFF solution can be tuned to optimize the trade off between hold time and transparent window size without the minimum pulse-width constraints of pulsed-clock designs, and widespread use of pulsed-clock latches will make the design more prone to variability-induced functional failure, as described in the next section.

In addition to the use of soft cycle boundaries, many chips are being designed with programmable clock edges which can also be used to mitigate the effects of variability and uncertainty [28]. Delaying specific clock edges can mitigate the longest timing paths found in the hardware. It is expected that techniques such as this will become more prevalent in the future, as variability continues to increase, and as the improvement in technology decreases from generation to generation. One might expect to see more automated techniques used to optimize clock control settings on a chip-specific basis, or the use of adaptive or autonomous techniques, moving from the more global adaptive deskew techniques in use today [55] (also described in Chaps. 2 and 7), to more fine-grained adjustments in the future.

3.6.2 Variability-Induced Functional Failures

In contrast to variability-induced frequency degradation, a race path or other functional hazard may involve only a small number of devices, so that local random variations may play a big role in determining the margins required to protect against such failures. For specific circuit configurations within certain CSE topologies, it will be necessary to carry out a complete statistical analysis at the desired conditions for use, and at various process corners, to make sure that the design is robust enough against any probable statistical variation [6]. However, in the general case, it will be impossible to analyze all race conditions between all CSEs to that level of detail without a global statistical timing methodology. Also, since the amount of variability will depend on the detailed circuit parameters (for example channel widths involved, device types, the types and amounts of parasitic resistance and capacitances, etc.) it will become extremely difficult to guardband every potential race path against the worst-case combination of statistical fluctuations. Thus it is likely that advanced CSE designs and the need to protect against statistical-variability-induced failures will drive the need for an increasingly sophisticated true statistical timing methodology.

In fact, the need for such a methodology is arguably even greater here than for the operating frequency analysis discussed in the preceding section.

Although there are many sources of variability, any one of which may lead to failure, it is expected that future designs will continue to push towards lower supply voltages, while device dimensions continue to shrink. This means that factors leading to threshold voltage variability are likely to play an increasingly significant role [56]. For this reason, statistical analyses of race conditions will be needed at the minimum operating voltage, and this low-voltage corner will often impose more stringent restrictions on the design than will the high-voltage corner, especially if the write of the latch must overcome a "weak" keeper device. Recent effort has been reported on measurement of hold time variability, using structures designed to look at race conditions under realistic design conditions, showing directly, in a hardware-based measurement, the increase in margins needed as the technology is scaled to finer dimensions [57], and/or as the operating voltage is lowered [58].

Pulsed static, or pulsed semi-dynamic designs may be particularly susceptible to fluctuation-induced writeability failures. For this reason, designers have added local pulse-width controls to the pulse generators used for these latches [6, 28], for test purposes and also as an emergency option in case of unexpected hardware problems. Such an example is shown in Fig. 3.20. Another technique mentioned earlier in the context of power-reduction features is to configure the clock drivers for MSFF latches such that they can provide either regular clocks, or a pulsed clock to the slave with the master clock held at a constant high value [5, 22, 28]. This scheme maintains a fail-safe option in case of either writeability issues or data races. In principle, with techniques like this it should be possible to cut down on the margins normally needed for guaranteeing race-free operation; registers containing latches which are seen to fail in the hardware with pulsed-mode clocking could be configured to run in MSFF mode.

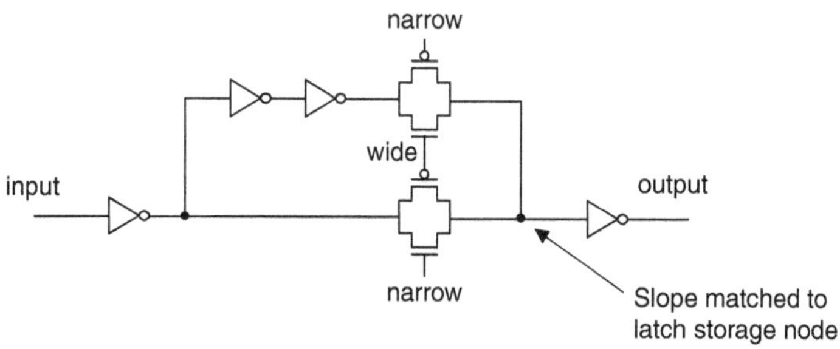

Fig. 3.20. Programmable delay line defining the trailing edge of a local clock pulse. Transmission gate structure is designed to match that in the latch. Reproduced with permission from [6], ©2008 IEEE

Techniques like these may be used not only to guard against unexpected problems, but may also help with test and debug, and one may imagine that in the future, designs may be automatically configured with the optimal pulse-width settings for simultaneously maximizing cycle stealing across soft cycle boundaries while maintaining a safe margin against race failures.

3.7 Reliability Issues

While issues related to process variability or random parameter fluctuations can, in principle, be screened or protected against by proper testing of the part, reliability issues can be much more difficult to protect against. It will be no surprise to the reader that adequate reliability of CSEs is becoming increasingly difficult to guarantee, as the number of CSEs on a single chip continues to increase and as the technology feature size continues to shrink. This section will examine two types of reliability issues, soft error upsets which can disturb the data in the CSEs, and wear out mechanisms which may cause failure after a prolonged period of use.

3.7.1 Soft Error Rate Considerations

Several factors combine to make SER robustness an increasingly important design concern in future systems. Although many designers may consider SER to be important only for situations where extremely high reliability is required, or a problem only for large, high-density SRAMs, recent work has shown that this assumption is no longer true, and as a result, soft errors in CSEs have been getting an increasing amount of attention in recent years.

As CMOS technology has continued to scale well into the sub-100nm dimensions, the combination of feature size reduction, and especially power supply voltage reduction has resulted in a steadily increasing susceptibility of individual latches and flip-flops to soft error upset [59–61], and by some accounts the SER rate in a typical latch or flip-flop has already eclipsed the SRAM SER rate in unprotected rate-per-bit comparisons [62]. Furthermore, error correction codes may be used to help improve the SER in SRAMs, with a relatively low overhead, but no such low-overhead solution is currently available for the collection of CSEs in a typical microprocessor. The voltage factor is especially important, and as systems become ever more adept at lowering chip voltage whenever possible to cut back on power dissipation, there will be a price to be paid in terms of SER. At the same time, the microprocessor industry has seen a shift away from steady clock frequency increases for system performance improvements, towards increased parallelism, increased functionality, and increased integration levels. Therefore, it is expected that the number of CSEs per chip will continue to grow rapidly. At the same time, the SER per element is also growing, leading to a looming "SER crisis" if no action is taken.

To improve the SER situation, in addition to specific CSE design techniques, there are various system, error checking/detection, and technology options which have been discussed in the literature [41, 60, 63–66]. Improvements in processes and materials may also help to reduce contaminants and/or provide increased immunity. However, this section will focus on local CSE design options, and methods therein to improve SER robustness. Perhaps the first approach to try might be to target improvements in SER by careful layout optimization. However, studies of the SER as a function of the CSE layout details have not shown any significant dependencies [62], and there appears to be little success reported in this regard. Another possibility is to improve the SER robustness of individual latches or flip-flops by simply adding additional capacitance to sensitive nodes [67], or by selectively increasing the size of devices holding sensitive nodes [68]. These techniques have been shown to help, and overall area cost may not be too excessive in some situations [69]. However, these techniques are more useful for specific design areas where high reliability is needed, and operating frequency/power is not a big issue. These hardening techniques are not very suitable for general use in microprocessor design. Moreover, these SER mitigation strategies will not scale very well to future technologies.

Future strategies for improving the soft error characteristics of CSEs are more likely to involve topological modifications to the circuits themselves. It is expected that the industry is likely to make more use of circuit related hardening techniques such as the Dual Interlocked storage Cell (DICE) latch [70]. The DICE latch uses redundant and interconnected storage nodes such that the cell cannot be flipped by the upset of a single device. An example [71] is shown in Fig.3.21. This design is resistant to single device upsets, although it can be seen that certain nodes may be left floating for a short period of time after a strike occurs. Depending on the operating voltage, experimental data on this cell showed a factor of 30× to 100× improvement in SER robustness with the observed errors likely caused by multiple upsets. Since the failure mechanism for this type of CSE involves charge collection on multiple nodes, care should be taken during layout to separate critical devices in order to reduce the probability of an error caused in this fashion [71].

DICE latches have been employed in a recent high-end microprocessor design [6], illustrating the combination of pulsed-clock techniques for power reduction and soft cycle boundaries, scan capability for test and debug, and special design tech-

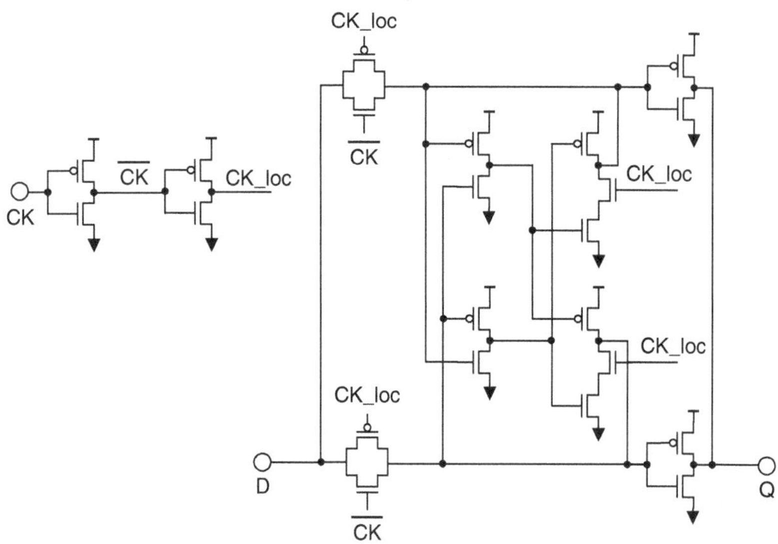

Fig. 3.21. DICE latch topology. Reproduced with permission in a form similar to that in [71], ©2004 IEEE

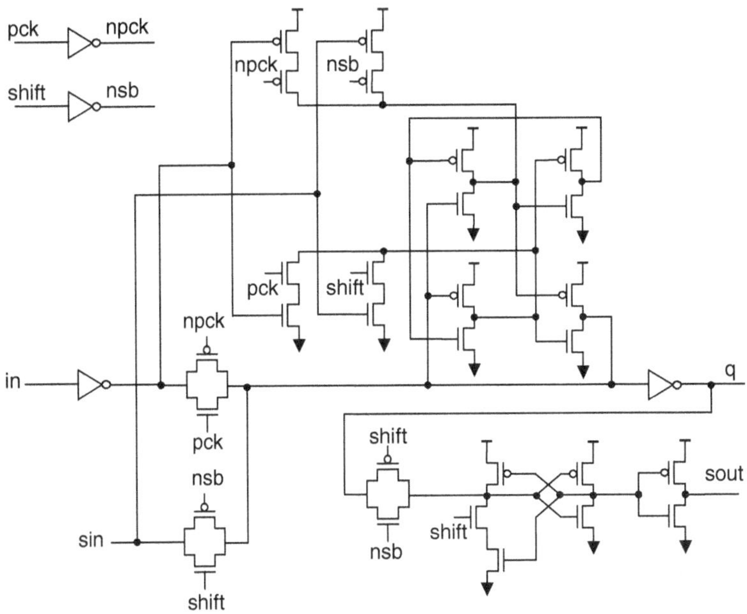

Fig. 3.22. Scannable pulsed-clock DICE latch. Reproduced with permission in a form similar to that in [6], ©2008 IEEE

niques for SER robustness. This scannable, pulsed-clock DICE latch is shown in Fig.3.22. The DICE method is expensive; an overall latch area increase of about 35% and power increase of about 25% has been reported [6]. However, there will be many situations where error checking may not be practical, or the overhead may be too high (much higher than that of switching to DICE elements), and it is expected that techniques such as these will become more widespread in the future as designers are forced to turn their attention to this problem.

3.7.2 End of Life Considerations for CSE Design

There are a number of wear out mechanisms which affect the operation of silicon CMOS circuits, some of which may result in relatively sudden catastrophic failures, while others result in a more gradual parametric device degradation over time, leading eventually to failure. The former type of problem is difficult to deal with at the single element level; usually higher level checking or redundancy schemes are needed, which are beyond the scope of this chapter. However, the second type of problem is more amenable to local circuit solutions, and some of these solutions will be described in this section.

Two major phenomena which may lead to gradual circuit degradation are hot carrier injection (HCI) and negative bias temperature instability (NBTI) [72]. HCI has long been recognized as a potential problem for CMOS circuits [73, 74], but as modern CMOS technology has moved to lower voltages, degradation due to HCI has tended to diminish in terms of its importance. This is especially true for digital CMOS circuits where the current flow through the devices is very transient in nature, and devices are not biased in a way that would subject them to hot carrier stress for long periods of time. Thus for CSE designs with reasonable signal slews, HCI is unlikely to cause significant degradation, although with the advent of new materials this an area that will bear watching for the future [75]. Both HCI and NBTI are addressed in more detail in Sect. 8.3.5

For CSE designs in today's technologies, NBTI is a more important concern. NBTI is specific to PFETs, increasing the magnitude of the threshold voltage, and decreasing carrier mobility over time, depending on the stress conditions. The customary manner of treating this issue is to design test sequences such that all parts which pass are guaranteed to have adequate margin under all operating conditions against the impact of any future degradation. Random collections of parts may be stressed over periods of time to assess their reliability, given a particular set of screening tests at time zero. However, given the intrinsic statistical nature of the phenomenon involved, the shrinking of the device geometries with the resultant growing variability, and the increasing numbers of devices integrated on a single chip, it may be necessary to ask whether the margins needed during test will remain within reasonable bounds in the future, and, given the power/performance cost of maintaining such margins, whether there are CSE design techniques which could be used to reduce this overhead.

Most of the work on this subject is focused on the issue of frequency degradation, or the slow down of components which occurs over time. One method that has been studied is the so called Razor technique, originally proposed [76] to allow aggressive dynamic voltage scaling, but also applicable to wearout-induced circuit degradation as well. A Razor MSL is shown in Fig. 3.23. The clock to the shadow latch is delayed enough to guarantee that the incoming data is successfully latched, even when the main latch fails to capture the correct data. In the event that the two latches contain different data, an error signal is registered, and it is possible for the

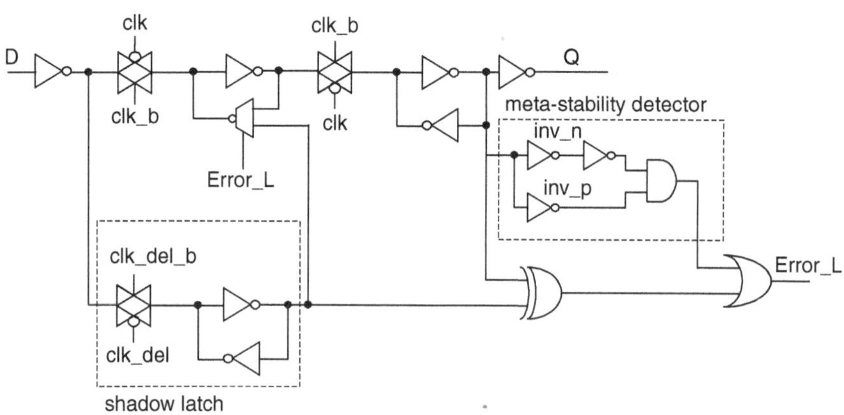

Fig. 3.23. Razor master–slave latch. Reproduced with permission from [76], ©2003 IEEE

system to recover by later swapping in the correct data from the shadow latch (at the cost of a few cycles, depending on the exact recovery scheme used). An error monitor could take appropriate action to maintain a reasonably low error rate, for example by adjusting the processor voltage or frequency, thereby avoiding excessive performance loss. The optimal error rate could then represent a trade off between the performance degradation caused by the overhead of error correction and the benefits of a higher frequency or lower operating voltage. The Razor technique applies considerable overhead to a typical MSL (not to mention the recovery logic overhead) but would not be needed on all CSEs. The Razor MSFF is itself vulnerable to hold time issues, and in fact the delayed clock to the shadow latch significantly increases the overall hold time for the CSE. Another issue of such a scheme would be that either the power or performance of the system could change over time, as adjustments were made for NBTI degradation (or simply in response to environmental changes), perhaps requiring re-instatement of some of the guardbands that were to be avoided in the first place. Sections 7.4.1–7.4.3 describe the Razor methodology in more detail from the viewpoint of addressing process variation through resiliency.

Other research has focused on trying to predict impending errors, and take action before the error actually occurs [78]. In this case, a transition detector watches for transitions which are arriving very late at the capturing CSE, and triggers if these transitions fall within the defined danger window. Chip monitoring hardware or software may then take action before any errors occur. This eliminates the logic complexity and overhead associated with the error recovery mechanisms. Such transition detectors may also be integrated into pulsed latches and used as error detectors [77, 79] in a similar fashion to the original Razor design, but with less overhead inside the CSE. An example of such a scheme is shown in Fig. 3.24. A more detailed description of these techniques appears in Sect. 7.4. One drawback of these transition detector schemes is that, in order to reliably signal the presence of an error (or an impending error), such circuits will need enough built-in margin to work reliably; it has to be guaranteed that the transition is always detected before an error actually occurs. This built-in margin will tend to lower the achievable operating frequency.

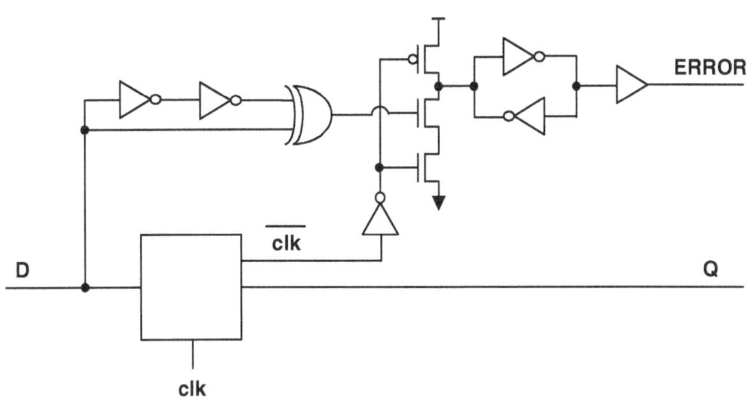

Fig. 3.24. Transition detection scheme. Reproduced with permission from [77], ©2008 IEEE

As a "last resort," redundant logic techniques have also been studied, where the CSE has the ability to automatically swap out a whole block of logic on sensing an impending fail [80], replacing it with an equivalent set of logic gates. Of course, the overhead here is extremely high. Regardless of whether or not techniques like these ever become adopted in a widespread fashion by the industry, it is likely that future microprocessors will require more advanced techniques to guard against wear out-induced reliability failures, either locally at the CSE-level, as described above, or by using more global monitoring and checking algorithms.

Finally, throughout all of the above, it has been assumed that race conditions, pulse-width and/or latch writeability margins can be ensured through a combination of design margin and test conditions, without too much overhead. While this latter assumption may still hold true for some time in the future, usage of large numbers of pulsed-clock components in future technologies are likely to drive the need for more advanced testing techniques, including both race path and pulse-width stressing using some of the special clock pulse width/and/or clock edge control features described earlier. In addition, that fact that race path or hold time failures generally involve a small number of logic gates means that they will tend to be more sensitive to variabilities inherent in the various device degradation mechanisms. As this variability increases, it may no longer be possible to ensure adequate margin through specific test voltage and temperature conditions alone.

3.8 Conclusion

In light of the ongoing power crisis in modern microprocessors, tomorrow's high-performance processors are likely to continue the push towards aggressive use of pulsed-static latches, which require only a single clock and provide for a soft cycle boundary. Improved analysis tools will be needed to guarantee robust operation of a large collection of such circuits across the full PVT space, and especially to be able to handle the ever-increasing impact of random local fluctuations. To ensure the highest quality, reliability, and system performance, future designs will use an increasingly sophisticated collection of special features for test, debug, and chip optimization. Finally, SER-related reliability will become a key issue for CSE design in the future. Tomorrow's latch and flip-flop designers will need to consider not only the usual power/performance aspects of their solutions, but will also need to design for enhanced testability, robustness against statistical variations, and high levels of reliability, and SER immunity.

Acknowledgements

The author would like to acknowledge the feedback and comments from Leon Sigal and Thucydides Xanthopoulos.

References

[1] S. Unger and C.-J. Tan, "Clocking schemes for high-speed digital systems," *IEEE Trans. Comput.*, vol. 35, no. 10, pp. 880–895, Oct. 1986.

[2] K. Wagner, "Clock system design," *IEEE Des. Test Comput.*, vol. 5, no. 5, pp. 9–27, Oct. 1988.

[3] V. Oklobdzija, V. Stojanovic, D. Markovic, and N. Nedovic, *Digital System Clocking*. Wiley-IEEE Press, New York, 2003.

[4] V. Oklobdzija, "Clocking and clocked storage elements in a multi-gigahertz environment," *IBM J. Res. Dev.*, vol. 47, no. 5/6, pp. 567–583, September/November 2003.

[5] J. Warnock, D. Wendel, T. Aipperspach, E. Behnen, R. Cordes, S. Dhong, K. Hirairi, H. Murakami, S. Onishi, D. Pham, J. Pille, S. Posluszny, O. Takahashi, and H. Wen, "Circuit design techniques for a first-generation Cell broadband engine processor," *IEEE J. Solid-State Circuits*, vol. 41, no. 8, pp. 1692–1706, Aug. 2006.

[6] D. Krueger, E. Francom, and J. Langsdorf, "Circuit design for voltage scaling and SER immunity on a quad-core Itanium® processor," in *Digest of Technical Papers IEEE International Solid-State Circuits Conference (ISSCC 2008)*, 2008, pp. 94–95.

[7] J. Friedrich, B. McCredie, N. James, B. Huott, B. Curran, E. Fluhr, G. Mittal, E. Chan, Y. Chan, D. Plass, S. Chu, H. Le, L. Clark, J. Ripley, S. Taylor, J. Dilullo, and M. Lanzerotti, "Design of the POWER6 microprocessor," in *Digest of Technical Papers IEEE International Solid-State Circuits Conference (ISSCC 2007)*, 2007, pp. 96–97.

[8] S. Naffziger, G. Colon-Bonet, T. Fischer, R. Riedlinger, T. Sullivan, and T. Grutkowski, "The implementation of the Itanium 2 microprocessor," *IEEE J. Solid-State Circuits*, vol. 37, no. 11, pp. 1448–1460, Nov. 2002.

[9] C. Giacomotto, N. Nedovic, and V. Oklobdzija, "The effect of the system specification on the optimal selection of clocked storage elements," *IEEE J. Solid-State Circuits*, vol. 42, no. 6, pp. 1392–1404, June 2007.

[10] V. Zyuban, D. Brooks, V. Srinivasan, M. Gschwind, P. Bose, P. Strenski, and P. Emma, "Integrated analysis of power and performance for pipelined microprocessors," *IEEE Trans. Comput.*, vol. 53, no. 8, pp. 1004–1016, Aug. 2004.

[11] V. Stojanovic and V. Oklobdzija, "Comparative analysis of master-slave latches and flip-flops for high-performance and low-power systems," *IEEE J. Solid-State Circuits*, vol. 34, no. 4, pp. 536–548, April 1999.

[12] D. Markovic, B. Nikolic, and R. Brodersen, "Analysis and design of low-energy flip-flops," in *Proceedings of the Low Power Electronics and Design, International Symposium*, 6–7 Aug. 2001, pp. 52–55.

[13] V. Zyuban, "Optimization of scannable latches for low energy," *IEEE Trans. VLSI Syst.*, vol. 11, no. 5, pp. 778–788, Oct. 2003.

[14] J. Tschanz, S. Narendra, Z. Chen, S. Borkar, M. Sachdev, and V. De, "Comparative delay and energy of single edge-triggered and dual edge-triggered pulsed flip-flops for high-performance microprocessors," in *Proceedings of the Low*

Power Electronics and Design, International Symposium, 6–7 Aug. 2001, pp. 147–152.

[15] M. Hamada, H. Hara, T. Fujita, C. K. Teh, T. Shimazawa, N. Kawabe, T. Kitahara, Y. Kikuchi, T. Nishikawa, M. Takahashi, and Y. Oowaki, "A conditional clocking flip-flop for low power H.264/MPEG-4 audio/visual codec LSI," in *Proceedings of the IEEE Custom Integrated Circuits Conference (CICC 2005)*, 18–21 Sept. 2005, pp. 527–530.

[16] S. DasGupta, E. Eichelberger, and T. Williams, "LSI chip design for testability," in *Digest of Technical Papers IEEE International Solid-State Circuits Conference (ISSCC 1978)*, 1978, pp. 216–217.

[17] G. Gerosa, S. Gary, C. Dietz, D. Pham, K. Hoover, J. Alvarez, H. Sanchez, P. Ippolito, T. Ngo, S. Litch, J. Eno, J. Golab, N. Vanderschaaf, and J. Kahle, "A 2.2 W, 80 MHz superscalar RISC microprocessor," *IEEE J. Solid-State Circuits*, vol. 29, no. 12, pp. 1440–1454, Dec. 1994.

[18] R. Ho, K. Mai, and M. Horowitz, "The future of wires," *Proceedings of the IEEE*, vol. 89, no. 4, pp. 490–504, April 2001.

[19] Y. Suzuki, K. Odagawa, and T. Abe, "Clocked CMOS calculator circuitry," *IEEE J. Solid-State Circuits*, vol. 8, no. 6, pp. 462–469, Dec 1973.

[20] J. Warnock, J. Keaty, J. Petrovick, J. Clabes, C. Kircher, B. Krauter, P. Restle, B. Zoric, and C. Anderson, "The circuit and physical design of the POWER4 microprocessor," *IBM J. Res. Dev.*, vol. 46, no. 1, pp. 27–51, January 2002.

[21] D. Lackey, "Efficient latch and clock structures for system-on-chip test flexibility," in *Proceedings of the IEEE International Test Conference ITC '06*, Oct. 2006, pp. 1–7.

[22] B. Stolt, Y. Mittlefehldt, S. Dubey, G. Mittal, M. Lee, J. Friedrich, and E. Fluhr, "Design and implementation of the POWER6 microprocessor," *IEEE J. Solid-State Circuits*, vol. 43, no. 1, pp. 21–28, Jan. 2008.

[23] I. Lin, J. Ludwig, and K. Eng, "Analyzing cycle stealing on synchronous circuits with level-sensitive latches," in *Proceedings of the ACM/IEEE Design Automation Conference*, 8–12 June 1992, pp. 393–398.

[24] E. Shriver, D. Hall, N. Nassif, N. Rahman, N. Rethman, G. Watt, and J. Farrell, "Timing verification of the 21264: A 600 MHz full-custom microprocessor," in *Proceedings of the International Conference on Computer Design: VLSI in Computers and Processors ICCD '98*, 5–7 Oct. 1998, pp. 96–103.

[25] K. Sakallah, T. Mudge, and O. Olukotun, "$checkT_c$ and $minT_c$: Timing verification and optimal clocking of synchronous digital circuits," in *Proceedings of the IEEE International Conference on Computer-Aided Design ICCAD-90. Digest of Technical Papers*, 11–15 Nov. 1990, pp. 552–555.

[26] T. Szymanski and N. Shenoy, "Verifying clock schedules," in *Proceedings of the IEEE/ACM International Conference on Computer-Aided Design ICCAD-92. Digest of Technical Papers*, 8–12 Nov. 1992, pp. 124–131.

[27] T. Burks, K. Sakallah, and T. Mudge, "Identification of critical paths in circuits with level-sensitive latches," in *Proceedings of the IEEE/ACM International Conference on Computer-Aided Design ICCAD-92. Digest of Technical Papers*, 8–12 Nov. 1992, pp. 137–141.

[28] B. Curran, "Power-constrained high-frequency circuits for the IBM POWER6 microprocessor," *IBM J. Res. Dev.*, vol. 51, no. 6, pp. 715–731, November 2007.
[29] M. Matsui, H. Hara, Y. Uetani, L.-S. Kim, T. Nagamatsu, Y. Watanabe, A. Chiba, K. Matsuda, and T. Sakurai, "A 200 MHz 13 mm^2-2D DCT macrocell using sense-amplifying pipeline flip-flop scheme," *IEEE J. Solid-State Circuits*, vol. 29, no. 12, pp. 1482–1490, Dec. 1994.
[30] J. Montanaro, R. Witek, K. Anne, A. Black, E. Cooper, D. Dobberpuhl, P. Donahue, J. Eno, W. Hoeppner, D. Kruckemyer, T. Lee, P. Lin, L. Madden, D. Murray, M. Pearce, S. Santhanam, K. Snyder, R. Stehpany, and S. Thierauf, "A 160-MHz, 32-b, 0.5-W CMOS RISC microprocessor," *IEEE J. Solid-State Circuits*, vol. 31, no. 11, pp. 1703–1714, Nov. 1996.
[31] P. Gronowski, W. Bowhill, R. Preston, M. Gowan, and R. Allmon, "High-performance microprocessor design," *IEEE J. Solid-State Circuits*, vol. 33, no. 5, pp. 676–686, May 1998.
[32] B. Nikolic, V. Oklobdzija, V. Stojanovic, W. Jia, J. K.-S. Chiu, and M. Ming-Tak Leung, "Improved sense-amplifier-based flip-flop: design and measurements," *IEEE J. Solid-State Circuits*, vol. 35, no. 6, pp. 876–884, June 2000.
[33] Y. Zhang, H. Yang, and H. Wang, "Low clock-swing conditional-precharge flip-flop for more than 30% power reduction," *Electron. Lett.*, vol. 36, no. 9, pp. 785–786, 2000.
[34] T. Darwish and M. Bayoumi, "Reducing the switching activity of modified SAFF flip-flop for low power applications," in *Proceedings of the 14th International Conference on 2002 – ICM Microelectronics*, 11–13 Dec. 2002, pp. 96–99.
[35] J.-C. Kim, Y.-C. Jang, and H.-J. Park, "CMOS sense amplifier-based flip-flop with two $N-C^2MOS$ output latches," *Electron. Lett.*, vol. 36, no. 6, pp. 498–500, 16 March 2000.
[36] A. Strollo, D. De Caro, E. Napoli, and N. Petra, "A novel high-speed sense-amplifier-based flip-flop," *IEEE Trans. VLSI Syst.*, vol. 13, no. 11, pp. 1266–1274, Nov. 2005.
[37] H. Partovi, R. Burd, U. Salim, F. Weber, L. DiGregorio, and D. Draper, "Flow-through latch and edge-triggered flip-flop hybrid elements," in *Digest of Technical Papers IEEE International Solid-State Circuits Conference (ISSCC 1996)*, 1996, pp. 138–139.
[38] N. Nedovic and V. Oklobdzija, "Dynamic flip-flop with improved power," in *Proceedings of the 26th European ESSCIRC Solid-State Circuits Conference '00*, 19–21 Sept. 2000, pp. 376–379.
[39] N. Nedovic and V. Oklobdzija, "Hybrid latch flip-flop with improved power efficiency," in *Proceedings of the 13th Symposium on Integrated Circuits and Systems Design*, 18–24 Sept. 2000, pp. 211–215.
[40] F. Klass, C. Amir, A. Das, K. Aingaran, C. Truong, R. Wang, A. Mehta, R. Heald, and G. Yee, "A new family of semidynamic and dynamic flip-flops with embedded logic for high-performance processors," *IEEE J. Solid-State Circuits*, vol. 34, no. 5, pp. 712–716, May 1999.

[41] C. Webb, C. Anderson, L. Sigal, K. Shepard, J. Liptay, J. Warnock, B. Curran, B. Krumm, M. Mayo, P. Camporese, E. Schwarz, M. Farrell, P. Restle, I. Averill, R.M., T. Slegel, W. Houtt, Y. Chan, B. Wile, T. Nguyen, P. Emma, D. Beece, C.-T. Chuang, and C. Price, "A 400-MHz S/390 microprocessor," *IEEE J. Solid-State Circuits*, vol. 32, no. 11, pp. 1665–1675, Nov. 1997.

[42] R. Heald, K. Aingaran, C. Amir, M. Ang, M. Boland, P. Dixit, G. Gouldsberry, D. Greenley, J. Grinberg, J. Hart, T. Horel, W.-J. Hsu, J. Kaku, C. Kim, S. Kim, F. Klass, H. Kwan, G. Lauterbach, R. Lo, H. McIntyre, A. Mehta, D. Murata, S. Nguyen, Y.-P. Pai, S. Patel, K. Shin, K. Tam, S. Vishwanthaiah, J. Wu, G. Yee, and E. You, "A third-generation SPARC V9 64-b microprocessor," *IEEE J. Solid-State Circuits*, vol. 35, no. 11, pp. 1526–1538, Nov. 2000.

[43] W. Belluomini, D. Jamsek, A. Martin, C. McDowell, R. Montoye, T. Nguyen, H. Ngo, J. Sawada, I. Vo, and R. Datta, "An 8 GHz floating-point multiply," in *Digest of Technical Papers IEEE International Solid-State Circuits Conference (ISSCC 2005)*, 2005, pp. 374–375, 604.

[44] J. Parkhurst, J. Darringer, and B. Grundmann, "From single core to multi-core: Preparing for a new exponential," in *Proceedings of the IEEE/ACM International Conference on Computer-Aided Design ICCAD '06*, 2006, pp. 67–72.

[45] B. Stackhouse, B. Cherkauer, M. Gowan, P. Gronowski, and C. Lyles, "A 65nm 2-billion-transistor quad-core Itanium® processor," in *Digest of Technical Papers IEEE International Solid-State Circuits Conference (ISSCC 2008)*, 2008, pp. 92–93, 598.

[46] C.-G. Hwang, "New paradigms in the silicon industry," in *Proceedings of the International Electron Devices Meeting IEDM '06*, 11–13 Dec. 2006, pp. 1–8.

[47] D. Josephson, S. Poehlman, V. Govan, and C. Mumford, "Test methodology for the McKinley processor," in *Proceedings of the International Test Conference*, 30 Oct.–1 Nov. 2001, pp. 578–585.

[48] R. Molyneaux, T. Ziaja, H. Kim, S. Aryani, S. Hwang, and A. Hsieh, "Design for testability features of the SUN Microsystems Niagara2 CMP/CMT SPARC chip," in *Proceedings of the IEEE International Test Conference ITC 2007*, 21–26 Oct. 2007, pp. 1–8.

[49] S. DasGupta, R. Walther, T. Williams, and E. Eichelberger, "An enhancement to LSSD and some applications of LSSD in reliability, availability, and serviceability," in *Proceedings of the Twenty-Fifth International Symposium on Fault-Tolerant Computing, ' Highlights from Twenty-Five Years'*, 1995, p. 289.

[50] K.-T. Cheng, S. Devadas, and K. Keutzer, "A partial enhanced-scan approach to robust delay-fault test generation for sequential circuits," in *Proceedings of the International Test Conference*, 26–30 Oct 1991, p. 403.

[51] J. Savir and S. Patil, "On broad-side delay test," *IEEE Trans. VLSI Syst.*, vol. 2, no. 3, pp. 368–372, Sept. 1994.

[52] J. Savir and S. Patil, "Scan-based transition test," *IEEE Trans. Comput.-Aided Design Integr. Circuits Syst.*, vol. 12, no. 8, pp. 1232–1241, Aug. 1993.

[53] A. Asenov, "Simulation of statistical variability in nano MOSFETs," in *Proceedings of the IEEE Symposium on VLSI Technology*, 12–14 June 2007, pp. 86–87.
[54] C. Visweswariah, "Death, taxes and failing chips," in *Proceedings of the Design Automation Conference*, 2–6 June 2003, pp. 343–347.
[55] E. Fetzer, "Using adaptive circuits to mitigate process variations in a microprocessor design," *IEEE Design & Test Comput.*, vol. 23, no. 6, pp. 476–483, June 2006.
[56] G. Roy, A. Brown, F. Adamu-Lema, S. Roy, and A. Asenov, "Simulation study of individual and combined sources of intrinsic parameter fluctuations in conventional Nano-MOSFETs," *IEEE Trans. Electron. Devices*, vol. 53, no. 12, pp. 3063–3070, Dec. 2006.
[57] G. Neuberger, F. Kastensmidt, R. Reis, G. Wirth, R. Brederlow, and C. Pacha, "Statistical analysis of systematic and random variability of flip-flop race immunity in 130nm and 90nm CMOS technologies," in *Proceedings of the IFIP International Conference on Very Large Scale Integration VLSI - SoC 2007*, 15–17 Oct. 2007, pp. 78–83.
[58] F. Klass, A. Jain, G. Hess, and B. Park, "An all-digital on-chip process-control monitor for process-variability measurements," in *Digest of Technical Papers IEEE International Solid-State Circuits Conference (ISSCC 2008)*, 3–7 Feb. 2008, pp. 408–409, 623.
[59] T. Karnik and P. Hazucha, "Characterization of soft errors caused by single event upsets in CMOS processes," *IEEE Trans. Dependable Secure Comput.*, vol. 1, no. 2, pp. 128–143, April–June 2004.
[60] R. Baumann, "Radiation-induced soft errors in advanced semiconductor technologies," *IEEE Trans. Device Mater. Rel.*, vol. 5, no. 3, pp. 305–316, Sept. 2005.
[61] H. Fukui, M. Hamaguchi, H. Yoshimura, H. Oyamatsu, F. Matsuoka, T. Noguchi, T. Hirao, H. Abe, S. Onoda, T. Yamakawa, T. Wakasa, and T. Kamiya, "Comprehensive study on layout dependence of soft errors in CMOS latch circuits and its scaling trend for 65 nm technology node and beyond," in *Proceedings of the Digest of Technical Papers VLSI Technology 2005 Symposium*, 14–16 June 2005, pp. 222–223.
[62] T. Heijmen, P. Roche, G. Gasiot, K. Forbes, and D. Giot, "A comprehensive study on the soft-error rate of flip-flops from 90-nm production libraries," *IEEE Trans. Device Mater. Rel.*, vol. 7, no. 1, pp. 84–96, March 2007.
[63] F. Wang and V. D. Agrawal, "Single event upset: An embedded tutorial," in *Proceedings of the 21st International Conference on VLSI Design VLSID 2208*, 4–8 Jan. 2008, pp. 429–434.
[64] M. Nicolaidis, "Design for soft error mitigation," *IEEE Trans. Device Mater. Rel.*, vol. 5, no. 3, pp. 405–418, Sept. 2005.
[65] P. Meaney, S. Swaney, P. Sanda, and L. Spainhower, "IBM z990 soft error detection and recovery," *IEEE Trans. Device Mater. Rel.*, vol. 5, no. 3, pp. 419–427, Sept. 2005.

[66] S. Mitra, M. Zhang, N. Seifert, T. Mak, and K. S. Kim, "Built-in soft error resilience for robust system design," in *Proceedings of the IEEE International Conference on Integrated Circuit Design and Technology ICICDT '07*, May 30 2007–June 1 2007, pp. 1–6.

[67] Y. Dhillon, A. Diril, A. Chatterjee, and A. Singh, "Analysis and optimization of nanometer CMOS circuits for soft-error tolerance," *IEEE Trans. VLSI Syst.*, vol. 14, no. 5, pp. 514–524, May 2006.

[68] Q. Zhou and K. Mohanram, "Gate sizing to radiation harden combinational logic," *IEEE Trans. Comput.-Aided Design Integr. Circuits Syst.*, vol. 25, no. 1, pp. 155–166, Jan. 2006.

[69] A. KleinOsowski, E. Cannon, M. Gordon, D. Heidel, P. Oldiges, C. Plettner, K. Rodbell, R. Rose, and H. Tang, "Latch design techniques for mitigating single event upsets in 65 nm SOI device technology," *IEEE Trans. Nucl. Sci.*, vol. 54, no. 6, pp. 2021–2027, Dec. 2007.

[70] T. Calin, M. Nicolaidis, and R. Velazco, "Upset hardened memory design for submicron CMOS technology," *IEEE Trans. Nucl. Sci.*, vol. 43, no. 6, pp. 2874–2878, Dec. 1996.

[71] P. Hazucha, T. Karnik, S. Walstra, B. Bloechel, J. Tschanz, J. Maiz, K. Soumyanath, G. Dermer, S. Narendra, V. De, and S. Borkar, "Measurements and analysis of SER-tolerant latch in a 90-nm dual-Vt CMOS process," *IEEE J. Solid-State Circuits*, vol. 39, no. 9, pp. 1536–1543, Sept. 2004.

[72] D. Schroder and J. Babcock, "Negative bias temperature instability: Road to cross in deep submicron silicon semiconductor manufacturing," *J. Appl. Phys.*, vol. 94, p. 1, 2003.

[73] T. Ning, P. Cook, R. Dennard, C. Osburn, S. Schuster, and H. Yu, "1 μm MOSFET VLSI technology: Part IV – hot-electron design constraints," *IEEE Trans. Electron. Devices*, vol. 26, no. 4, pp. 346–353, Apr 1979.

[74] C. Hu, S. C. Tam, F.-C. Hsu, P.-K. Ko, T.-Y. Chan, and K. Terrill, "Hot-electron-induced MOSFET degradation – model, monitor, and improvement," *IEEE J. Solid-State Circuits*, vol. 20, no. 1, pp. 295–305, Feb 1985.

[75] J. McPherson, "Reliability trends with advanced CMOS scaling and the implications for design," in *Proceedings of the IEEE Custom Integrated Circuits Conference CICC '07*, 16–19 Sept. 2007, pp. 405–412.

[76] D. Ernst, N. S. Kim, S. Das, S. Pant, R. Rao, T. Pham, C. Ziesler, D. Blaauw, T. Austin, K. Flautner, and T. Mudge, "Razor: a low-power pipeline based on circuit-level timing speculation," in *Proceedings of the 36th Annual IEEE/ACM International Symposium on MICRO-36 Microarchitecture*, 2003, pp. 7–18.

[77] K. Bowman, J. Tschanz, N. S. Kim, J. Lee, C. Wilkerson, S.-L. Lu, T. Karnik, and V. De, "Energy-efficient and metastability-immune timing-error detection and instruction-replay-based recovery circuits for dynamic-variation tolerance," in *Digest of Technical Papers IEEE International Solid-State Circuits Conference (ISSCC 2008)*, 3–7 Feb. 2008, pp. 402–403, 623.

[78] M. Agarwal, B. Paul, M. Zhang, and S. Mitra, "Circuit failure prediction and its application to transistor aging," in *Proceedings of the 25th IEEE VLSI Test Symposium*, 6–10 May 2007, pp. 277–286.

[79] D. Blaauw, S. Kalaiselvan, K. Lai, W.-H. Ma, S. Pant, C. Tokunaga, S. Das, and D. Bull, "Razor II: In situ error detection and correction for PVT and SER tolerance," in *Digest of Technical Papers IEEE International Solid-State Circuits Conference (ISSCC 2008)*, 2008, pp. 400–401, 622.

[80] T. Nakura, K. Nose, and M. Mizuno, "Fine-grain redundant logic using defect-prediction flip-flops," in *Digest of Technical Papers IEEE International Solid-State Circuits Conference (ISSCC 2007)*, 2007, pp. 402–403, 611.

4

Exploiting Inductance

Nestoras Tzartzanis

Fujitsu Laboratories of America

In this chapter, the benefits of using inductance for generating and distributing clocks are explored. Starting with the implementation and modeling of spiral inductors and transmission lines the focus, subsequently, shifts into LC and transmission-line oscillators for generating two or more phases with low phase noise. Finally, resonant clock distribution methods including rotary traveling-wave oscillator arrays, standing wave oscillators and grids, and inductor-based clock grids are presented.

4.1 Introduction

Although inductive components have been used as off-chip components for several decades, their first practical implementation on silicon was reported in early nineties in silicon bipolar and BiCMOS processes [1, 2]. There are two reasons for this delay. First, the low Q obtained for inductors in early silicon processes with only two or three metal layers compared to off-chip inductors was prohibitive for their use in any practical system. Second, the low-frequency requirements for the inductors made system-level integration in a single die very difficult. Since inductor size is inversely proportional to the square of the frequency, the inductors required at the system level were too large to accommodate for on-chip implementations.

Following the evolution of CMOS processes, new applications and new challenges emerged. CMOS has been the technology of choice for large scale applications such as microprocessors in which large integration is the primary goal. However, the rapid increase of CMOS transition (cut-off) frequency (f_T) coupled with its high energy efficiency and high integration scale compared to competing technologies (e.g., SiGe, GaAs) makes CMOS the primary choice for high frequency wireless, wireline, and optical communication applications. Utilizing on-chip inductors to build oscillators became attractive for communication applications as their operating frequency increased to GHz range. Furthermore, one of the drawbacks of CMOS technology is its large process variation which translates into uneven phase generation for ring oscillators and clock skew for clock distribution networks. Oscillator phase deviation can be improved by using LC-based oscillators instead

T. Xanthopoulos (ed.), *Clocking in Modern VLSI Systems*, Integrated Circuits and Systems, DOI 10.1007/978-1-4419-0261-0_4,

of ring oscillators. Also, LC-based oscillators have lower phase noise compared to ring oscillators since they exhibit better supply noise rejection. The clock skew issue gets worse for increasing clock load since it is proportional to the number of buffer stages required to drive the load [3]. To make things even worse, clock skew takes a larger fraction of the clock cycle as frequency increases, even if the clock distribution network and load remain the same. Clock skew can be addressed by replacing conventional buffer-based clock networks with transmission-line-based and LC-based clock distribution.

Spiral inductors and transmission lines are the two dominant structures for forming on-chip inductive components. In this chapter, both spiral inductors and transmission lines are described including implementation and modeling methods. Subsequently, LC-based Voltage-Controlled Oscillators (VCO) for generating two or more phases as well as transmission-line based distributed VCOs are described. In Sect. 4.4, clock distribution methods based on resonance are presented. Finally, Sect. 4.5 summarizes the chapter.

4.2 Monolithic Inductance

4.2.1 Spiral Inductors

The implementation of spiral inductors depends mostly on the target application and the process parameters. For instance, the inductor quality (Q) factor is a key parameter for spiral inductors designed for LC VCOs. Maximizing the Q of the inductor around the VCO operating frequency results in decreasing the VCO phase noise. Furthermore, to meet the VCO frequency range, it is necessary to accurately predict the inductance L of the spiral inductor. For other applications (e.g., shunt and series peaking [4], etc.), inductor area is more important than its Q or inductance accuracy. Finally, for practical verification purposes, an inductor model should be readily available. Inductor models vary depending on the application. Narrowband models are necessary for LC VCOs since they operate around a well-determined frequency whereas broadband models are required for other applications. In this chapter, we focus on inductor implementation and modeling for LC VCOs.

Spiral inductors can be implemented in various shapes. Figure 4.1 shows examples of 2-turn square, octagonal, and symmetrical octagonal spiral inductors. Other common shapes include hexagonal and circular spiral inductors. Typically, high-Q spiral inductors are implemented using low sheet resistance, high-level metal layers. It is also possible to strap together multiple metal layers in order to reduce the inductor resistance. Using high-level metals results in low inductor capacitance and substrate losses as we see next.

The inductance L of spiral inductors can be estimated using Greenhouse's method [5]. However, simpler analytical models are also available [4, 6, 7] based on the inductor geometry (shape, number of turns, radius, metal width, metal spacing, and metal thickness). The inductance model presented in [6] is based on inductor segment decomposition. The length, l, of square spiral inductors is given by:

$$l = (4n+1)\,d_{\text{in}} + (4N_{\text{i}}+1)\,N_{\text{i}}\,(w+s)\,, \tag{4.1}$$

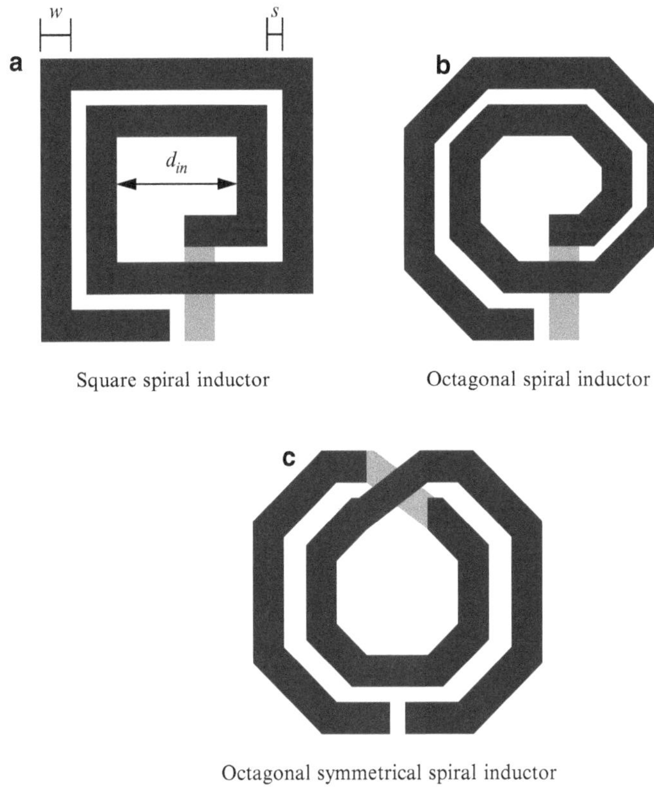

Fig. 4.1. Common 2-turn spiral inductor shapes

where n is the number of turns, N_i is the integer part of n, d_in is the inner diameter of the inductor, w is the metal width, and s is the metal spacing.

The inductance L is estimated taking into consideration the self inductance of the spiral inductor, the negative mutual inductance between segments, and the positive mutual inductance between segments. The total inductance of the square spiral inductance is given by [6]:

$$L = \frac{\mu_0}{2\pi} l \left[\begin{array}{l} \ln \frac{l}{n(w+t)} - 0.2 - 0.47n \\ + (n+1) \left(\ln \left(\sqrt{1 + \left(\frac{l}{4nd_\mathrm{av}} \right)^2} + \frac{l}{4nd_\mathrm{av}} \right) - \sqrt{1 + \left(\frac{4nd_\mathrm{av}}{l} \right)^2} + \frac{4nd_\mathrm{av}}{l} \right) \end{array} \right], \tag{4.2}$$

where μ_0 is the permeability of vacuum, l and n are the inductor length and number of turns, w and t are the metal width and thickness, and d_av is the average segment distance for the positive mutual inductance given by (4.3).

$$d_\mathrm{av} = (w+s) \frac{(3n - 2N_\mathrm{i} - 1)(N_\mathrm{i} + 1)}{3(2n - N_\mathrm{i} - 1)}. \tag{4.3}$$

Inductance analytical models can be derived for different shapes similar to square inductors. In [6], (4.2) is modified for octagonal spiral inductors.

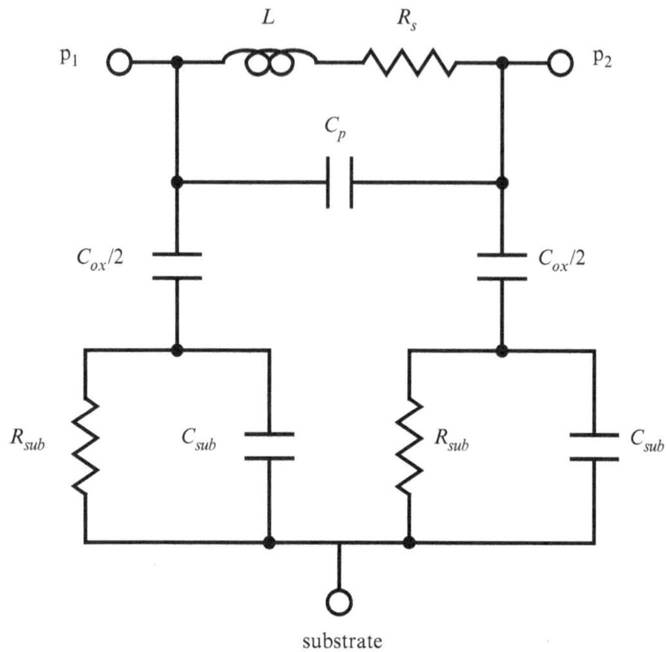

Fig. 4.2. Spiral inductor narrowband lumped model

Deriving lumped element models for spiral inductors is of utmost importance. First, they facilitate circuit simulations (especially in time domain). Second, they provide a well-defined description of spiral inductors that can be used to improve their design. Figure 4.2 shows a widely used lumped element model for spiral inductors [4, 8, 9], which is more appropriate for narrowband applications. A broadband inductor model has also been proposed [10], but such models are beyond the scope of this chapter. The narrowband model has two signal ports p_1 and p_2 and the substrate port which typically connects to ground. It includes the inductance L connected to a series resistance R_s, a shunt capacitance C_p and the inductor to substrate capacitance C_ox. Finally, R_sub and C_sub model substrate resistance and capacitance respectively.

The series resistance R_s models losses due to the metal resistivity and the skin effect. The latter term is used to describe that at high frequency, the current tends to flow at the surface of a conductor causing its resistance to increase. Therefore, R_s is frequency dependent and is given by [4, 9]:

$$R_\mathrm{s} = \frac{\rho l}{w\delta\left(1 - \mathrm{e}^{-t/\delta}\right)}, \tag{4.4}$$

where ρ is the metal resistivity at DC, l is the inductor length, w is the metal width, t is the metal thickness and δ is the metal skin depth obtained by (4.5):

$$\delta = \sqrt{\frac{2\rho}{\omega\mu_0}}. \tag{4.5}$$

Other losses not modeled with (4.4) and (4.5) include proximity effect losses and eddy currents losses to the substrate. The proximity effect term describes the phenomenon of the currents of two parallel conductors runing along the side of their surface that is close to each other [4]. This effect is caused by the magnetic field between the two conductors. Substrate losses due to eddy currents depend on the substrate doping and can be neglected for lightly doped substrates.

Capacitance C_p models the overlap between the metal layer used for the inductor and the underpass metal to connect to the other port:

$$C_p = nw^2\frac{\varepsilon_{oxM}}{t_{oxM}}, \tag{4.6}$$

where n is the number of turns, w is the metal width, ε_{oxM} and t_{oxM} are the oxide permittivity and the oxide thickness between the two metal layers respectively.

C_{ox} is the capacitance between the inductor metal layers and the substrate modeled with:

$$C_{ox} = wl\frac{\varepsilon_{ox}}{t_{ox}}, \tag{4.7}$$

where ε_{ox} and t_{ox} are the oxide permittivity and the oxide thickness between the inductor and the substrate, respectively.

Resistance R_{sub} and capacitance C_{sub} model losses to the substrate and are fitting parameters based on measurements. R_{sub} models loss due to current capacitively coupled to substrate and is given by [4, 9]:

$$R_{sub} = \frac{2}{lwG_{sub}}, \tag{4.8}$$

where G_{sub} is the substrate conductance per unit area. C_{sub} is given by [4, 9]:

$$C_{sub} = \frac{lwC_{subu}}{2}, \tag{4.9}$$

where C_{subu} is the substrate capacitance per unit area.

The analytical expressions presented in this section can be used as first-order approximations to develop inductor models. There are effects that cannot be accounted for by using the analytical equations. For instance dummy metal fills are inserted in modern chemical mechanical polishing (CMP) processes which impact at least the inductor to substrate capacitance. More accurate inductor models are based on hardware characterization and parameter fitting for the components shown in Fig.4.2. Generally, this task should be performed for each inductor used in a circuit. However, it is possible to develop scalable inductor models for varying inductor radius if the other parameters such as shape, number of turns, width, and space are kept constant. Although this restriction imposes limitations to the designers, it can speed up the inductor modeling process. Only a few inductors with different radius must be characterized whereas a full scalable model can be developed through extrapolation from the measured data. In addition, several EM analysis tools [11, 12] are available as an option between simple analytical models and hardware characterization. There is a trade-off between their accuracy and the inductor analysis time and they are more effective when their process settings are calibrated based on hardware measurements to produce accurate models for fast or modest analysis time.

Besides the inductance, the Q factor is the other key parameter for inductors used for clock generation. Analytical models for inductor Q factor based on the inductor lumped model have been developed [9, 13, 14]. However, the Q factor which is a function of frequency can be extracted through hardware measurements for better accuracy. In general the Q of an inductor can be improved by reducing losses. Certain processes provide options to boost the Q of inductors. For example, low-resistivity thick high-level metal layers can be available for inductor implementation. If such option is not available, multiple metal layers can be strapped together to form the inductor. Another practical issue that can impact the inductor Q factor is the necessity of dummy metal fills in modern processes which increase inductor coupling to the substrate. If it is not possible to exclude the dummy metal fills below the inductor, they can be added manually. This serves two purposes. First, only the minimum amount of dummy metal fills is inserted to satisfy process requirements. Second, it resolves the issue of random dummy metal fills which depend on the location of the inductor in the die. Another technique to improve inductor Q is the insertion of patterned ground shields [14] using a low metal layer. Shielding the inductor from the substrate reduces inductor coupling to the substrate as well as noise that reaches the inductor from the substrate. The drawback is the increase of the inductor capacitance since the shield is closer to the inductor than the substrate.

4.2.2 Transmission Lines

As the integrated circuit operating frequency increases, long interconnects behave as distributed instead of lumped elements. This happens when the length of the interconnect approaches the wavelength of the signal highest frequency. In this mode of operation, long interconnects behave as transmission lines and they cannot be modeled using simple RC-based models. Figure 4.3 shows a lumped model for such interconnect. Components L, R, C, and G represent inductance, resistance, capacitance, and conductance per unit length respectively. The inductance L and capacitance C model the magnetic and electric energy-stored components, respectively. The resistance and conductance model energy losses in the transmission line. Specifically, R models losses due to the series resistance of the transmission line including the skin effect, whereas G models losses due to the dielectric conductance.

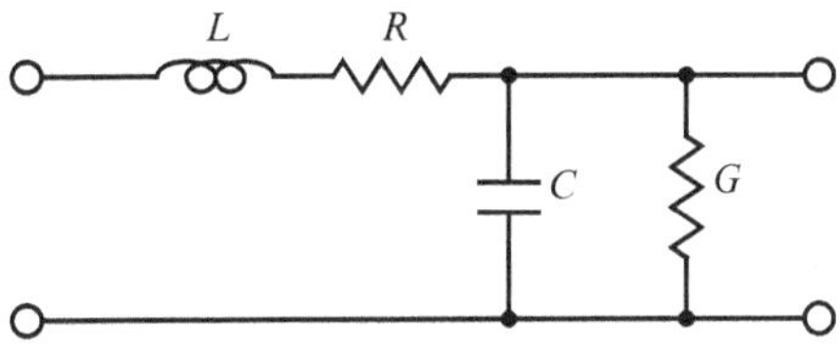

Fig. 4.3. Transmission line lumped model

The characteristic impedance Z_0 of a transmission line determines the voltage to current ratio of a signal propagating along its length and is given by:

$$Z_0 = \sqrt{\frac{R + j\omega L}{G + j\omega C}}. \tag{4.10}$$

If the losses (i.e., R and G) are negligible or if RC=GL, then (4.10) simplifies to:

$$Z_0 = \sqrt{\frac{L}{C}}. \tag{4.11}$$

Another important parameter of a transmission line is its propagation constant, γ, that determines the signal attenuation through the line. The voltage V at any point in the transmission line that is in distance z from its beginning, is given by:

$$V(z) = V_0 \mathrm{e}^{-\gamma z}, \tag{4.12}$$

where V_0 is the voltage at $z = 0$.

The propagation constant, γ, is given by:

$$\gamma = \sqrt{(R + j\omega L)(G + j\omega C)}, \tag{4.13}$$

According to (4.13), γ can be generalized into the following complex form [4]:

$$\gamma = \alpha + j\beta, \tag{4.14}$$

where α determines the attenuation in the line due to the distance increase and β has only phase contribution.

If the losses are negligible, (4.13) reduces to:

$$\gamma = j\omega\sqrt{LC}. \tag{4.15}$$

As expected, there is no attenuation to signal amplitude, only its phase changes proportionally to its frequency. As shown in [4], the delay of a lossless line is a constant independent from frequency, but proportional to $\sqrt{LC}$. At the same time, the bandwidth in the lossless transmission line is infinite because there is no signal attenuation. As long as R and G can be neglected, increasing L and C causes the transmission line delay to increase, but has no effect on its bandwidth.

If RC=GL, then (4.13) simplifies to:

$$\gamma = \sqrt{RG} + j\omega\sqrt{LC}. \tag{4.16}$$

In this case, the transmission line has losses proportional to $\sqrt{RG}$, and therefore limited bandwidth. However, its delay is still independent from the frequency and its characteristic impedance is still given by (4.11) similar to the lossless line. Therefore, the transmission line is distortionless which improves the transmission of broadband signals. This property of the transmission line was first recognized by Heaviside [15] and Pupin [16] and led to the improvement of the telephony industry through the periodic insertion of coils so that the RC and GL products match. The same approach was also used recently [17, 18] to implement distortionless differential transmission lines by adding periodic resistances to increase their shunt conductance G in order for RC and GL products to match. Using distortionless transmission lines is beneficial for transmitting broadband signals because it reduces their intersymbol interference (ISI) since their loss and group delay are frequency independent. However, such an improvement is not necessary for transmitting clock signals because their frequency is constant. Although not directly related to the potential distortionless nature of transmission lines, [19] has demonstrated that on-chip interconnect transmission-line effects can be used to correct duty cycle deviations as well as sensitivity to PVT variations.

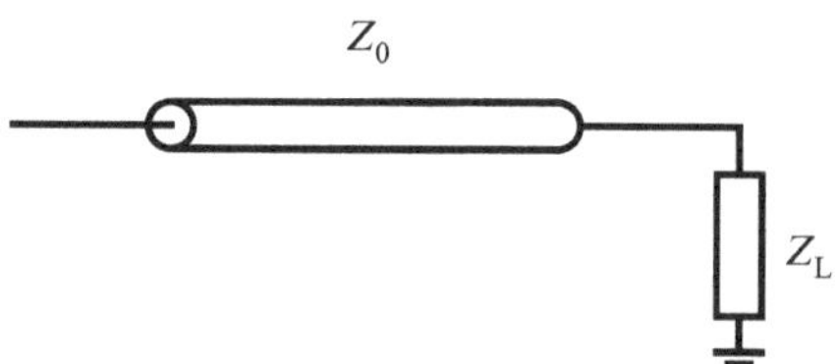

Fig. 4.4. Finite-length transmission line with termination

The transmission line characteristics presented above apply to transmission lines with infinite length. In practice, transmission lines have a finite length and are used to transmit a signal between two points. In order to ensure that transmission-line characteristics resemble those of an infinite line, a termination impedance must be inserted at the receiving end with impedance Z_L, equal to the characteristic impedance Z_0 of the line (Fig.4.4). If the termination impedance does not match the line characteristic impedance, there is a reflection propagating in the opposite direction of the transmitted signal. Let Γ_L denote the reflection coefficient defined as the ratio of reflected to incident voltage and current at the receiving end of the line. Then Γ_L is given by:

$$\Gamma_L = \frac{Z_L - Z_0}{Z_L + Z_0}. \tag{4.17}$$

The reflection coefficient absolute value can vary between 0, when Z_L perfectly matches Z_0, and ± 1, when termination is an open or short. As discussed in Sects. 4.3 and 4.4, certain circuit topologies for clock generation and distribution are based on circular transmission lines eliminating the need for impedance termination.

Similar to spiral inductors, transmission lines are implemented in silicon using low-resistive, thick, high-level metals. In general there are two major forms of transmission lines: microstrips and coplanar waveguides. Microstrips are implemented as metal lines on the top of substrate or a ground shield using a low-level metal layer (Fig. 4.5). Coplanar waveguides include side ground shields for ground current returns and can also run either on top of the substrate or a ground shield (Fig. 4.6). Coplanar waveguides are better shielded from surrounding circuits and other interconnects at the expense of area overhead for the side ground shields. Since different width and spacing settings can give the same characteristic impedance Z_0, there are several different parameters that need to be considered for their implementation:

- Low-resistivity substrate with epi layer have higher losses compared to high-resistivity substrate. Therefore, bottom ground metal shield may be used to reduce losses in the former case.
- Using ground metal shield reduces losses to the substrate as well as substrate noise, but at the same time it increases the transmission-line capacitance since

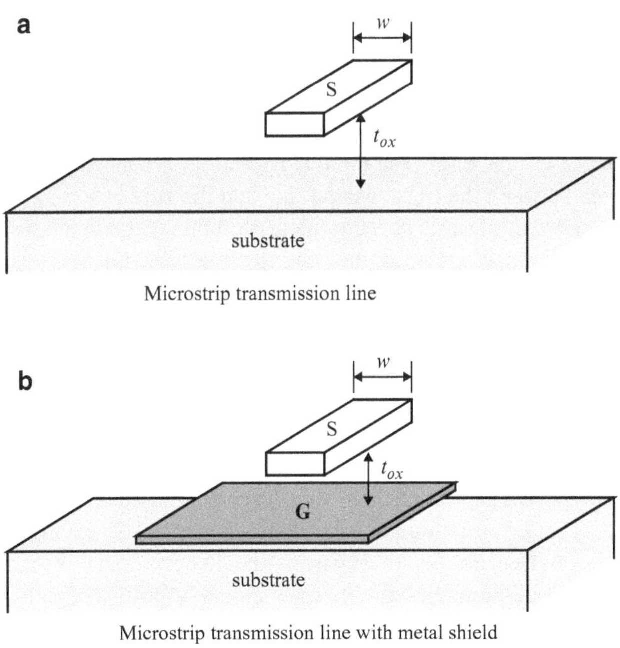

Fig. 4.5. Microstrip transmission line with and without ground shield

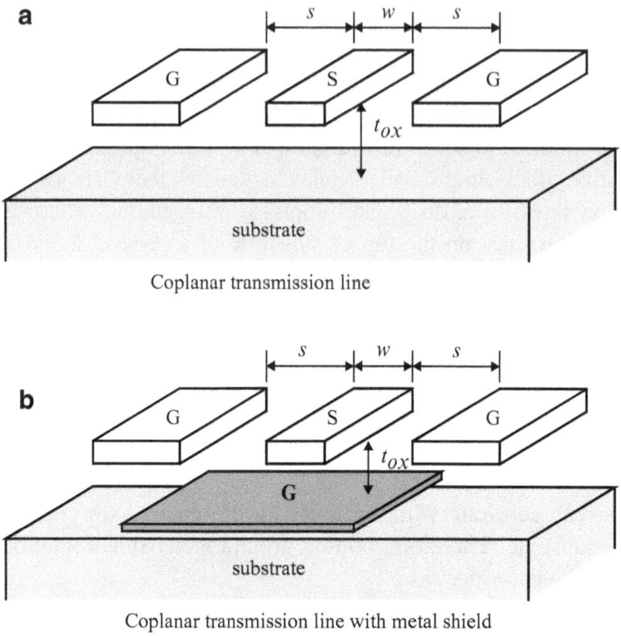

Fig. 4.6. Coplanar transmission line with and without ground shield

t_{ox} is reduced. Another issue with bottom ground shield is that transmission line current will tend to run along its bottom side due to the proximity and skin effects.

- Increasing the width of the transmission line reduces resistive losses but it increases the capacitance. In case there is no bottom ground shield, depending on the distance between the transmission line and the substrate, more electric field lines penetrate the substrate as the transmission line width increases.
- In case of coplanar waveguides, the space between the transmission line and the side ground lines affects both the inductance and capacitance.
- Dummy metal insertion required by certain processes can change both the characteristic impedance and the losses of a transmission line. The dummy metal effect can be reduced if these metals are placed manually.

Trade-offs of these parameters are studied in [20] through hardware measurements. Transmission lines can be also analyzed through EM software [11, 12]. Figure 4.7 shows a 60Ω U-shaped coplanar waveguide used for a distributed VCO in 90nm CMOS [21]. The bottom ground shield is in metal 2, leaving metal 1 available for feed-throughs under the transmission line. Dummy metal is manually placed to meet the process minimum requirements. Figure 4.8 shows a coplanar differential waveguide. The space between the two differential signals is twice the space between each line and the side ground line.

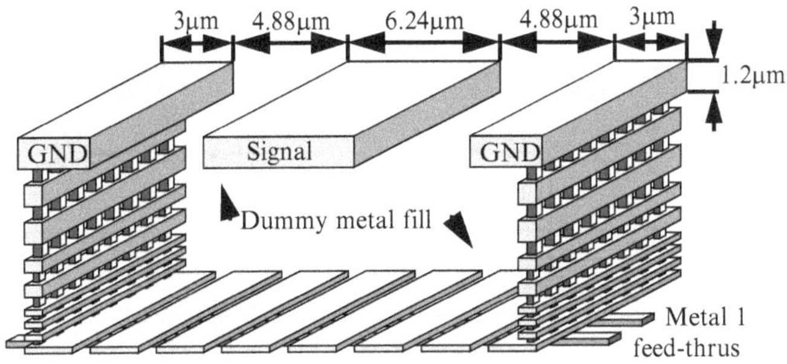

Fig. 4.7. U-shaped coplanar waveguide. Reproduced with permission from [21], © 2006 IEEE

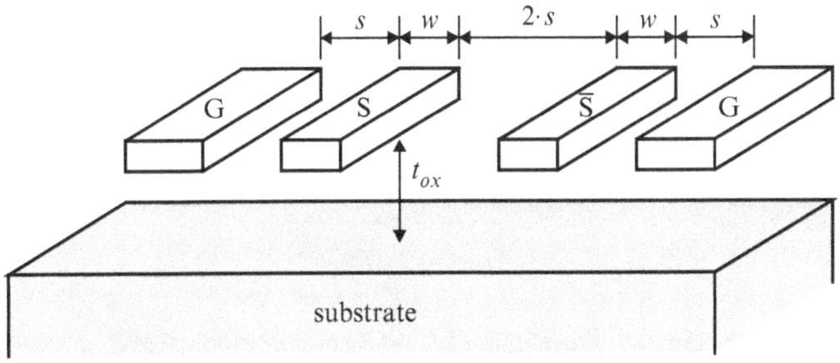

Fig. 4.8. Coplanar differential transmission line without ground shield

4.3 Inductor-Based Clock Generation

4.3.1 Differential LC VCO

As was noted before, LC VCO implementation was one of the first applications of on-chip spiral inductors [2]. In general, LC oscillators consist of two main components: the LC tank and the negative resistance. The latter is an active circuit necessary to start and maintain the oscillation [22]. Figure 4.9 shows commonly used differential LC VCO circuit topologies in CMOS. The first two circuits are similar with the main difference being the bias configuration (top biased vs. bottom biased). They both rely on cross-coupled NFETs MN_0 and MN_1 to form the negative resistance. The third circuit uses cross-coupled NFETs and PFETs for negative resistance with the advantage of requiring a single inductor for both LC tanks. The other two topologies require two inductors or one center-tapped inductor. In any case, the physical placement and wiring is greatly simplified when the VCO topology requires a single inductor. Another advantage for the third VCO topology is that the clock phases do not exceed the supply voltage. In contrast, it is possible especially for the clock phases of the first topology to surpass the supply voltage, which can cause process

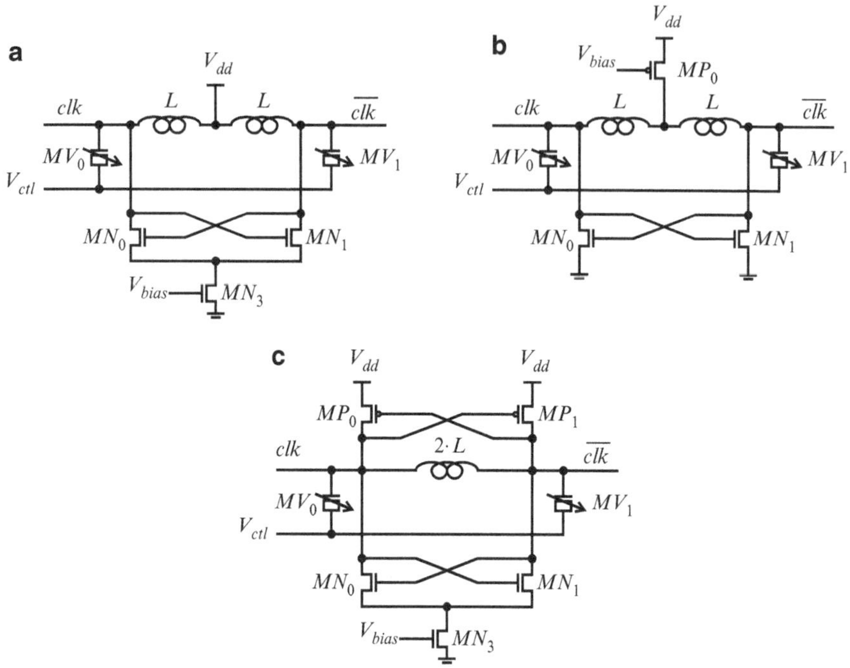

Fig. 4.9. Differential LC VCO circuit topologies

break down assuming the VCO supply voltage is the same as the process maximum allowed voltage. On the other hand, the common-mode voltage of the first VCO is equal to the supply voltage allowing for more voltage headroom.

In LC VCOs, the frequency is controlled by varying the voltage across the varactors. There are two types of varactors: PN junctions and MOSFET [23, 24]. MV_0 and MV_1 represent MOSFET varactors in Fig. 4.9 while V_{ctl} denotes an analog control voltage. The frequency of the VCO is given by:

$$f = \frac{1}{2\pi\sqrt{LC}}, \tag{4.18}$$

where L and C denote the VCO inductance and capacitance respectively. The capacitance consists of a fixed part (transistor gate, drain, and parasitics) and a varying part (varactors) with the latter determining the frequency range. Another method to adjust the VCO frequency is to use magnetic tuning through transformers [25] which effectively changes the inductance factor in (4.18).

Besides the VCO frequency and range, its phase noise is the most important characteristic, especially for communication applications in which low clock jitter is necessary in order to meet system jitter specifications for transmitting and receiving data. Starting with the generic Leeson model for oscillator phase noise [26], several theories have been developed recently specifically for LC oscillators [27–30]. Leeson's equation for oscillator phase noise due to thermal noise is:

$$L(\omega_m) = F\frac{1}{V^2}\frac{kT}{C}\frac{\omega_0}{Q}\frac{1}{\omega_m^2}, \tag{4.19}$$

where ω_0 and ω_m are the center and offset frequency respectively, kT/C is the thermal noise density, V is the voltage amplitude, Q is the resonator quality factor, and F is a circuit-dependent noise factor. Extending Leeson's linear equation in order to more accurately model the inherently nonlinear behavior of the LC oscillators, it has been shown that for the VCO topology of Fig. 4.9a the noise factor F is given by [30, 31]:

$$F = 1 + \frac{4\gamma IR}{\pi V} + \gamma\frac{4}{9}g_{mtail}R, \tag{4.20}$$

where γ is the FET channel noise coefficient, I is the tail current, V is the voltage amplitude, R is the tank resistance, and g_{mtail} is the transconductance of the current-source FET. Based on the phase noise equations above, it appears that increasing the VCO voltage amplitude reduces phase noise since it improves signal to noise ratio. However, in practice there is a limit on how large the voltage amplitude can be before in fact the phase noise starts increasing. As the voltage amplitude increases, the differential pair FETs start operating in the linear region for part of the cycle time. Depending on the circuit topology and process parameters, it is possible for the differential pair FETs to completely turn off. Furthermore, the current-source FET may also go out of saturation into the linear region of operation. In fact, there are two modes of operation for LC VCOs [31–33]: current-limited and voltage-limited mode. In current-limited mode operation, the clock voltage amplitude is determined by the tail current and the equivalent tank resistance. In voltage-limited mode operation, the amplitude approaches or exceeds the supply voltage for the VCOs shown in Fig. 4.9c and Fig. 4.9a, respectively.

One of the main contributors to phase noise is the tail current source noise. This noise is both high frequency and low frequency. The former is mainly due to the second harmonic of the VCO frequency since even harmonics flow into the common-mode path. This noise can be filtered by placing a large capacitor parallel to the current source to short the noise frequencies around the second harmonic and an inductor between the current source and the tail sized properly to provide high impedance at the second harmonic [31]. Low-frequency noise in the tail current source can also be suppressed using inductive degeneration and capacitive filtering which require off chip inductor and capacitor, respectively [34].

To ensure that the VCO frequency range has enough margin to meet its specification despite process, voltage, and temperature variations, extra varactor banks are typically included. The problem is exacerbated in applications that require operation in multiple frequencies [35]. The VCO gain is proportional to the varactor size. Furthermore, the control voltage usually varies around half of the supply voltage where the VCO gain is close to its maximum. Any VCO noise (control voltage as well as supply and substrate noise) gets converted into phase noise in proportion with its gain [36]. Since the gain of the VCO is minimized when the control voltage is set to ground or to the supply voltage, phase noise can be reduced by operating the VCO at a lower gain. This can be accomplished by splitting the varactor banks into digitally-controlled and analog-controlled [36] with the majority of them being controlled digitally. VCOs with hybrid analog/digital control complicate the design at the system-level (e.g., PLL, CDR, etc.) since they require both coarse and fine tuning. There is also the extreme case of completely digitally-controlled oscillators (DCO) [37].

4.3.2 Quadrature LC VCO

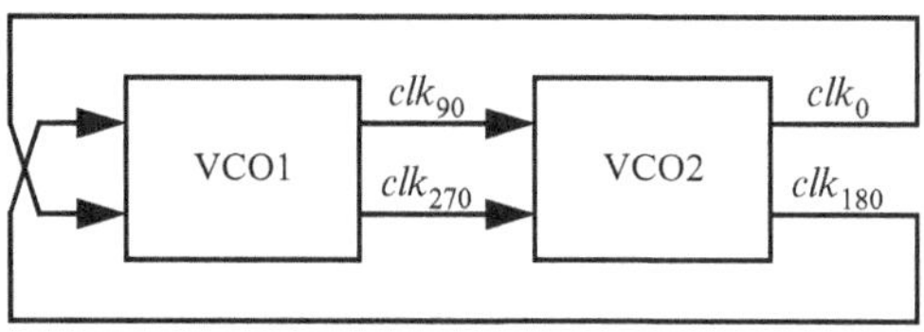

Fig. 4.10. Quadrature LC VCO block diagram

Several communication applications require four clock phases with 90° phase difference between consecutive phases. Such applications include half-rate CDR architectures [38, 39] and wireless transceivers [40]. To meet strict phase deviation and phase noise requirements imposed by these applications at multigigahertz frequency range, using LC VCOs is the primary choice especially in CMOS. A quadrature LC VCO can be formed by coupling two differential LC VCOs (Fig.4.10). Starting with a typical differential LC VCO (Fig.4.9a), coupling can be introduced either in parallel or in series with the cross-coupled pair transistors. Figure 4.11a shows a parallel-coupled quadrature LC VCO [41]. Varactors are omitted for clarity. MN_4, MN_5, MN_6, and MN_7 are the coupling transistors connected in parallel with the stage internal cross-coupled switch transistors MN_0, MN_1, MN_2, and MN_3 respectively. Assuming that coupling and switch transistors have the same length, the coupling strength, α, is defined as the ratio between the width of the coupling transistors to the width of the switch transistors [42]. Coupling strength α has a value between 0 and 1. It has been shown [42] that in case of parallel coupling both the phase noise and phase deviation depend on α in an inverse fashion. As α increases, phase deviation decreases whereas phase noise increases. Figure 4.11b shows an alternative parallel-coupled quadrature VCO in which the coupling parallel path is biased using a different tail-current transistor. In this case, α is determined by the bias current ratio between the coupling path and the switch path.

Figure 4.12 shows two series-coupled quadrature LC VCOs [42, 43]. In both cases, the coupling transistors in each stage are connected in series with the cross-coupled transistors. In Fig.4.12a, the coupling transistors MN_4–MN_7 are on the

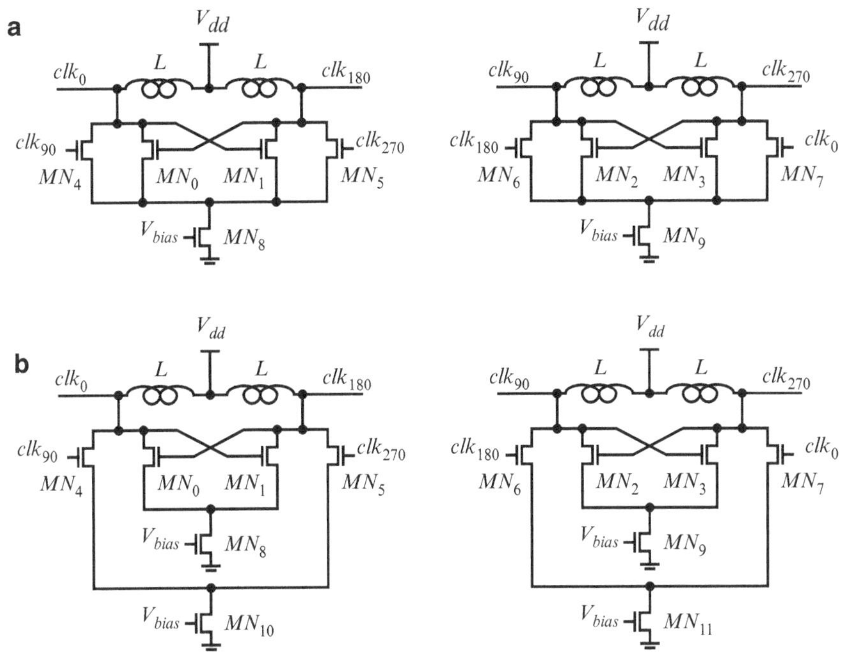

Fig. 4.11. Parallel-coupled quadrature LC VCOs

top whereas in Fig.4.12b they are on the bottom. As has been reported [42, 43], the phase deviation in series-coupled quadrature LC VCOs is independent of the coupling strength for all reasonable values (e.g., between 1/3 and 1/2). Therefore, the coupling strength can be optimized for phase noise without affecting phase deviation between the four clock phases. In studies comparing parallel- and series-coupled quadrature VCOs, it has been shown [42, 43] that series-coupled VCOs exhibit better phase noise characteristics compared to equivalent parallel-coupled VCOs. Phase noise and oscillation stability of parallel-coupled quadrature VCOs can be improved by shifting the coupling phases by 90° (Fig.4.13) [44–46]. The outputs of the phase shifters that are used to couple the VCO stages are in phase with their respective cross-coupled phases that drive the other branch of the parallel structure. Phase shifter circuits can be based on a differentiator or an integrator [45] or a passive RC-CR filter [46]. As shown in [46], the accuracy of the phase shift is not essential since same results can be obtained with smaller than 90° phase shift.

Other quadrature LC VCO architectures have been proposed that require two inductors instead of four [47, 48]. In the first design, the coupling is introduced using PFETs. The second design introduces both pull-down and pull-up parallel coupling networks using NFETs and PFETs, respectively.

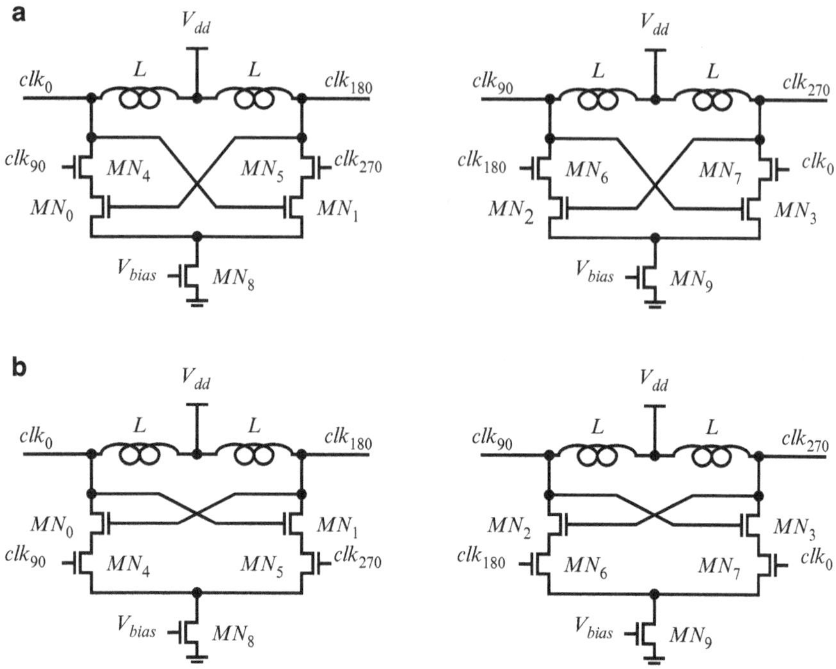

Fig. 4.12. Series-coupled quadrature LC VCOs

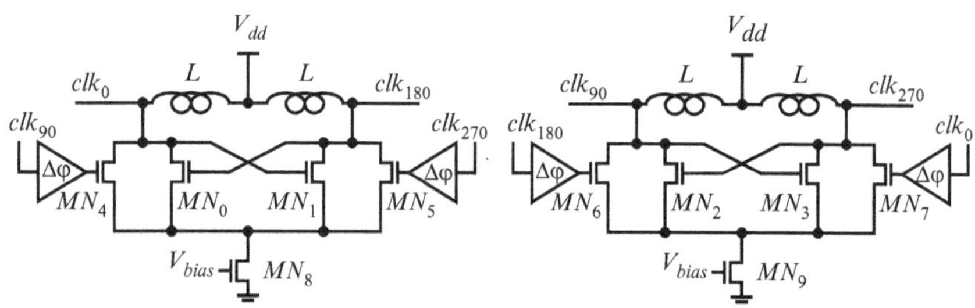

Fig. 4.13. Parallel-coupled quadrature LC VCO with 90° phase shift in the coupling phase

4.3.3 Distributed VCO

Distributed amplifiers are used in applications with high bandwidth requirements. Despite the physical challenges associated with distributed amplifier implementation, they can achieve high bandwidth because parasitic capacitance (including transistor gate and drain) is absorbed in the transmission lines. Distributed oscillators can be implemented based on distributed amplifiers by feeding the output of the amplifier back to its input [49, 50]. A 3-stage distributed oscillator is shown in Fig. 4.14. Com-

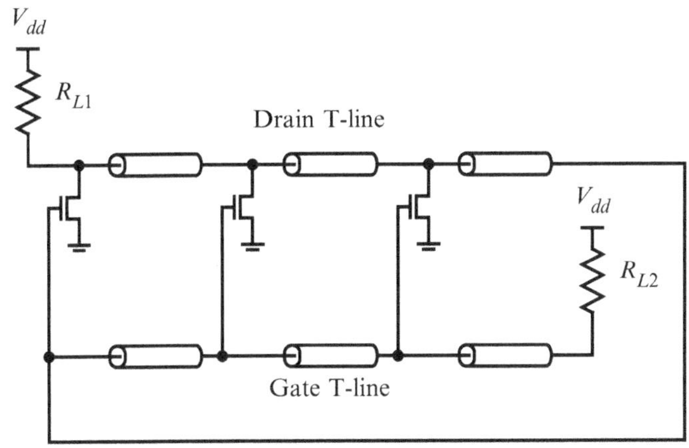

Fig. 4.14. Distributed oscillator

pared to lumped LC oscillators, distributed oscillators can be operated at frequencies close to the f_T of the process. As reported in [50], using conventional varactors for frequency tuning can result in loading the transmission line excessively which would cause the oscillator operating frequency to be reduced. An alternative frequency tuning method is proposed in [50]. Specifically, the oscillator frequency is tuned by effectively changing the length of the transmission line. This can be achieved by shortening or lengthening the path of the current either in the drain or the gate transmission line. A variation of the distributed oscillator shown in Fig. 4.14 has also been implemented using discrete components [51–53]. In this design, the drain line connects to the gate line at the drain line load.

There are several issues associated with the implementation of conventional distributed oscillators. First, the voltage amplitude across both the drain and gate lines varies depending on the distance to the respective load resistors. Furthermore, the termination resistors are sources of thermal noise that affect the oscillator phase noise. Finally, it is difficult to implement differential distributed oscillators since they require two gate and two drain lines.

4.3.4 Poly-Phase Circularly Distributed VCO

Communication applications such as quarter-rate and/or oversampled CDRs require eight [54] or more [18] clock phases equally-spaced in time. Distributed oscillators offer the ability to generate multiple clock phases. However, as was discussed before conventional distributed oscillators have several drawbacks that make their use difficult. Another type of distributed oscillators can be formed using circular transmission line [21, 54] as shown in Fig. 4.15 which are based on rotary traveling waves [55, 56]. The distributed oscillator shown in Fig. 4.15 produces eight clock phases with 45° phase difference between any two consecutive phases. The transmission line is folded so that clock phases are delivered as four differential pairs (clk$_i$ and

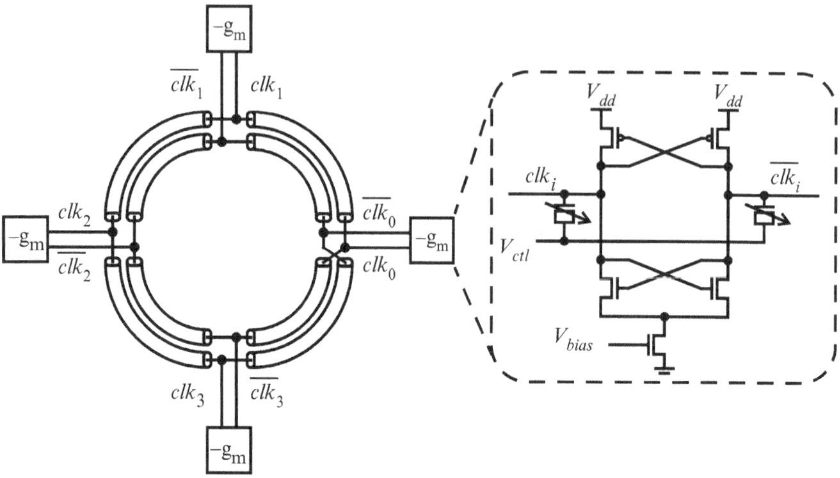

Fig. 4.15. Circular distributed oscillator. Reproduced with permission from [21], ©2006 IEEE

$\overline{\mathrm{clk}_i}$, $i = 1,..., 3$). Four negative-g_m circuits connected to the differential pairs start and sustain the oscillation. The negative-g_m circuits can be cross-coupled NFET/PFET pairs. Varactors are included so that the oscillator frequency can be adjusted through a control voltage, V_ctl. The circular distributed oscillator overcomes all issues related to the amplifier-based distributed oscillator. All clock phases have the same amplitude regardless of their position. The circular transmission line has virtually infinite length, and therefore eliminates the need for termination resistors. Finally, the clock phases are inherently delivered as differential pairs.

A distributed oscillator is useful only if the direction of the clock phases is controllable. Circular distributed oscillators are symmetric resulting in uncertainty for the clock direction. The clock phases can propagate either clockwise or counter-clockwise. After power up clk_i can either lead or trail clk_{i-1} depending on several physical design factors such as the location of the pads, the supply distribution network, and the physical arrangement of the transmission lines. In [21], a method to control the direction of the clock phases has been demonstrated on a 24-phase closed-loop distributed VCO. The method is outlined in Fig. 4.16. In each stage two NFETs are added that conditionally pull down clk_i and $\overline{\mathrm{clk}_i}$. When on, these NFETs add positive resistance that cancels out the negative-g_m circuits. Their gates are controlled by signal ccw/cw starting with clk_0 and ending with $\overline{\mathrm{clk}_3}$ or the other way around, following the VCO loop. Pulling up the control signal stops the VCO from oscillating and resets the clock phases to a DC voltage determined by the strength of the FETs in the negative-g_m circuit and the positive-resistance NFETs. After all phases are reset, the control signal is pulled down either from the ccw side (clk_0 leads) or from the cw side ($\overline{\mathrm{clk}_3}$ leads), releasing the clock phases in a predetermined order.

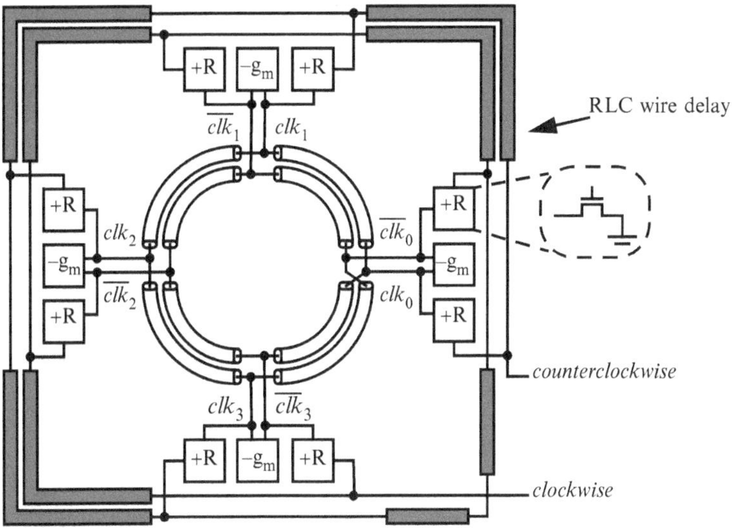

Fig. 4.16. Circular distributed oscillator with clock direction control. Reproduced with permission from [21], ©2006 IEEE

4.4 Clock Distribution Using Inductance

4.4.1 Rotary Traveling-Wave Oscillator Arrays

Rotary traveling-wave oscillator arrays are formed using multiple rotary traveling-wave oscillators [55, 56] as shown in Fig. 4.17. Assuming all loops are symmetrically matched, each loop has a traveling wave. The cross-coupled inverters are used to start and sustain the operation due to losses in the transmission-line loops. They also serve as latches that amplify the differential signal and generate square waveform clocks instead of sinusoidal. The individual loops are synchronized at their junctions through a phase-locking phenomenon. In any four port junction there are two pulses that are arriving simultaneously and are propagated to the output ports if the impedances are matched. The clock frequency is determined by the characteristics of a single loop and is given by [56]:

$$f = \frac{1}{2\sqrt{L_{\text{lp}} C_{\text{lp}}}}, \tag{4.21}$$

where L_{lp} and C_{lp} are the total loop inductance and capacitance, respectively. The factor of 2 is inserted because the phase wave has to go around the loop twice per clock cycle. As is the case with other oscillators, the frequency can be tuned by placing varactor banks along the transmission line.

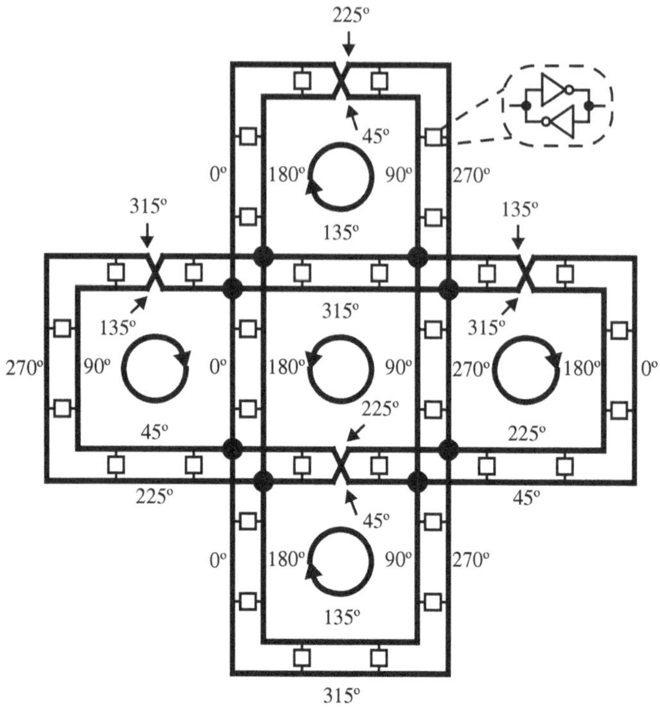

Fig. 4.17. Rotary traveling-wave arrays. Reproduced with permission from [55], ©2001 IEEE

As shown in [56], rotary traveling-wave oscillator arrays have low skew even in the presence of mismatches between the various loops. Furthermore, they operate adiabatically [57] when the losses in the transmission line and the gate resistance of the cross-coupled inverters are small. Finally, as suggested by Fig.4.17, rotary traveling-wave oscillator arrays are expandable since more loops can be connected together to form larger arrays. Rotary traveling-wave clocks operate well with digital logic that requires two phases. In addition to conventional logic, a new two-phase adiabatic logic style operating with such a clock has been demonstrated [58]. However, careful consideration must be given at the system level since the phase of the clocks depends on their location within each loop. For instance, moving data globally may require retiming to a different clock domain. Another issue is coupling between the clock transmission lines and signal interconnects that need to be routed underneath the transmission lines.

4.4.2 Standing Wave Oscillator and Grid

Another type of oscillators based on transmission line signal propagation is standing wave oscillators [59]. The oscillating signals are formed in the transmission line by the incident and the reflected waves that travel in opposite directions. A standing

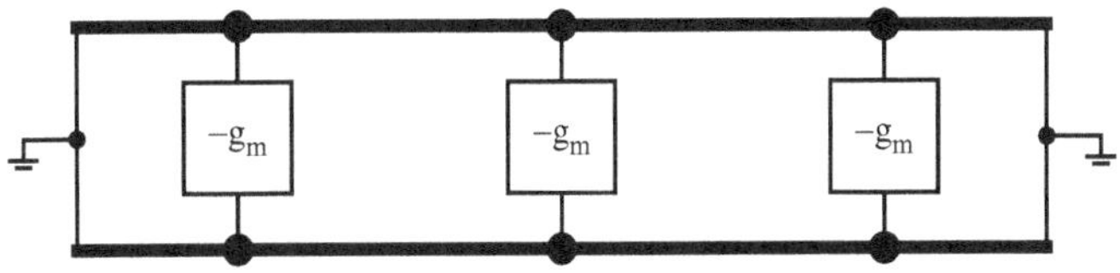

Fig. 4.18. Standing wave oscillator

wave oscillator (Fig.4.18) consists of a differential transmission line and several negative-g_m cells distributed along the line. The transmission lines are shorted to ground at both ends creating a half-wave ($\lambda/2$) resonator, where λ is the wavelength. The negative-g_m cells used in [59, 60] contain cross-coupled NFETs to compensate for gain losses in the transmission lines and diode-connected PFETs for loads to set the common mode voltage. Unlike other LC-based oscillators, standing-wave oscillators may require clock injection in order to oscillate. Using diode connected loads simplifies clock injection as shown in [60]. Compared to rotary traveling-wave oscillators, standing wave oscillator clocks have the same phase irrespective of their position along the line. However, clock amplitude is position-dependent and varies sinusoidally along the transmission line. Specifically, clock amplitude is high in the middle of the transmission line and low at the two ends. Therefore, clock phases close to the transmission line ends require amplification to be usable. As shown in [60], a clock buffer consisting of a limiting amplifier and a sine-to-square converter is required especially for digital applications. The standing wave oscillator oscillates at a frequency f if the following equations are satisfied [59]:

$$g < \frac{1}{n}\frac{R_{tl}C_{tl}}{L_{tl}} \tag{4.22}$$

and

$$l = \frac{1}{2f\sqrt{L_{tl}C_{tl}}}, \tag{4.23}$$

where R_{tl}, C_{tl}, and L_{tl} are the resistance, capacitance, and inductance of the distributed transmission line, l is the length of the transmission line, n is the number of negative-g_m cells along the transmission line, and g is the transconductance of each one of the negative-g_m cells. A more generic analytical model for the operation of standing wave oscillators is derived in [60].

Multiple standing wave oscillators can be coupled into clock grids as shown in Fig.4.19 [59, 60]. A single standing wave oscillator is diagrammed in the top right corner of the grid. The ends of standing wave oscillators are folded outside the grid since the low clock amplitude in these segments makes the clocks unusable. Each oscillator has five negative-g_m cells. Although not shown in Fig.4.19, each oscillator can contain varactors in order to adjust its resonance frequency so that it matches the injected clock frequency. The clock injection point is also indicated in Fig.4.19. Similar to rotary traveling-wave oscillators, standing wave oscillators exhibit a phase averaging effect when connected in grids resulting in overall low-jitter and low-skew clock distribution.

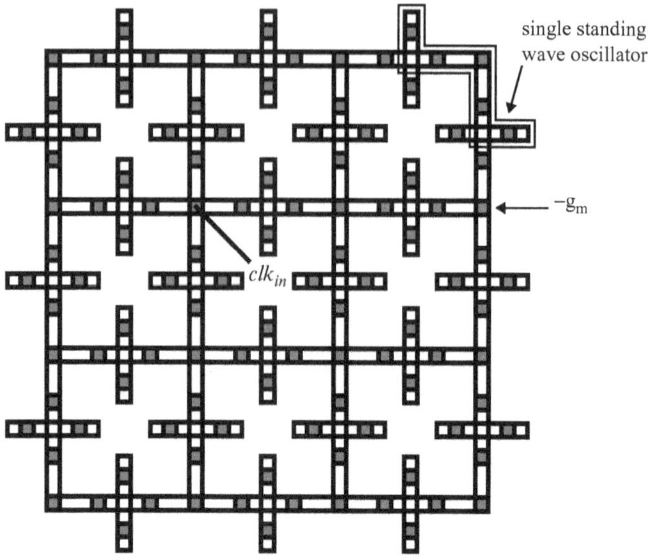

Fig. 4.19. Coupled standing wave oscillators used for clock distribution network. Reproduced with permission from [60], ©2003 IEEE

One of the drawbacks of the standing wave oscillator shown in Fig.4.18 is the low voltage amplitude for a rather large portion of the transmission line length. To overcome this issue, spiral inductors can be placed at the two ends of the differential transmission line (Fig.4.20a) [61, 62]. The resulting oscillator can have the same frequency while exhibiting a small voltage variation along its transmission line. Furthermore, its resonance frequency does not depend on the transmission line length as is the case for the half-wave oscillator. For instance, the transmission line length is 3.6mm for a 20GHz oscillator when fixed to $\lambda/2$ [61]. Since the transmission line is relatively short, only two negative-g_m cells are placed at both ends of the line. In [61], the negative-g_m cell consists of cross-coupled NFETs and PFETs and a single inductor is placed between the differential signals at each end of the line (i.e., a total of two inductors). In [62], the negative-g_m cell consists of cross-coupled NFETs and two inductors are placed at each end connected to the supply voltage. The oscillator frequency is given by [62]:

$$f = \frac{1}{2\pi}\frac{Z_0}{L_{ld}}\tan(\pi - \beta l), \tag{4.24}$$

where Z_0 is the transmission line characteristic impedance, L_{ld} is the inductance of the load, β is the transmission line phase constant, and l is the transmission line length.

In [62], multiple standing wave oscillators with inductive loads are coupled forming a clock distribution grid (Fig.4.20b). After power up, all the oscillators are synchronized due to the inductive coupling between adjacent oscillators. The clock

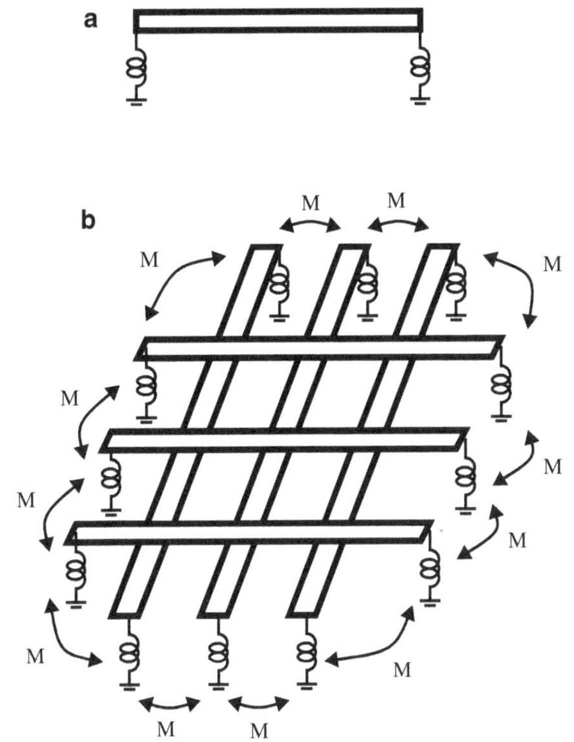

Fig. 4.20. (**a**) Standing-wave oscillator with inductive loads and (**b**) Multiple oscillators magnetically coupled. Reproduced with permission from [62], ©2007 IEEE

frequency is also given by (4.24). However, in this case mutual inductance must be taken into account, and thus L_{ld} equals $(1+2k)L$ where L is the oscillator load inductance and $k = M/L$ with M being the mutual inductance.

Another standing wave oscillator topology is presented in [63]. In this oscillator, the differential transmission line is only terminated in one end, whereas a negative-g_{m} cell is placed in the other end resulting in a $\lambda/4$ standing wave oscillator. The voltage amplitude is position-dependent having a maximum value at the end where the negative-g_{m} cell is located and being reduced sinusoidally along the transmission line until it becomes 0 at the other end where the short is located. To remedy the voltage amplitude reduction along the transmission line, a tapered line is used instead of a uniform line resulting in lower phase noise.

In [64], a circular standing wave oscillator is introduced (Fig.4.21). In this standing wave oscillator the lines are not shorted, but they are circularly connected. Two negative-g_{m} cells are placed in two opposite sides. The circular standing wave oscillator requires that voltage $V(\phi)$ for any angle ϕ be equal to $V(\phi + 2\pi)$. Therefore, the length l of the circumference is an integer multiple n of the wavelength λ of the mode. To suppress the even mode, the differential nodes from the two negative-g_{m}

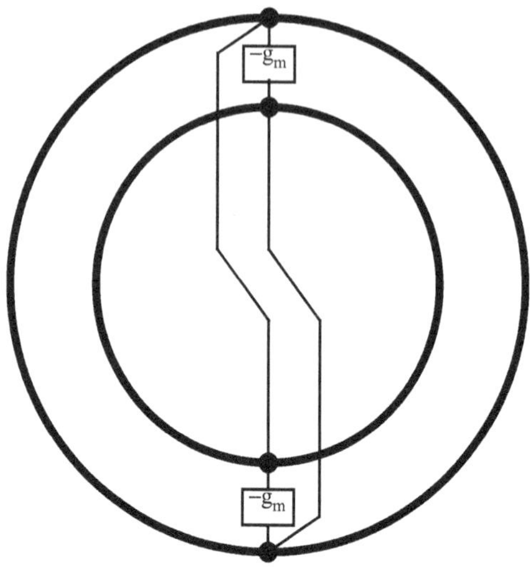

Fig. 4.21. Circular standing wave oscillator. Reproduced with permission from [64], ©2004 IEEE

cells are connected as shown in Fig. 4.21. To be able to make these connections, the circular differential transmission line is formed so that the two negative-g_m cells are physically placed close to each other. The voltage amplitude is also position-dependent with the negative-g_m cell nodes having the highest amplitude.

4.4.3 Inductor-Based Resonant Global Clock Distribution

Resonant global clock distribution was primarily used as a means to significantly reduce power dissipation through adiabatic charging and energy recovery [57]. The resonantly generated clocks power both the clock network as well as circuit nodes. A differential LC clock generator and driver [65] is used to produce two non-overlapping clock phases. The active components of the clock generator are integrated in the same chip with the rest of the circuit. Furthermore, in [66] the active components are distributed along the clock grid forming a distributed clock network. However, the two inductors are placed off chip [66–68]: First, the Q of on-chip inductors was too small for these CMOS processes. Second, the inductors would be too large to integrate since their size is inversely proportional to the operating frequency. As CMOS processes advanced, both the inductor Q increased and clock frequency reached the GHz-range, allowing for on-chip inductor integration. In [69], an adiabatic logic family is demonstrated to operate from a fully-integrated resonant clock driver.

Other LC-based clocking methods target power reduction only for the clock network (not data nodes) as well as low clock jitter and skew. Figure 4.22 illustrates one of those methods [70] which is based on conventional clock distribution. Specifically, the clock network is divided into sectors with a clock buffer chain driving each sector. One such sector is shown in Fig. 4.22. The clock sector consists of a global H-tree with the ends of the tree driving a global grid. Local clock buffers connect to the clock grid and deliver the clock to local blocks. To enhance clock distribution, spiral inductors are attached to certain nodes in the H-tree. Decoupling capacitors are placed in series to the inductors so that one port of the inductors connects to the clock tree whereas the other port connects to the decoupling capacitors. Figure 4.23 shows a simplified lumped circuit for a clock sector [70]. Capacitors C_{clk} and C_{dec} indicate the clock and decoupling capacitances respectively, L indicates the spiral inductance, and R_{clk} and R_{ind} represent loads in the clock network and the inductor parasitic resistance respectively. The decoupling capacitance must be an order of magnitude larger than the clock capacitance so that the corresponding pole introduced does not

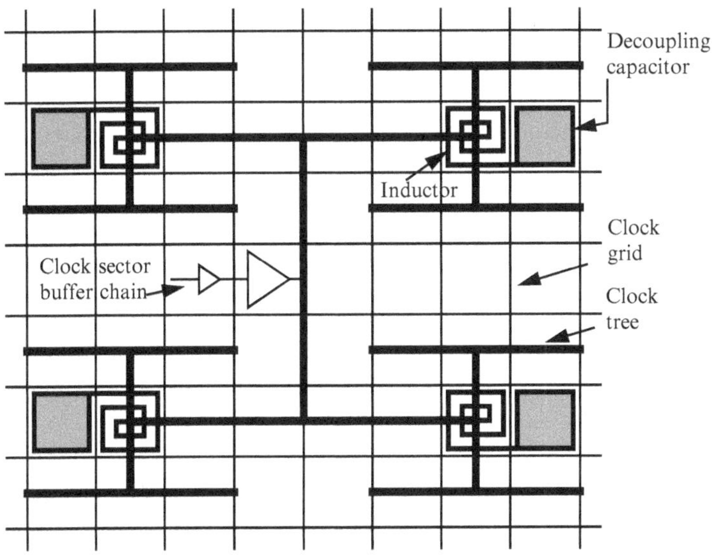

Fig. 4.22. Global clock distribution with resonant load. Reproduced with permission from [70], ©2005 IEEE

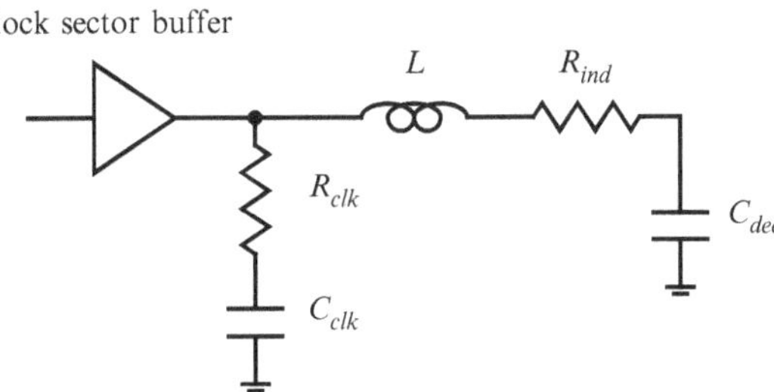

Fig. 4.23. Lumped circuit model of the clock sector. Reproduced with permission from [70], ©2005 IEEE

interfere with the resonant frequency of this circuit. When the clock frequency is approximately equal to the resonant frequency (i.e., $1/\left(2\pi\sqrt{LC_{\mathrm{clk}}}\right)$), the inductive reactance of the spiral inductor cancels the capacitive reactance of the clock load resulting in significantly lower power dissipation compared to an equivalent conventional clock network [70]. In addition to the energy being recovered and re-used through the LC tank, the sector clock buffers do not need to be as strong as in the conventional case. The reduction of clock latency in the distribution network combined with the resonance causes both clock skew and jitter to be reduced as well. To further demonstrate the viability of this method, the conventional clock distribution of a commercial microprocessor was modified to implement this clocking method [71]. Specifically, a thick top metal layer was added to accommodate the spiral inductors without affecting the existing microprocessor design.

Another LC-based clock distribution (Fig. 4.24) uses differential clocks [72]. The clock network consists of H-trees that drive a global clock grid similar to the previous method. The clock network is divided into tiles. Each tile contains an H-tree driven by a differential LC-oscillator formed by a negative-g_{m} cell and a spiral inductor. Each tile contains varactors to adjust its resonance frequency. A single buffer is used for clock injection from an external clock. Local clock buffers tap into the clock grid. These buffers convert the differential sinusoidal clocks into a single-ended square clock that drives local blocks. As shown in [72], both clock power and jitter are reduced by an order of magnitude compared to an equivalent conventional clock distribution network.

In [73], a central LC-based clock generator [65] delivers differential clocks that are distributed through a clock grid. The active devices of the clock generator are

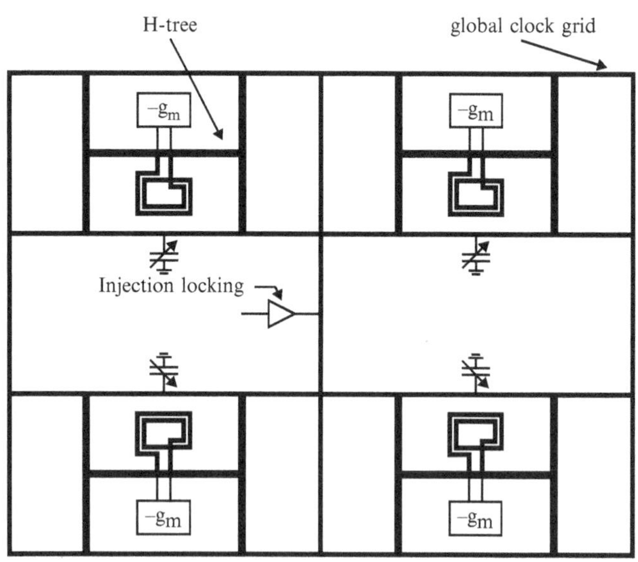

Fig. 4.24. Distributed differential global clock network. Reproduced with permission from [72], ©2006 IEEE

distributed in the clock grid similar to [66]. Latches that can operate from sinusoidal clocks are used as clocked storage elements similar to [68]. In [74], a hybrid injection-locked oscillator and LC-based buffer is used to drive the clock load. This technique can perform an interesting trade-off between input clock and clock buffer jitter.

4.5 Conclusion

In this chapter both the implementation and use of monolithic inductance were discussed. Most common monolithic inductors are spiral inductors and transmission lines. In addition to the implementation of spiral inductors and transmission lines, their accurate modeling is of utmost importance to circuit designers. A narrowband model for spiral inductors was presented. Modeling of transmission lines is simpler than spiral inductors. Their characteristics (i.e., inductance, capacitance, resistance, and conductance) are proportional to their length and can be modeled with simpler extraction methods.

The second part of the chapter focused on clock generation. LC VCOs are commonly used for high-speed wireless, wireline, and optical communications due to their low phase noise, small phase deviation, and low-power dissipation compared to ring oscillators. LC VCOs can efficiently generate up to four phases. Although in theory, more LC tanks can be coupled to generate more than four phases, this is not practically possible due to the physical arrangement of the LC tanks, coupling connections between the LC tanks, and potentially asymmetric clock phase routing to their destination. For all these reasons, distributed oscillators become the primary choice for generating more than four phases.

The third part of the chapter covered resonant-based methods to improve global clock distribution. Chapter 2 has demonstrated that conventional global clock distribution methods are more and more challenged to meet clock skew and jitter requirements for increasing clock load and clock frequency. Moreover, a large portion of total power dissipation is typically attributed to clock networks. Resonant-based methods can remedy these problems. Transmission-line-based methods such as rotary traveling-wave oscillator arrays and standing wave grids were presented. Furthermore, transmission lines can be efficient for transmitting high-frequency clocks without buffering as shown in [75]. The section concludes with LC-based global clock grids which are extensions of conventional global grids with the addition of spiral inductors.

Acknowledgments

I would like to thank Fujitsu Laboratories of America for providing me the time to complete this manuscript and W. Walker for discussions about spiral inductor and transmission line modeling. Last but not least, I would like to thank my wife, Malina, for support.

References

[1] N. Nguyen and R. Meyer, "Si IC-compatible inductors and LC passive filters," *IEEE J. Solid-State Circuits*, vol. 25, no. 4, pp. 1028–1031, 1990.
[2] N. Nguyen and R. Meyer, "A 1.8-GHz monolithic LC voltage-controlled oscillator," *IEEE J. Solid-State Circuits*, vol. 27, no. 3, pp. 444–450, 1992.
[3] P. Restle, T. McNamara, D. Webber, P. Camporese, K. Eng, K. Jenkins, D. Allen, M. Rohn, M. Quaranta, D. Boerstler, C. Alpert, C. Carter, R. Bailey, J. Petrovick, B. Krauter, and B. McCredie, "A clock distribution network for microprocessors," *IEEE J. Solid-State Circuits*, vol. 36, no. 5, pp. 792–799, 2001.
[4] T. Lee, *The design of CMOS radio-frequency integrated circuits, 2nd Edition*. Cambridge, United Kingdom: Cambridge University Press, 2004.
[5] H. Greenhouse, "Design of planar rectangular microelectronic inductors," *IEEE Trans Parts, Hybrids, and Packaging*, vol. 10, no. 2, pp. 101–109, June 1974.
[6] S. Jenei, B. Nauwelaers, and S. Decoutere, "Physics-based closed-form inductance expression for compact modeling of integrated spiral inductors," *IEEE J. Solid-State Circuits*, vol. 37, no. 1, pp. 77–80, Jan. 2002.
[7] S. Mohan, M. del Mar Hershenson, S. Boyd, and T. Lee, "Simple accurate expressions for planar spiral inductances," *IEEE J. Solid-State Circuits*, vol. 34, no. 10, pp. 1419–1424, Oct. 1999.
[8] K. Ashby, I. Koullias, W. Finley, J. Bastek, and S. Moinian, "High Q inductors for wireless applications in a complementary silicon bipolar process," *IEEE J. Solid-State Circuits*, vol. 31, no. 1, pp. 4–9, Jan. 1996.
[9] C. Yue, C. Ryu, J. Lau, T. Lee, and S. Wong, "A physical model for planar spiral inductors on silicon," in *Proceedings of International Electron Devices Meeting*, 8–11 Dec. 1996, pp. 155–158.
[10] Y. Cao, R. Groves, X. Huang, N. Zamdmer, J.-O. Plouchart, R. Wachnik, T.-J. King, and C. Hu, "Frequency-independent equivalent-circuit model for on-chip spiral inductors," *IEEE J. Solid-State Circuits*, vol. 38, no. 3, pp. 419–426, March 2003.
[11] "Sonnet user's guide," Sonnet High Frequency Electromagnetic Software.
[12] A. Niknejad, "ASITIC." [Online]. Available: http://rfic.eecs.berkley.edu/niknejad/asitic.html
[13] J. Long and M. Copeland, "The modeling, characterization, and design of monolithic inductors for silicon RF IC's," *IEEE J. Solid-State Circuits*, vol. 32, no. 3, pp. 357–369, March 1997.
[14] C. Yue and S. Wong, "On-chip spiral inductors with patterned ground shields for Si-based RF ICs," *IEEE J. Solid-State Circuits*, vol. 33, no. 5, pp. 743–752, May 1998.
[15] O. Heaviside, "Electromagnetic induction and its propagation," *The Electrician*, vol. 19, no. 40, pp. 79–81, 1887.
[16] M. Pupin, "Art of reducing attenuation of electrical apparatus therefore," U.S. Patent no. 652,230, June 1900.

[17] H. Chen, R. Shi, C.-K. Cheng, and D. Harris, "Surfliner: a distortionless electrical signaling scheme for speed of light on-chip communications," in *Proceedings of the IEEE International Conference on Computer Design: VLSI in Computers and Processors ICCD 2005*, 2–5 Oct. 2005, pp. 497–502.
[18] N. Nedovic, N. Tzartzanis, H. Tamura, F. Rotella, M. Wiklund, Y. Mizutani, Y. Okaniwa, T. Kuroda, J. Ogawa, and W. Walker, "A 40-to-44 Gb/s 3x oversampling CMOS CDR/1:16 DEMUX," *IEEE J. Solid-State Circuits*, vol. 42, no. 12, pp. 2726–2735, Dec. 2007.
[19] M. Thomson, P. Restle, and N. James, "A 5GHz duty-cycle correcting clock distribution network for the POWER6 microprocessor," in *Digest of Technical Papers IEEE International Solid-State Circuits Conference (ISSCC 2006)*, 2006, pp. 384–385.
[20] B. Kleveland, C. Diaz, D. Vook, L. Madden, T. Lee, and S. Wong, "Exploiting CMOS reverse interconnect scaling in multigigahertz amplifier and oscillator design," *IEEE J. Solid-State Circuits*, vol. 36, no. 10, pp. 1480–1488, Oct. 2001.
[21] N. Tzartzanis and W. Walker, "A reversible poly-phase distributed VCO," in *Digest of Technical Papers IEEE International Solid-State Circuits Conference (ISSCC 2006)*, 2006, pp. 596–597.
[22] N. Nguyen and R. Meyer, "Start-up and frequency stability in high-frequency oscillators," *IEEE J. Solid-State Circuits*, vol. 27, no. 5, pp. 810–820, May 1992.
[23] P. Andreani and S. Mattisson, "A 1.8-GHz CMOS VCO tuned by an accumulation-mode MOS varactor," in *Proceedings of the ISCAS 2000 Geneva Circuits and Systems the 2000 IEEE International Symposium*, vol. 1, 28–31 May 2000, pp. 315–318.
[24] A.-S. Porret, T. Melly, C. Enz, and E. Vittoz, "Design of high-Q varactors for low-power wireless applications using a standard CMOS process," *IEEE J. Solid-State Circuits*, vol. 35, no. 3, pp. 337–345, March 2000.
[25] G. Cusmai, M. Repossi, G. Albasini, A. Mazzanti, and F. Svelto, "A magnetically tuned quadrature oscillator," *IEEE J. Solid-State Circuits*, vol. 42, no. 12, pp. 2870–2877, Dec. 2007.
[26] D. Leeson, "A simple model of feedback oscillator noise spectrum," *Proc. IEEE*, vol. 54, no. 2, pp. 329–330, Feb. 1966.
[27] A. Hajimiri and T. Lee, "A general theory of phase noise in electrical oscillators," *IEEE J. Solid-State Circuits*, vol. 33, no. 2, pp. 179–194, Feb. 1998.
[28] T. Lee and A. Hajimiri, "Oscillator phase noise: A tutorial," *IEEE J. Solid-State Circuits*, vol. 35, no. 3, pp. 326–336, March 2000.
[29] A. Demir, A. Mehrotra, and J. Roychowdhury, "Phase noise in oscillators: A unifying theory and numerical methods for characterization," *IEEE Trans. Circuits Syst. I*, vol. 47, no. 5, pp. 655–674, May 2000.
[30] J. Rael and A. Abidi, "Physical processes of phase noise in differential LC oscillators," in *Proc. IEEE Custom Integrated Circuits Conference (CICC 2000)*, 21–24 May 2000, pp. 569–572.

[31] E. Hegazi, H. Sjoland, and A. Abidi, "A filtering technique to lower LC oscillator phase noise," *IEEE J. Solid-State Circuits*, vol. 36, no. 12, pp. 1921–1930, Dec. 2001.
[32] A. Hajimiri and T. Lee, "Design issues in CMOS differential LC oscillators," *IEEE J. Solid-State Circuits*, vol. 34, no. 5, pp. 717–724, May 1999.
[33] D. Ham and A. Hajimiri, "Concepts and methods in optimization of integrated LC VCOs," *IEEE J. Solid-State Circuits*, vol. 36, no. 6, pp. 896–909, June 2001.
[34] P. Andreani and H. Sjoland, "Tail current noise suppression in RF CMOS VCOs," *IEEE J. Solid-State Circuits*, vol. 37, no. 3, pp. 342–348, March 2002.
[35] H. Nosaka, E. Sano, K. Ishii, M. Ida, K. Kurishima, S. Yamahata, T. Shibata, H. Fukuyama, M. Yoneyama, T. Enoki, and M. Muraguchi, "A 39-to-45-Gbit/s multi-data-rate clock and data recovery circuit with a robust lock detector," *IEEE J. Solid-State Circuits*, vol. 39, no. 8, pp. 1361–1365, Aug. 2004.
[36] T.-H. Lin and W. Kaiser, "A 900-MHz 2.5-mA CMOS frequency synthesizer with an automatic SC tuning loop," *IEEE J. Solid-State Circuits*, vol. 36, no. 3, pp. 424–431, March 2001.
[37] R. Staszewski, D. Leipold, C.-M. Hung, and P. Balsara, "A first digitally-controlled oscillator in a deep-submicron CMOS process for multi-GHz wireless applications," in *Proceedings of the IEEE Radio Frequency Integrated Circuits (RFIC) Symposium*, 8–10 June 2003, pp. 81–84.
[38] J. Hauenschild, C. Dorschky, T. Winkler von Mohrenfels, and R. Seitz, "A plastic packaged 10 Gb/s BiCMOS clock and data recovering 1:4-demultiplexer with external VCO," *IEEE J. Solid-State Circuits*, vol. 31, no. 12, pp. 2056–2059, Dec. 1996.
[39] M. Rau, T. Oberst, R. Lares, A. Rothermel, R. Schweer, and N. Menoux, "Clock/data recovery PLL using half-frequency clock," *IEEE J. Solid-State Circuits*, vol. 32, no. 7, pp. 1156–1159, July 1997.
[40] A. Rofougaran, G. Chang, J. Rael, J.-C. Chang, M. Rofougaran, P. Chang, M. Djafari, M.-K. Ku, E. Roth, A. Abidi, and H. Samueli, "A single-chip 900-MHz spread-spectrum wireless transceiver in 1-mm CMOS – Part I: Architecture and transmitter design," *IEEE J. Solid-State Circuits*, vol. 33, no. 4, pp. 515–534, April 1998.
[41] A. Rofougaran, J. Rael, M. Rofougaran, and A. Abidi, "A 900 MHz CMOS LC-oscillator with quadrature outputs," in *Digest of Technical Papers IEEE International Solid-State Circuits Conference (ISSCC 1996)*, 1996, pp. 392–393.
[42] P. Andreani, A. Bonfanti, L. Romano, and C. Samori, "Analysis and design of a 1.8-GHz CMOS LC quadrature VCO," *IEEE J. Solid-State Circuits*, vol. 37, no. 12, pp. 1737–1747, Dec. 2002.
[43] P. Andreani and X. Wang, "On the phase-noise and phase-error performances of multiphase LC CMOS VCOs," *IEEE J. Solid-State Circuits*, vol. 39, no. 11, pp. 1883–1893, Nov. 2004.
[44] P. Vancorenland and M. Steyaert, "A 1.57-GHz fully integrated very low-phase-noise quadrature VCO," *IEEE J. Solid-State Circuits*, vol. 37, no. 5, pp. 653–656, 2002.

[45] J. van der Tang, P. van de Ven, D. Kasperkovitz, and A. van Roermund, "Analysis and design of an optimally coupled 5-GHz quadrature LC oscillator," *IEEE J. Solid-State Circuits*, vol. 37, no. 5, pp. 657–661, 2002.

[46] A. Mirzaei, M. Heidari, R. Bagheri, S. Chehrazi, and A. Abidi, "The quadrature LC oscillator: A complete portrait based on injection locking," *IEEE J. Solid-State Circuits*, vol. 42, no. 9, pp. 1916–1932, 2007.

[47] C.-Y. Wu and H.-S. Kao, "A 1.8 GHz CMOS quadrature voltage-controlled oscillator (VCO) using the constant-current LC ring oscillator structure," in *Proceedings of the IEEE International Symposium on Circuits and Systems ISCAS '98*, vol. 4, pp. 378–381, 1998.

[48] M. Tiebout, "Low-power low-phase-noise differentially tuned quadrature VCO design in standard CMOS," *IEEE J. Solid-State Circuits*, vol. 36, no. 7, pp. 1018–1024, 2001.

[49] B. Kleveland, C. Diaz, D. Vock, L. Madden, T. Lee, and S. Wong, "Monolithic CMOS distributed amplifier and oscillator," in *Digest of Technical Papers IEEE International Solid-State Circuits Conference (ISSCC 1999)*, 1999, pp. 70–71.

[50] H. Wu and A. Hajimiri, "Silicon-based distributed voltage-controlled oscillators," *IEEE J. Solid-State Circuits*, vol. 36, no. 3, pp. 493–502, 2001.

[51] Z. Skvor, S. Saunders, and C. Aitchison, "Novel decade electronically tunable microwave oscillator based on the distributed amplifier," *Electron. Lett.*, vol. 28, no. 17, pp. 1647–1648, 1992.

[52] L. Divina and Z. Skvor, "Experimental verification of a distributed amplifier oscillator," in *Proceedings of the 25th European Microwave Conference*, vol. 2, 1995, pp. 1163–1167.

[53] L. Divina and Z. Skvor, "The distributed oscillator at 4 GHz," *IEEE Trans. Microw. Theory Tech.*, vol. 46, no. 12, pp. 2240–2243, 1998.

[54] J. Lee and B. Razavi, "A 40-Gb/s clock and data recovery circuit in 0.18-μm CMOS technology," *IEEE J. Solid-State Circuits*, vol. 38, no. 12, pp. 2181–2190, 2003.

[55] J. Wood, S. Lipa, P. Franzon, and M. Steer, "Multi-gigahertz low-power low-skew rotary clock scheme," in *Digest of Technical Papers IEEE International Solid-State Circuits Conference (ISSCC 2001)*, 2001, pp. 400–401, 470.

[56] J. Wood, T. Edwards, and S. Lipa, "Rotary traveling-wave oscillator arrays: A new clock technology," *IEEE J. Solid-State Circuits*, vol. 36, no. 11, pp. 1654–1665, 2001.

[57] W. Athas, L. Svensson, J. Koller, N. Tzartzanis, and E. Ying-Chin Chou, "Low-power digital systems based on adiabatic-switching principles," *IEEE Trans. VLSI Syst.*, vol. 2, no. 4, pp. 398–407, Dec. 1994.

[58] J. Wood, T. Edwards, and C. Ziesler, "A 3.5GHz rotary-traveling-wave-oscillator clocked dynamic logic family in 0.25μm CMOS," in *Digest of Technical Papers IEEE International Solid-State Circuits Conference (ISSCC 2006)*, 2006, pp. 390–391.

[59] F. O'Mahony, C. Yue, M. Horowitz, and S. Wong, "A 10-GHz global clock distribution using coupled standing-wave oscillators," in *Digest of Technical*

Papers IEEE International Solid-State Circuits Conference (ISSCC 2003), 2003, pp. 428–429.

[60] F. O'Mahony, C. Yue, M. Horowitz, and S. Wong, "A 10-GHz global clock distribution using coupled standing-wave oscillators," *IEEE J. Solid-State Circuits*, vol. 38, no. 11, pp. 1813–1820, Nov. 2003.

[61] M. Sasaki, M. Shiozaki, A. Mori, A. Iwata, and H. Ikeda, "17GHz fine grid clock distribution with uniform-amplitude standing-wave oscillator," in *Proceedings of Digest of Technical Papers VLSI Circuits 2006 Symposium*, 2006, pp. 124–125.

[62] M. Sasaki, M. Shiozaki, A. Mori, A. Iwata, and H. Ikeda, "12GHz low-area-overhead standing-wave clock distribution with inductively-loaded and coupled technique," in *Digest of Technical Papers IEEE International Solid-State Circuits Conference (ISSCC 2007)*, 2007, pp. 180–181.

[63] W. Andress and D. Ham, "Standing wave oscillators utilizing wave-adaptive tapered transmission lines," *IEEE J. Solid-State Circuits*, vol. 40, no. 3, pp. 638–651, Mar. 2005.

[64] D. Ham and W. Andress, "A circular standing wave oscillator," in *Digest of Technical Papers IEEE International Solid-State Circuits Conference (ISSCC 2004)*, 2004, pp. 380–381.

[65] W. Athas, L. Svensson, and N. Tzartzanis, "A resonant signal driver for two-phase, almost-non-overlapping clocks," in *Proceedings of the IEEE International Symposium on Circuits and Systems ISCAS '96, 'Connecting the World'*, vol. 4, 12–15 May 1996, pp. 129–132.

[66] W. Athas, N. Tzartzanis, L. Svensson, and L. Peterson, "A low-power microprocessor based on resonant energy," *IEEE J. Solid-State Circuits*, vol. 32, no. 11, pp. 1693–1701, Nov. 1997.

[67] N. Tzartzanis and W. Athas, "Clock-powered logic for a 50 MHz low-power RISC datapath," in *Digest of Technical Papers IEEE International Solid-State Circuits Conference (ISSCC 1997)*, 1997, pp. 338–339, 482.

[68] W. Athas, N. Tzartzanis, W. Mao, L. Peterson, R. Lal, K. Chong, J.-S. Moon, L. Svensson, and M. Bolotski, "The design and implementation of a low-power clock-powered microprocessor," *IEEE J. Solid-State Circuits*, vol. 35, no. 11, pp. 1561–1570, Nov. 2000.

[69] V. S. Sathe, J.-Y. Chueh, and M. C. Papaefthymiou, "Energy-efficient GHz-class charge-recovery logic," *IEEE J. Solid-State Circuits*, vol. 42, no. 1, pp. 38–47, Jan. 2007.

[70] S. Chan, K. Shepard, and P. Restle, "Uniform-phase uniform-amplitude resonant-load global clock distributions," *IEEE J. Solid-State Circuits*, vol. 40, no. 1, pp. 102–109, Jan. 2005.

[71] S. Chan, P. Restle, T. Bucelot, S. Weitzel, J. Keaty, J. Liberty, B. Flachs, R. Volant, P. Kapusta, and J. Zimmerman, "A resonant global clock distribution for the cell broadband-engine™ processor," in *Digest of Technical Papers IEEE International Solid-State Circuits Conference (ISSCC 2008)*, 2008, pp. 512–513.

[72] S. Chan, K. Shepard, and P. Restle, "Distributed differential oscillators for global clock networks," *IEEE J. Solid-State Circuits*, vol. 41, no. 9, pp. 2083–2094, Sept. 2006.
[73] V. Sathe, J. Kao, and M. Papaefthymiou, "Resonant-clock latch-based design," *IEEE J. Solid-State Circuits*, vol. 43, no. 4, pp. 864–873, April 2008.
[74] L. M. Lee and C.-K. K. Yang, "An adaptive low-jitter LC-based clock distribution," in *Digest of Technical Papers IEEE International Solid-State Circuits Conference (ISSCC 2007)*, 2007, pp. 182–183.
[75] F. O'Mahony, M. Mansuri, B. Casper, J. Jaussi, and R. Mooney, "A low-jitter PLL and repeaterless clock distribution network for a 20Gb/s link," in *Proceedings of the Digest of Technical Papers VLSI Circuits 2006 Symposium*, 2006, pp. 29–30.

5

Phase Noise and Jitter

Scott Meninger

Cavium Networks

5.1 Introduction

In this chapter we examine variations that occur in the edge locations of the clock signal in a synchronous system. These edge variations are referred to in the time domain as *jitter* and in the frequency domain as *phase noise*. We also describe the various mechanisms that can cause these non-idealities and present techniques to analyze their individual contributions to total jitter. We begin by defining precisely what we mean by jitter and relate the various types of jitter to one another. Next, we explore the relationship between the time domain representation of timing error as jitter and the frequency domain representation of timing error as phase noise. The fundamental relationship between phase jitter (also denoted as absolute jitter) and phase noise will provide a useful basis for analysis of all types of jitter via simple frequency domain filter functions.

Building on this frequency domain foundation, we explore the jitter behavior of the phase locked loop (PLL) system that is most often used to generate on-chip clock signals. We utilize a control system block diagram model [1] that allows for simple analysis to determine how the PLL dynamics filter the noise sources internal to the PLL (*intrinsic* noise) as well as how the PLL reacts to noise sources external to the PLL (*extrinsic* noise).

We then use this model to examine jitter performance and reference clock specification for the PLLs used by two different types of systems. The first is a microprocessor core clock PLL, where the key requirement is that the minimum clock period produced by the PLL does not violate a minimum cycle time requirement. The second system we examine is a serial link that has the clock embedded within the data stream and uses PLLs on both ends of the link to recover the clock local to their respective systems.

The overall goal for this chapter is to give the reader an understanding of the relationship between phase noise and jitter in PLL based systems and to see that by using phase noise as a basis for calculation, system jitter analysis can be performed using simple frequency domain filtering techniques.

T. Xanthopoulos (ed.), *Clocking in Modern VLSI Systems*, Integrated Circuits and Systems, DOI 10.1007/978-1-4419-0261-0_5,

5.2 Timing Error in the Time Domain: Jitter

The first task we face is to define what we mean by jitter. To do so, we use terms consistent with industry standards [2] and notation proposed by a high speed serial data (PCI Express) jitter subcommittee [3, 4].

Jitter can be caused by many different processes. Duty cycle distortion of a waveform can be thought of as a source of jitter, if both edges of the clock are used in the system. Power supply noise may induce jitter in a PLL either directly, by moving the edge crossing of the PLL's output signal by changing the delay through a buffer, or indirectly, by changing the oscillator control voltage inside the PLL and thereby changing the PLL output frequency. We will focus our discussion in this chapter on jitter mechanisms that are caused by noise processes within circuit elements that make up the PLL.

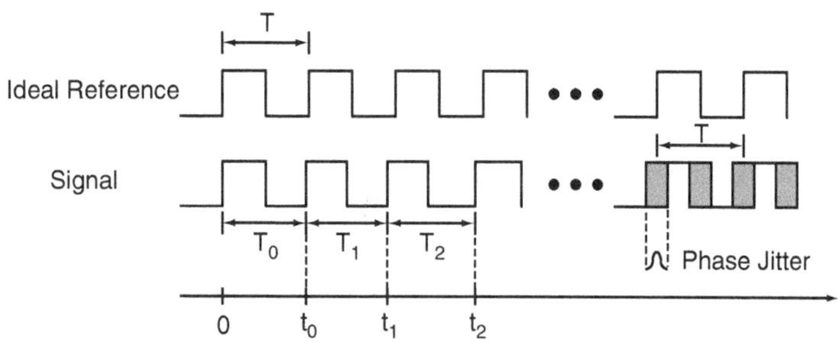

$$\text{Phase jitter} = \Phi[n] = t_n - nT$$

$$\text{Period jitter} = \Phi'[n] = (t_n - t_{n-1}) - T = T_n - T = \Phi[n] - \Phi[n-1]$$

$$\text{Cycle to cycle jitter} = \Phi''[n] = (t_n - t_{n-1}) - (t_{n-1} - t_{n-2}) = T_n - T_{n-1} = \Phi'[n] - \Phi'[n-1]$$

$$n = 0, 1, 2, \ldots.$$

Fig. 5.1. Jitter definitions

Figure 5.1 visually presents the definitions that we will use for the three different types of jitter commonly discussed in the literature [4]. These are phase jitter, period jitter, and cycle-to-cycle (c2c) jitter. In all cases, we use n in the standard way, namely as an index to indicate the edge sample (t_n) or period sample (T_n) under discussion. In general, jitter is described by statistical properties (rms or peak-to-peak values), which indicate that a large number of samples have been aggregated in the analysis. We will discuss the nature of jitter distributions in Sect. 5.6.

5.2.1 Phase Jitter

Phase jitter, or *absolute* jitter (sometimes also referred to as time interval error (TIE)) is defined as the difference between the edges of the clock signal being measured and the ideal locations where the edges would occur in the absence of jitter. In Fig. 5.1, we set up an ideal reference that allows us to determine the absolute jitter as the difference in edge locations between our signal and the ideal reference. We can express this phase jitter as:

$$\Phi[n] = t_n - nT \tag{5.1}$$

where t_n is the nth signal edge and nT is the nth period of the ideal reference. It is apparent that nT corresponds to the ideal n^{th} edge location, so Eq. (5.1) is consistent with our definition.

Phase jitter is the jitter type of interest in systems where absolute time difference between clock and signal matters. For example, in a serial data link, where the clock is recovered from the data, it is important to know the difference between the data edge and the recovered clock edge since data is being captured by the recovered clock. Phase jitter is, therefore, the jitter metric appropriate for serial data applications.

5.2.2 Period Jitter

Period jitter is defined as the difference between a given period of the signal, and the average period of the signal. In a synchronous system with no frequency offset, such as a PLL, the average period will be the same as the ideal period, T. The expression for period jitter is shown in Fig. 5.1 to be:

$$\Phi'[n] = (t_n - t_{n-1}) - T = T_n - T \tag{5.2}$$

which can also be expressed as:

$$\Phi'[n] = \Phi[n] - \Phi[n-1] \tag{5.3}$$

Equation (5.3) explains why the calculus "tick" notation is used to denote period jitter; namely, period jitter is simply the first difference of phase jitter. The importance of this relationship will become apparent in the analysis portion of this chapter, where the difference function will be utilized in the frequency domain as a filter function.

Period jitter is defined as the first difference of phase jitter, so we can say that phase jitter is a more fundamental jitter type. We also note that because period jitter can be derived from phase jitter through a difference operation, period jitter is a more high frequency phenomenon than phase jitter.[1]

[1] Finally, note that we lose some information in going from phase jitter to period jitter. That is to say that while we can construct a period jitter sequence from a phase jitter sequence, we cannot do the opposite because the starting phase is unknown. This is equivalent to saying that while we can differentiate a function without loss of information, we cannot integrate absolutely without knowing the initial condition (or offset).

Period jitter is the type of jitter of concern within a digital system where the minimum (or maximum) time period is of importance. For example, a microprocessor has a minimum cycle time that is allowed before its critical path undergoes a timing violation. In this case, knowledge of period jitter is critical since it will determine how much of the overall system timing budget will be occupied by the clock source.

5.2.3 Cycle-to-Cycle Jitter

Cycle-to-cycle (c2c) jitter is defined as the difference between successive periods of a signal. It is expressed as:

$$\Phi''[n] = T_n - T_{n-1} \tag{5.4}$$

which can also be written as:

$$\Phi''[n] = \Phi'[n] - \Phi'[n-1] \tag{5.5}$$

Just as period jitter is the first difference of phase jitter, so c2c jitter is the first difference of period jitter, or the second difference of phase jitter. Cycle-to-cycle jitter is, therefore, a very high frequency phenomenon. Once again, some information is lost in going from period jitter to c2c jitter, as the difference function removes any offset information.

5.3 Timing Error in the Frequency Domain: Phase Noise

The physical mechanisms that cause random jitter are noise processes, and these are best described and analyzed in the frequency domain. The frequency domain equivalent of jitter is phase noise.

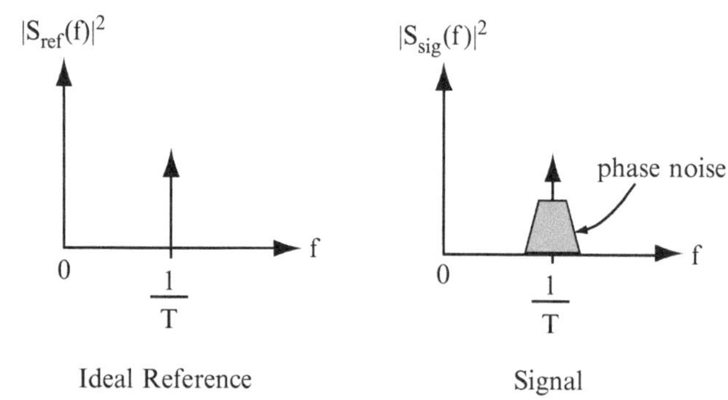

Fig. 5.2. Phase noise

Figure 5.2 depicts the spectra associated with the waveforms of Fig. 5.1. $|S_{\text{ref}}(f)|^2$ is the spectrum of the ideal reference, while $|S_{\text{sig}}(f)|^2$ is the spectrum of the signal[2].

[2] $S_{\text{ref}}(f)$ and $S_{\text{sig}}(f)$ are the power spectral densities of the time-domain signals $s_{\text{ref}}(t)$ and $s_{\text{sig}}(t)$, respectively. [26]

The ideal reference appears as an impulse at its frequency, $1/T$.[3] The signal spectrum, by contrast, has some degree of spreading over frequency. Phase noise is the term used to represent this spectral spreading. On average, the signal frequency is equal to $1/T$.[4] However, the instantaneous deviations from ideal behavior that appear as jitter in the time domain translate to phase noise in the frequency domain. The units of phase noise are dBc/Hz, and it represents the amount of noise power present in a 1 Hz bandwidth of spectrum.

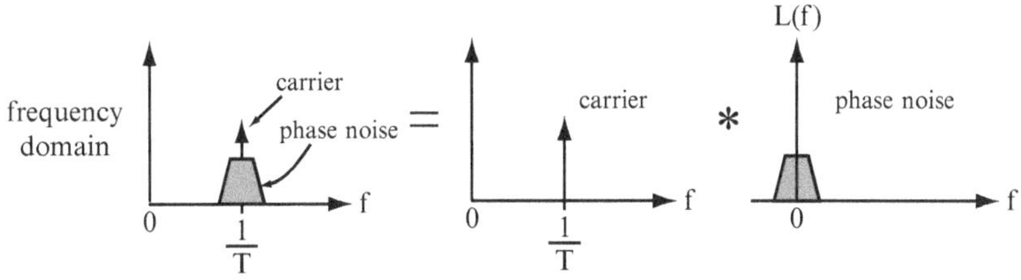

Fig. 5.3. Phase noise decomposition into carrier and L(f)

Since the system operates around a fundamental frequency $1/T$, it is possible to separate the noise spectrum from the carrier, as depicted in Fig.5.3. The phase noise spectrum is denoted as $L(f)$, which is centered around DC. Because $L(f)$ is symmetric, it is general practice to use a single-sided version of $L(f)$, wherein $L(f)$ is plotted only for positive frequencies.

5.3.1 Relationship Between Phase Noise and Jitter

In [5], arguments are made using Parceval's theorem to show the equivalence between the energy in the time domain jitter and frequency domain phase noise representations. We will simply present the result as:

$$\sigma_{\Phi}^2 = \left(\frac{T}{2\pi}\right)^2 \int_{-\infty}^{\infty} L(f)\mathrm{d}f \qquad (5.6)$$

Equation (5.6) is a powerful relationship. It states that the variance of the phase jitter of a signal is the integral over frequency of its phase noise spectrum, scaled by a factor that simply relates how much time corresponds to 2π radians of phase. Under the assumption that the signal jitter statistics can be described by a normal distribution, which is a good assumption under most PLL analysis cases[5], we can

[3] We have not plotted the impulses appearing at odd harmonics of the fundamental frequency caused by the square-wave nature of the ideal reference for simplicity.

[4] In the case of a PLL, which is the focus of our discussion, it is a fundamental property that the average output frequency is equal to the ideal output frequency.

[5] It is a property of systems with large numbers of independent noise processes that the overall noise distribution of the system will appear normal as described by the Central Limit Theorem [6].

directly calculate the rms and peak-to-peak values of phase jitter if we know the phase noise spectrum, $L(f)$. The rms value of the phase noise is just the square root of the variance given by Eq. (5.6). The peak-to-peak value is determined by how many standard deviations are required for a particular application. As we will show in Sect. 5.4.11, period jitter and cycle-to-cycle jitter may also be determined from $L(f)$ by applying the appropriate filter functions.

5.4 Frequency Domain Modeling of PLLs

Now that we have established the relationship between phase noise and phase jitter, we look at the phase noise properties of PLLs.

5.4.1 PLL Phase Noise

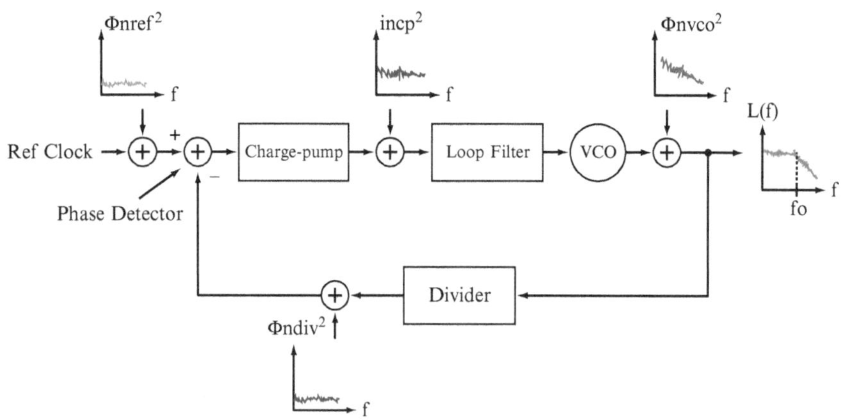

Fig. 5.4. PLL block diagram with noise sources included

Figure 5.4 presents a block diagram of a typical charge-pump type PLL along with the noise sources associated with each of the circuit elements that comprise the PLL. These noise sources are all contributors to the overall phase noise output of the PLL. Φn_{ref}^2 represents reference clock phase noise, in_{cp}^2 is a combination of phase detector, charge-pump and loop filter noise current, Φn_{vco}^2 is the output referred VCO phase noise, and Φn_{div}^2 is the feedback divider phase noise. The PLL closed loop response appears as a low-pass filter to charge-pump, loop filter, feedback divider, phase detector, and reference clock phase noise, and as a high-pass filter to VCO phase noise [1]. The overall PLL phase noise, $L(f)$, is the sum of all of the individual filtered components and has a low-pass nature, as depicted in Fig.5.4.

We divide the noise sources into two categories, *intrinsic* noise and *extrinsic* noise [7]. Intrinsic noise is associated with the components necessary to build the

PLL, namely the charge-pump, loop filter, VCO, feedback divider, and phase detector. Extrinsic noise is the noise that is processed by the PLL and is present either at the PLL input (i.e., the reference clock) or is due to undesired non-ideal system characteristics (such as poor supply noise rejection or duty cycle distortion). We will deal solely with reference clock noise as an extrinsic noise source in the analyses presented in this chapter. It is useful to go through the different noise sources in turn to develop a feel for how they impact the PLL output phase noise.

5.4.2 PLL Intrinsic Noise: VCO

VCO phase noise analysis is a rich field of research, with many different papers written on how to analyze the contributions of individual transistors to VCO output phase noise. Several interesting methods for VCO phase noise analysis are presented in [8–10], and [11] and [12] are journal paper collections with more references.

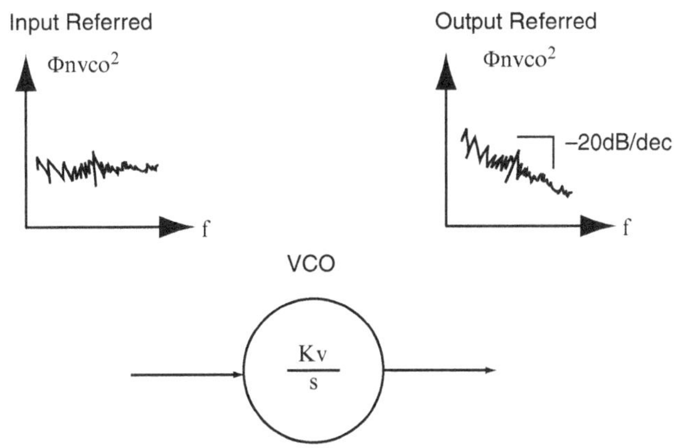

Fig. 5.5. VCO noise modeling

VCO noise behavior can be summarized by a few key characteristics. First, the VCO takes as its input a voltage[6] and outputs a specific frequency. The gain through the VCO will be denoted by K_v and has units of Hz/V or (Radians/s)/V. From the perspective of a PLL, which locks based on *phase*, the VCO transfer function from input to output looks like an ideal integrator, since phase is the integration of frequency. Second, while the VCO phase noise profile changes slope across frequency, it is generally a good assumption that in the PLL region of interest, namely around the PLL closed loop bandwidth, the VCO phase noise rolls off at −20 dB/dec.

[6] There are also current controlled oscillators (CCOs) and digitally controlled oscillators (DCOs), which have similar relationships between the control node or bus and VCO output phase. We will use VCOs in our analysis as they are still the most common oscillator type in use.

Since the VCO is modeled as an integrator, and in the frequency range of interest its phase noise rolls off at −20 dB/dec, its phase noise can be input referred as a white noise source that is then integrated, as depicted in Fig.5.5. This is a useful point to keep in mind when analyzing noise on the VCO control voltage input due to other sources, such as, for example, the loop filter. If the additional noise is white in the frequency range of interest, it is indistinguishable from the input referred VCO noise.

5.4.3 PLL Intrinsic Noise: Feedback Divider

As Fig.5.4 shows, the feedback divider is not easily distinguishable from the reference clock input in the block diagram. Fortunately, divider phase noise is generally very low, well below other system components that contribute phase noise in the same frequency band [12]. The nature of the divider phase noise will depend on its design. For example, a synchronous counter will typically exhibit less phase noise than an asynchronous cascade of divide-by stages because the synchronous design will have its output edge set directly by the VCO, whereas the asynchronous design will have its output edge set by a cascade of edges, each of which may contribute some jitter.

The challenge in modern PLLs, which operate in the multi-GHz range, is that it is difficult to create high speed synchronous dividers that do not consume significant power while operating. Therefore, lower power asynchronous dividers are very popular in high speed PLLs [12]. As phase noise requirements become more stringent, designers are developing techniques to re-synchronize the divider output to the VCO, effectively resetting its phase noise/jitter, and thereby minimizing its impact [13, 14].

In our analysis, we will assume that divider phase noise has a white profile and is below the reference clock phase noise magnitude. In such a case, we can approximate the divider phase noise as being zero.

5.4.4 PLL Intrinsic Noise: Phase Detector

Phase detectors (PD) are difficult to analyze for several reasons. First, they can be very non-linear in nature[7] [14]. The non-linearity is present, for example, in tri-state phase/frequency detectors (PFDs)[8] like the one depicted in Fig.5.6. Tri-state PFDs are of popular use in the charge-pump PLL, which is among the most popular PLL architectures in use. The PFD in Fig.5.6 compares the reference clock (REF) phase to the feedback divider (DIV) phase and creates output pulses UP or DOWN that are used to steer positive (UP) or negative (DOWN) current onto a capacitor. The more phase difference between the REF and DIV signals, the longer the current sources are on, and the more resultant positive charge Q_u or negative charge Q_d is integrated onto the capacitor.

[7] Which, of course, makes their inclusion in a linear model difficult.

[8] Tri-state PFDs offer the benefit that, in addition to phase detection, frequency detection is obtained "for free" meaning that harmonic false locking problems associated with pure phase detectors are eliminated [12].

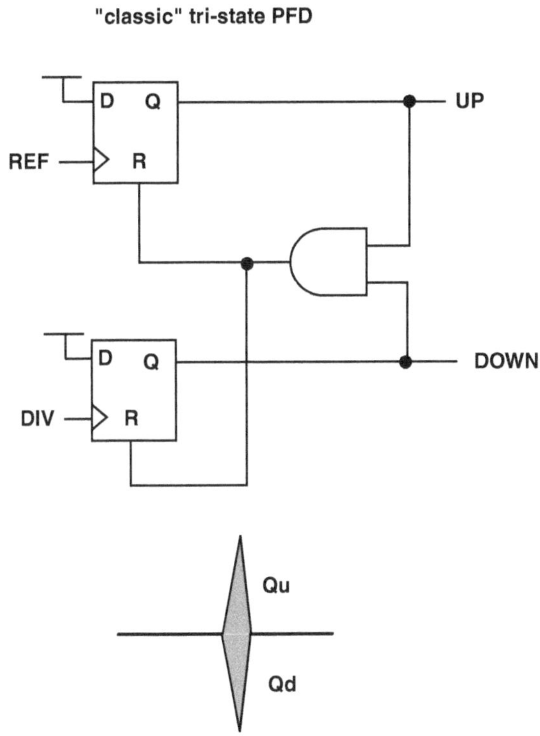

Fig. 5.6. Tri-state phase-frequency detector (PFD)

Under locked conditions, this PFD operates with very small phase differences between REF and DIV and, therefore, must produce either very short duration output pulses, or no pulses at all. Any time digital circuitry is asked to operate with narrow pulse inputs and produce narrow pulse outputs, there will be a nonlinearity present in the transfer from input pulse width to output pulse width. The simplest way to think of this is that finite edge rates in any real system will impact the pulse duration, and matching input and output edge rates perfectly is not practical. In Fig. 5.6, we show one possible steady-state locked condition, where both flip-flops fire for a short duration and produce charge packets that cancel. It is apparent that this is a situation ripe with non-linear possibilities, since either REF or DIV may occur first, and any edge-rate differences or delay differences in the reset path will allow the ordering and length of pulses to swap as the PLL feedback action seeks to correct any instantaneous errors.

There are linearization techniques to improve performance that mainly center around making the tri-state PFD operate with forced finite output pulse durations [15]. One possible configuration of this linearization technique is shown in Fig. 5.7 and results in the input phases being skewed apart enough so that the PFD operates linearly with respect to the input phase error. In this case, the cell that delays the reset signal to the REF flip-flop causes the Q_u charge packet to occur later in time than the

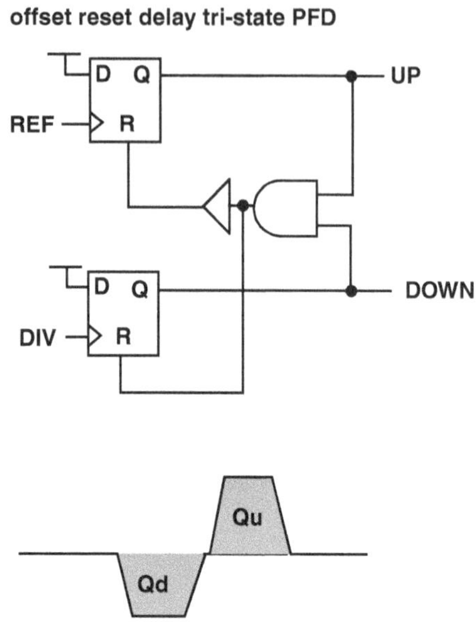

Fig. 5.7. Modified tri-state phase-frequency detector (PFD)

Q_d packet. This creates a constant (within noise limitations) positive charge packet Q_u that acts as a reference charge. REF is constrained to always occur after DIV, and no short pulses are generated. The system, therefore, behaves much more linearly under locked conditions than the PFD of Fig.5.6. The tradeoff is that, by having the PFD on longer, any noise associated with charge packet generation has a larger impact on overall PLL noise. In the case of the tri-state PFD, it is the charge-pump, which the PFD controls, that is the noise source of interest. This will be made clear in the next section.[9]

Generally speaking, the PFD itself is not considered as a noise source in phase noise analysis. This is because the phase comparison circuitry is synchronous with respect to the reference clock and feedback divider output. Any nonlinear behavior is best explored with behavioral simulators such as MATLAB or Cppsim [16]. It is mainly the way in which the PFD controls the charge-pump outputs that PLL noise is affected. Therefore, it is not uncommon to lump the charge-pump and PFD together for analysis.

5.4.5 PLL Intrinsic Noise: Charge Pump

The charge pump is used as a transducer that takes the phase difference information encoded in the PFD output pulse width and translates it into a correction signal

[9] It is also interesting to note that the tri-state PFD really has more than three states since there are two bits controlling the output. The "tri-state" name really comes into play when the charge-pump is considered. When we do this, we see that it is possible to have either a net negative charge, a net positive charge, or no charge integrated onto the loop filter (hence three states).

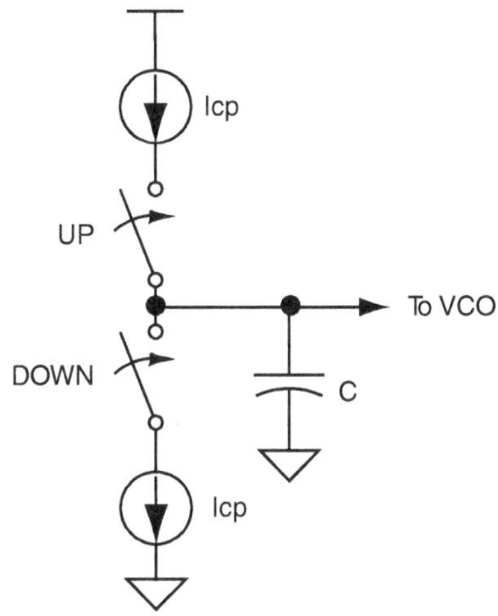

Fig. 5.8. Charge-pump architecture

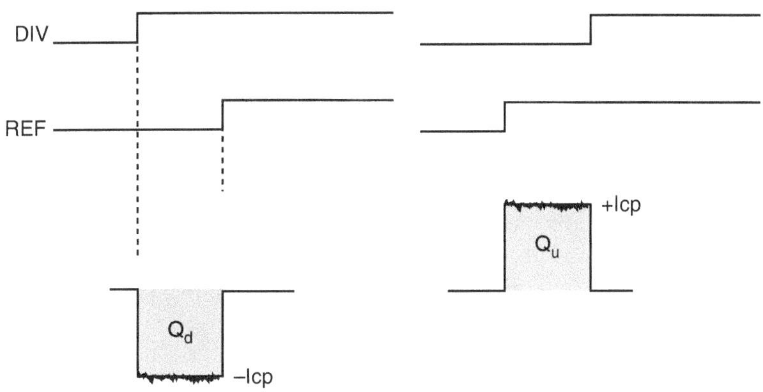

Fig. 5.9. Charge-pump operation

by integrating charge onto the loop filter capacitance. Figure 5.8 depicts a typical charge-pump used by a PLL. The UP and DOWN signals produced by the PFD close switches that cause either a positive or negative current to be routed to a loop filter, which in this case is represented by a simple capacitor.

Figure 5.9 shows the charge-pump of Fig. 5.8 in operation when controlled by a tri-state PFD under conditions where the PLL is not yet locked, and the charge-pump is on. The phase detector output turns on either a positive or negative current source that is integrated onto a loop filter capacitor. The corresponding correction charge, Q_u or Q_d, moves the VCO control voltage either down or up, depending on whether the divider phase leads or lags the reference clock. Thermal noise generated within the current sources supplying the charge appear in Q_u or Q_d and are depicted in Fig. 5.9 as the variations in the magnitude of $+Icp$ and $-Icp$. The noise current results in a noise voltage, which then produces output phase noise by modulating the VCO. It is apparent from the figure that the longer the charge-pump currents are on in steady-state, the more noise they contribute.

An offset tri-state PFD and charge-pump combination, such as the one shown in Fig. 5.7, produces a steady-state ripple on the control voltage due to the Q_d and Q_u charge packets continuously lowering and then raising the VCO control voltage back to its starting point when the PLL is locked. This periodic modulation of the VCO control voltage occurs at the reference frequency, and appears in the VCO output spectrum as a spur. This spur contains noise energy and, therefore, adds to PLL output jitter.

Sample-and-hold type loop filters can be used to dramatically reduce the spur by adding a sampling network to a tri-state PFD and charge-pump based PLL [14, 17]. The idea is to construct a circuit that samples the loop filter voltage after the up and down pulses have completed. Under a locked condition, the net excursion of the voltage due to up and down currents will be zero except for any charge-pump noise. The sampled voltage is, therefore, theoretically spur-free when presented to the VCO control input. Practical limitations in implementing the sample and hold circuitry will most likely result in reference spur suppression on the order of 20–40 dB, which is still a significant improvement.

Charge-pump noise itself is a device current noise. In MOS designs, it will, therefore, have a $1/f$ component and a thermal noise component.

5.4.6 PLL Intrinsic Noise: Loop Filter

Loop filters may be implemented in many different ways. Active or passive loop filters have different tradeoffs, as do split loop filters employed for faster acquisition of lock. The reader is referred to [11] and [12] for details on various loop filter structures.

The simplest loop filter is a totally passive RC network. The resistor elements will generate thermal noise that is filtered by the capacitor(s). Figure 5.10 depicts such a filter for use in a charge-pump PLL. The VCO is included in the figure for context.

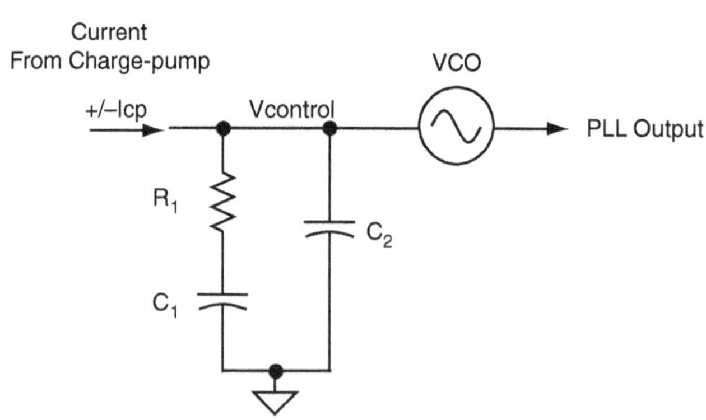

Fig. 5.10. Typical charge-pump PLL passive RC filter

Capacitor C_1 is large and acts as an integrator for the charge-pump current. R_1 adds a zero to the closed loop PLL response for stability reasons, and C_2 is a small capacitor that acts to add another pole to filter the control voltage ripple introduced by adding R_1. R_1 noise will appear directly at the VCO input and, therefore, should be considered when designing the PLL. Recall from the discussion of VCO noise modeling that the input referred phase noise profile of the VCO in the frequency range of interest appears white. The loop filter thermal noise power spectral density (PSD) will also be a white voltage source of magnitude $4kTR^2$ and is filtered by a single pole low-pass filter due to R_1 and the series combination of C_1 and C_2. It is, therefore, necessary to do a quick check of the resistor thermal noise and compare it to the VCO input referred noise and make sure that the overall phase noise is acceptable. Note that as the VCO is tuned to different voltages, its control voltage must change accordingly. Therefore, if the current sources internal to the charge-pump have finite output resistance, any change in control voltage could modulate the value of $+Icp$ or $-Icp$ and, thereby, change PLL behavior. In cases where exact control of Icp is important, an active loop filter may be employed. In an active loop filter implementation, a feedback op-amp is used to keep the charge-pump output node at some desired voltage while the loop filter output changes to the value that is required by the VCO. A downside to such an architecture is that the op-amp becomes another noise generator appearing at the VCO input in the frequency range where the VCO input referred noise should be dominant. See [14] for an example of an active loop filter implementation along with the relevant analysis.

In the analyses presented in this chapter, we will combine the loop filter noise with the charge-pump noise for simplicity, since they can be both approximated by white noise sources over a large frequency range, and they add in that range.

5.4.7 PLL Extrinsic Noise: Reference Clock

We finally come to the first extrinsic noise source of interest, namely reference clock phase noise. Reference clock phase noise is a key design parameter in PLL based systems, yet is often not known a-priori by the PLL designer, since the reference clock is picked by the system designer (the PLL end user). As we will show in Sect. 5.4.10, the PLL filters the reference clock phase noise as it proceeds to the PLL output. Therefore, the PLL filters the reference clock jitter, potentially reducing it significantly.[10]

The key concept to keep in mind is that filter analysis is much more easily performed in the frequency domain, so it is necessary that we know the frequency response of the PLL as well as the phase noise profile of the reference clock to perform this kind of analysis. Simply specifying an rms or peak-to-peak jitter number is not enough, since it does not tell us anything about the shape of the noise content of the reference clock, and, therefore, we can not predict precisely how the PLL will filter the noise.

[10] One PLL application is to be a so-called "clean-up" or "jitter filter" PLL, where a jittery system clock is passed through a PLL solely for the purpose of jitter reduction.

Reference clock phase noise can have different properties, depending on what the clock source is. For example, a crystal oscillator has a complex low frequency (sub kHz) response that levels out at higher frequencies to a white noise floor profile over the frequency range of interest to the PLL designer. The reference clock phase noise of a crystal can, therefore, often be approximated as white for analysis purposes.

Alternatively, the reference clock may be supplied by a PLL instead of a crystal oscillator. In this case, the PLL generating the reference clock will have its own intrinsic noise and filter response[11]. Knowledge of the frequency response and intrinsic phase noise of the clock generator PLL is, therefore, necessary for such a configuration.

5.4.8 PLL Extrinsic Noise: Supply Noise

We mention here that supply noise also impacts overall PLL performance. In particular, supply noise coupling to the VCO control voltage node will directly modulate the VCO output phase. Because this noise appears inside the PLL loop, it has the same impact as an intrinsic noise source and, therefore, must be minimized. The transfer of supply noise to the VCO input is highly dependent on VCO and loop filter architecture, so we will ignore supply noise in our analysis, assuming that the design has been constructed to minimize its impact.

5.4.9 PLL Extrinsic Noise: Buffer Delay and Noise

Supply noise outside the loop also impacts overall PLL noise performance. For example, the PLL output may be buffered before it is distributed throughout the chip to its various loads. Supply noise will impact the switching point of these buffers and modulate the buffer delay, adding to jitter. One can argue that this noise is outside the context of the PLL and, therefore, a different problem, but for the system clock designer, it is a phenomenon that has to be considered. In general, this type of added noise scales with the total delay through the buffers. Simply put, the more delay you create in initially buffering the clock, the more buffer jitter you will have. This statement has been quantified to first order in Eq. (2.9).

In our analysis, we focus on the PLL itself and, therefore, don't account for buffer phase noise/jitter. In a well designed system, buffer induced jitter will be very small compared to PLL jitter, especially in the context of phase jitter[12]. Section 6.8, presents an analytical model for clock buffer supply noise-induced jitter. The modeling focuses on period jitter ($\Phi'[n]$) validating the claim that the effect on phase jitter is limited.

---

[11] The clock generator PLL will also require its own reference, meaning there is probably a crystal oscillator somewhere in the system.

[12] This is because buffer jitter only impacts the clock edge on which the noise event occurs, whereas intrinsic noise sources in a PLL affect not only the current edge, but all edges afterward, since the VCO integrates noise at its input.

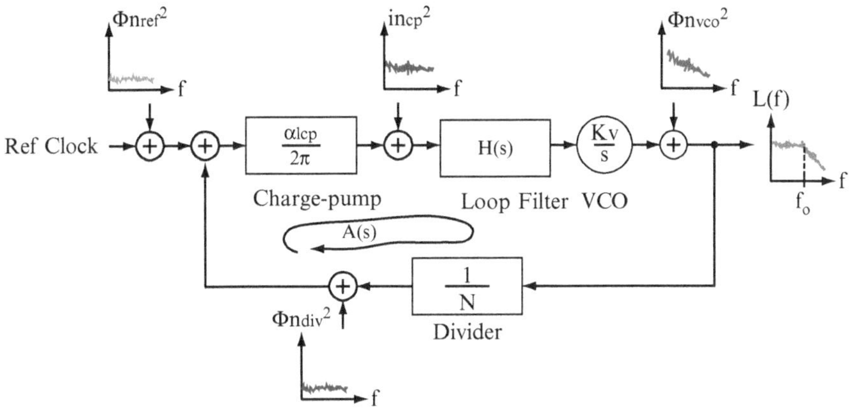

Fig. 5.11. PLL noise analysis model proposed in [1]

5.4.10 PLL Phase Noise Filtering

We now explore how the PLL filters noise. We use as our basis a simplified version of the noise model proposed in [1], which is depicted in Fig.5.11. The individual circuit blocks are replaced by their frequency domain models. The feedback divider is represented by its phase gain, which is $1/N$, where N is the PLL multiplier. The charge-pump and PFD are combined into a single block and comprise a gain term $(\alpha Icp)/(2\pi)$, where α is the charge-pump phase gain ($\alpha = 1$ for a tri-state PFD [1]), and I_{cp} is the charge-pump current magnitude. The loop filter is represented by its Laplace transform, $H(s)$, and its noise has been lumped together with the charge-pump noise. The VCO is depicted as an integrator with gain Kv, and its phase noise is output referred.

In [1], a closed loop function $G(s)$ is constructed from the loop gain, $A(s)$, where:

$$A(s) = \frac{\alpha I_{cp} K_v H(s)}{2\pi N s} \tag{5.7}$$

and

$$G(s) = \frac{A(s)}{1 + A(s)} \tag{5.8}$$

$G(s)$ has a low-pass nature with a bandwidth of f_o Hz and represents the closed loop dynamics of the PLL. Also, in [1], the block diagram is manipulated such that the transfer functions from each noise source to the PLL output are calculated in terms of $G(s)$ and the individual gain terms comprising $A(s)$. This is depicted in Fig.5.12[13]. We use the notation that the overall phase noise profile $L(f)$ is composed of the sum of the intrinsic phase noise $L_{\text{int}}(f)$ and the extrinsic phase noise $L_{\text{ext}}(f)$.

[13] As is noted in Fig.5.12, the transfer functions must be squared when processing the noise PSDs.

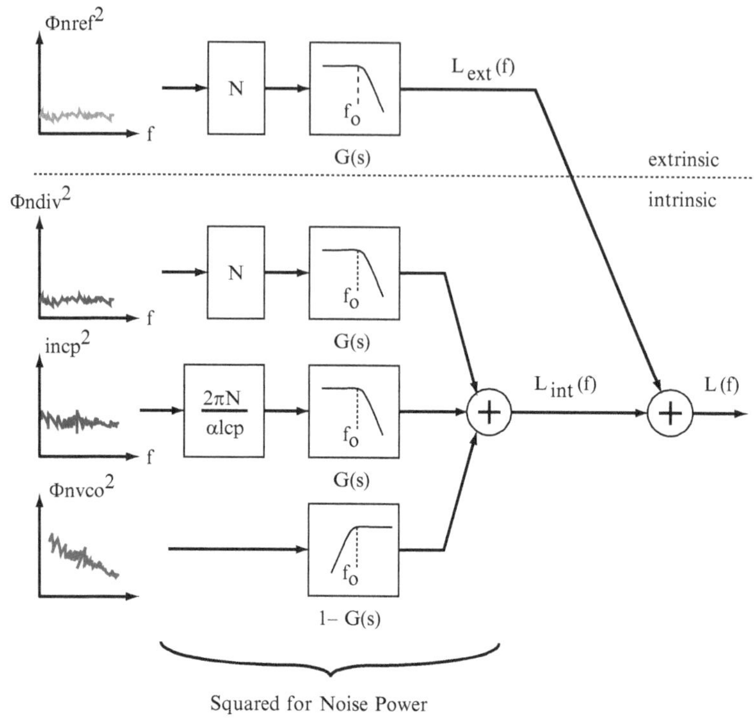

Fig. 5.12. PLL noise analysis model proposed in [1]

There are two key observations that we make from Fig.5.12. First, all intrinsic phase noise sources are low-pass filtered by the PLL with the exception of VCO phase noise, which is high-pass filtered. Second, the extrinsic reference clock phase noise is low-pass filtered by the PLL with a noise gain of N^2. The phase noise transfer for the reference clock noise power is

$$L_{\mathrm{ext}}(f) = \Phi n_{\mathrm{ref}}^2(f)\left(N^2|G(f)|^2\right) \tag{5.9}$$

where we have substituted $s = j2\pi f$ in $G(s)$. We can also calculate the intrinsic phase noise as

$$L_{\mathrm{int}}(f) = N^2|G(f)|^2\Phi n_{\mathrm{div}}^2(f) + \left(\frac{2\pi N}{\alpha Icp}\right)^2|G(f)|^2 in_{\mathrm{cp}}^2(f) + |1-G(f)|^2\Phi n_{\mathrm{vco}}^2(f) \tag{5.10}$$

The total phase noise is just the linear sum of the noise powers[14] and is expressed as

$$L(f) = L_{\mathrm{int}}(f) + L_{\mathrm{ext}}(f) \tag{5.11}$$

[14] The linear sum assumes that there is no correlation between the various noise sources, which is reasonable.

Once we have phase noise, we can convert to rms jitter using Eqs. (5.6) and (5.11):

$$\sigma_{\Phi}^2 = \left(\frac{T}{2\pi}\right)^2 \int_{-\infty}^{\infty} (L_{\text{int}}(f) + L_{\text{ext}}(f))\,\mathrm{d}f \tag{5.12}$$

where, to be consistent with earlier discussions, T is the PLL output period. Since integration is a linear operation, we can separate the components in Eq. (5.12) and express the total jitter variance as

$$\sigma_{\Phi}^2 = \sigma_{\Phi,\text{int}}^2 + \sigma_{\Phi,\text{ext}}^2 \tag{5.13}$$

The rms PLL phase jitter is just the square root of the variance[15]:

$$\sigma_{\Phi} = \sqrt{\sigma_{\Phi,\text{int}}^2 + \sigma_{\Phi,\text{ext}}^2} \tag{5.14}$$

Using Eq. (5.14), the designer can calculate the rms phase jitter resulting from his or her particular PLL design from the intrinsic and extrinsic phase noise sources. While intrinsic noise is within the control of the PLL designer, extrinsic reference clock induced phase noise is determined by the system designer's choice of reference clock. It is, therefore, necessary to properly specify the limitations on the reference clock so that total PLL output jitter is within the requirements of the desired application.

Some Intuition on Reference Clock Phase Noise (or Jitter) Filtering

As mentioned above, it is often necessary for the PLL designer to specify how much extrinsic noise is allowed at the PLL input for the PLL output to meet a required specification. Historically, this has been done by specifying a jitter requirement for the reference clock. The preceding analysis in this chapter has shown that specifying a jitter number alone does not really address the problem. The reason is that jitter, which is the result of noise processes, is filtered by the PLL according to the PLL closed loop response and reference clock phase noise profile. Jitter is the integration of this noise over frequency. So, specifying a scalar jitter quantity as an input specification does not allow us to determine how the PLL will filter the reference clock phase noise. For example, if the reference clock jitter has a phase noise profile that is entirely within the PLL bandwidth, the PLL does not filter it at all and merely passes the reference clock jitter through. If, on the other hand, the reference clock jitter has a phase noise profile with frequency components outside the PLL bandwidth, then the high frequency components are filtered by the PLL, and the reference clock contribution to total PLL jitter is reduced. Finally, there is the case where the reference clock is generated by a PLL with a bandwidth very similar to the PLL under

[15] Note that the intrinsic noise itself may be expanded and that $\sigma_{\Phi,\text{int}} = \sqrt{\sigma_{\text{vco}}^2 + \sigma_{\text{div}}^2 + \sigma_{\text{cp}}^2}$.

analysis. If both PLLs have peaking in their respective $G(f)$'s, then reference clock jitter (the output from the first PLL) may actually be *amplified* by the second PLL![16]

The PLL output jitter variance component due to the reference clock phase noise can be found using Eqs. (5.6) and (5.9) and is expressed as:

$$\sigma_{\Phi,\text{ext}}^2 = \left(\frac{T}{2\pi}\right)^2 \int_{-\infty}^{\infty} N^2 \Phi n_{\text{ref}}^2(f)|G(f)|^2 \text{d}f \tag{5.15}$$

Equation (5.15) states that the reference clock phase noise is multiplied by N^2 and filtered as it passes through the PLL. The reason for the N^2 factor has to do with the fact that the reference clock phase noise is referred to the reference clock period, T_{ref}. The PLL output phase, however, is referred to the PLL output period, T, which is N times smaller than T_{ref}. Any noise variations in the reference clock edge at the PLL input that get through the PLL and pass to the PLL output occupy the same amount of time duration, but N times more phase at the output than at the input. Noise calculations are done in noise power, so the multiplicative factor for noise analysis becomes N^2.

We can substitute $T = T_{\text{ref}}/N$ in Eq. (5.15) to get

$$\sigma_{\Phi,\text{ext}}^2 = \left(\frac{T_{\text{ref}}}{2\pi}\right)^2 \int_{-\infty}^{\infty} \Phi n_{\text{ref}}^2(f)|G(f)|^2 \text{d}f \tag{5.16}$$

What Eq. (5.16) shows is that it is possible to perform analysis on the extrinsic reference clock phase noise contribution to jitter at the PLL *output* using the reference clock phase noise profile present at the PLL *input* without explicitly taking into account the N^2 factor, as long as we remember to scale the phase noise at the PLL output by T_{ref} rather than T. From this viewpoint, the PLL is truly just a phase noise filter, $G(f)$, from input to output. Figure 5.13 depicts the simplification in the model, where the N^2 factor has been be factored out.

5.4.11 Phase Noise to Period Jitter and Phase Noise to C2C Jitter

In Sect. 5.2, we described phase jitter as being the most fundamental jitter type, with period and c2c jitter defined to be the first and second difference of phase jitter, respectively. Using this information, we can derive a frequency domain expression for period and c2c jitter based on phase jitter.

A discrete time difference function $y[n] = x[n] - x[n-1]$ is described in the frequency domain by its z transform:

$$Y(z) = X(z)(1 - z^{-1}) \tag{5.17}$$

[16] This is why it is a good practice to make sure that, when cascading PLLs, the bandwidths of the PLLs in question are "sufficiently different", where "sufficiently different" means that the two PLL bandwidths are far enough apart that any peaking does not get amplified when their closed loop responses are multiplied.

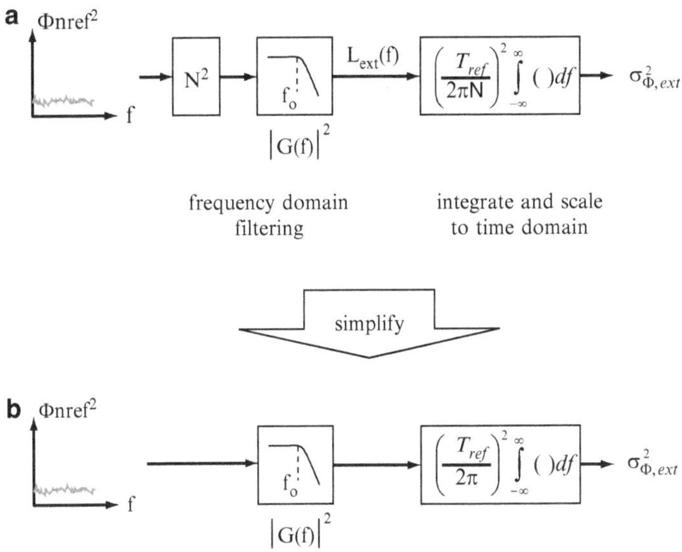

Fig. 5.13. Reference clock phase noise to jitter modeling

The transfer function from x to y, which we will call $D(z)$, is:

$$D(z) = \frac{Y(z)}{X(z)} = 1 - z^{-1} \tag{5.18}$$

Making the substitution $z = e^{j2\pi fT}$, where T is the discrete time sample period, we arrive at:

$$D(f) = 1 - e^{-j2\pi fT} \tag{5.19}$$

The log magnitude of Eq. (5.19) is plotted in Fig.5.14 assuming $T = 1$ ns to be consistent with examples that will be presented later on. $D(f)$ is periodic in frequency with nulls appearing at multiples of $1/T$. It is the "filter function" we apply to the PLL phase noise output, $L(f)$, in order to determine period jitter from phase noise. For c2c jitter, we apply $D(f)^2$ as the filter function since we know it is the second difference of phase jitter. Note that the nature of $D(f)$ is high-pass, as we expected from discussions in Sect. 5.2. We only show one "period" of $D(f)$ (up to the first null) for a reason that is fundamental to PLLs. $D(f)$ is periodic at multiples of the PLL output frequency, $1/T$. For reasons of stability [15], PLLs are designed with bandwidths significantly lower than their output frequency. A common rule of thumb is that the PLL bandwidth be set to at least $10\times$ lower than the reference clock frequency, which is N times lower than the output frequency. This implies that the PLL bandwidth is at least $10N$ times lower than $1/T$. Since the PLL is low-pass in

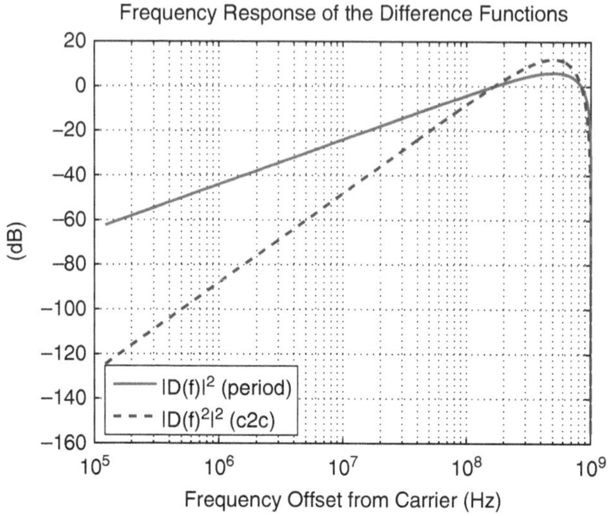

Fig. 5.14. Phase to period and C2C jitter difference functions

nature, it greatly attenuates the impact of $D(f)$ and/or $D(f)^2$ at high frequencies where they have significant magnitude.[17]

To determine period jitter from phase noise, we use Eq. (5.20):

$$\sigma_{\Phi'}^2 = \left(\frac{T}{2\pi}\right)^2 \int_{-\infty}^{\infty} L(f)|D(f)|^2 \mathrm{d}f \tag{5.20}$$

To determine c2c jitter from phase noise, we use Eq. (5.21):

$$\sigma_{\Phi''}^2 = \left(\frac{T}{2\pi}\right)^2 \int_{-\infty}^{\infty} L(f)|D(f)|^4 \mathrm{d}f \tag{5.21}$$

To determine the rms jitter contribution of any individual phase noise source to either period or c2c jitter, we simply substitute the appropriate transfer function for $G(f)$ (as depicted in Fig.5.11) and the individual phase noise profile for $L(f)$ in Eqs. (5.20) and (5.21), respectively. We will give examples in Sects. 5.5 and 5.7 regarding calculating the reference clock contribution to PLL output jitter.

[17] Here, we make the assumption that the PLL is of sufficiently high order that it rolls off fast enough to attenuate $D(f)$ or $D(f)^2$ significantly at high frequencies, which is reasonable because PLLs are typically designed to be 2nd order closed loop systems, which will roll off at the same rate that the $D(f)^2$ function rises. Unavoidable high frequency parasitic poles that are well outside the PLL closed loop bandwidth (typically 4× higher or more in frequency) will further attenuate any impact of the $D(f)$ and $D(f)^2$ filter functions.

5.4.12 Phase, Period, and C2C Jitter Examples

Before moving on to practical examples of jitter analysis, we will first use the information obtained thus far to calculate the various jitter types from a phase noise profile for an example PLL. In order to verify the frequency domain calculations, we will also explicitly measure phase, period, and c2c jitter from the time domain edge crossings of the PLL output and use the jitter definitions presented in Sect. 5.2.

To carry out the PLL simulations presented in this chapter, the CppSim behavioral modeling tool was used [1, 16]. CppSim is a constant time step C++ based simulator that is particularly useful for PLL analysis. CppSim's constant time step approach makes performing frequency domain analysis on the time domain waveforms simple via the use of FFTs.

Phase Jitter

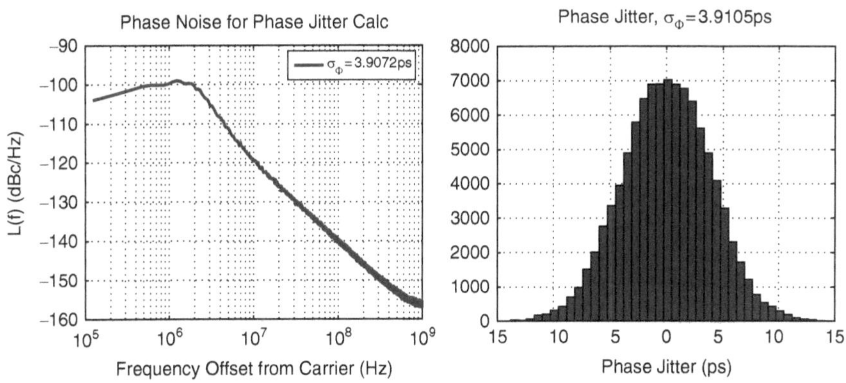

Fig. 5.15. Reference clock to phase jitter model

Figure 5.15 presents the PLL simulation results. The left plot is the PLL output $L(f)$, which includes contributions from all of the intrinsic and extrinsic phase noise sources in the system. Also shown on the plot is the calculated value of rms phase jitter using Eq. (5.6) and the simulated profile for $L(f)$. The PLL output frequency was set to 1 GHz, therefore $T = 1$ ns. The right plot shows a histogram of time domain phase jitter calculated according to the definition in Eq. (5.1). To perform this calculation, an ideal non-jittery timing reference at 1 GHz was synchronized to a PLL output edge once the PLL locked, and then the difference in edge locations between the ideal reference and the PLL output were computed over one million cycles. This large sample was taken so that the rms value calculated from the distribution would be based on a significant number of samples. Also, to a first order, we can think of this process as capturing noise frequency information down to about 1 KHz, since we have one million periods of 1 ns average length, which corresponds to 1 ms of time information, or down to 1 KHz of frequency information. The FFT in the plot does not go down to the KHz range in frequency purely for computational reasons in generating such a large FFT. This is acceptable because the phase noise components in the very low frequency range do not contribute significantly to total PLL noise power.

The frequency domain, phase noise based, rms jitter calculation is 3.9 ps. The direct measurement from the time domain simulation using the time domain definition for phase jitter is also 3.9 ps, showing excellent agreement between the two methods. This result validates our earlier discussions and Eq. (5.6). Note also that the shape of the time domain jitter measurement looks Gaussian, in agreement with our earlier assertion.

Period Jitter

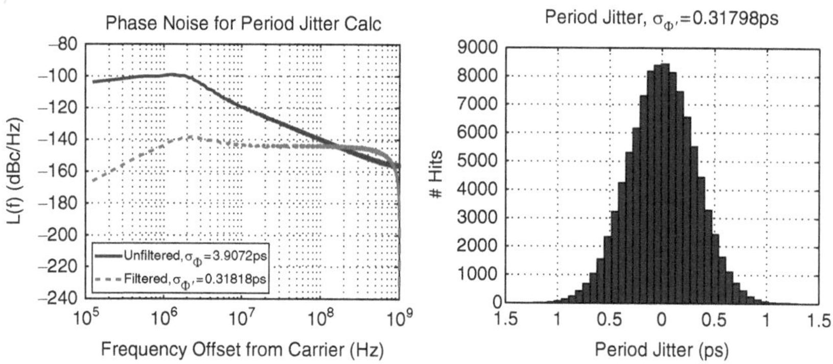

Fig. 5.16. Reference clock to period jitter model

Figure 5.16 shows the result of applying $D(f)$ from Fig.5.14 to the PLL output $L(f)$ of Fig.5.15. The unfiltered $L(f)$ is also shown in the plot as a reference for comparison. We see that the high-pass nature of $D(f)$ removes a significant amount of low frequency noise in the period jitter calculation. The low-pass nature of $G(f)$ attenuates the high frequency component of $D(f)$[18]. The right plot shows a direct measurement of period jitter as the difference between each period and the average period (1 ns).

We once again have excellent agreement between the rms jitter value based on the frequency domain analysis and the time domain measurement. This experiment validates Eqs. (5.19) and (5.20).

C2C Jitter

Finally, for completeness, we show the results in Fig.5.17 for the c2c jitter calculation based on both frequency and time domain analysis. They both agree very

[18] Also, as mentioned previously, in a real PLL, there will be additional high frequency poles in $G(f)$ in the many MHz to GHz range that will further attenuate $D(f)$. These poles were left out in the simulation for simplicity.

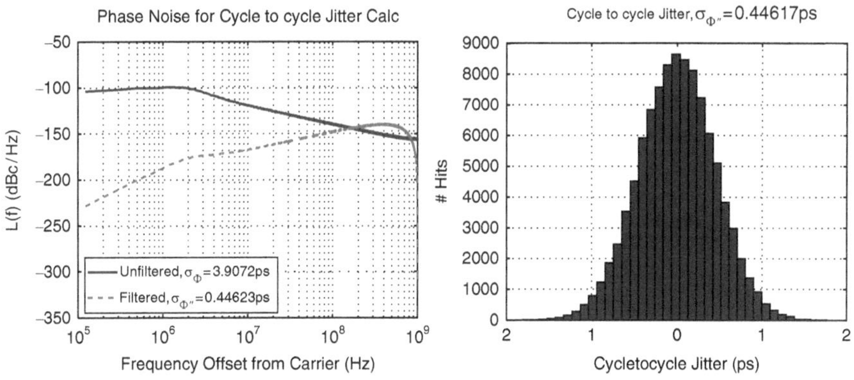

Fig. 5.17. Reference clock to C2C jitter model

well, which is to be expected since the phase and period jitter calculations were in agreement.

Having verified through simulation that our analysis methodology and jitter definitions are in agreement, we proceed with two examples of reference clock phase noise impact on system design.

5.5 Reference Clock Jitter Transfer Example: Microprocessor

In this section, we analyze the contribution of a reference clock to the output period jitter of a microprocessor core clock generator PLL. Since the microprocessor operates in its own time domain, we are primarily concerned with period jitter because it impacts the maximum operating frequency. Typically, the clock has been routed throughout the chip in a well controlled fashion (see Chap. 2) with minimum skew. The primary constraint for this exercise is to ensure that the clock period generated at the PLL output does not shrink substantially from the nominal value so that there will be a critical path violation and a catastrophic failure.

5.5.1 A Proposed Core Clock Methodology Using Mean Time Between Failures (MTBF)

Using all of the information presented thus far, we employ the model shown in Fig.5.18 to calculate the impact of reference clock phase noise on the PLL output jitter. The reference clock phase noise, Φn_{ref}^2, passes through the PLL filter function, $G(f)$, and then is processed by the period jitter difference function, $D(f)$, before it is integrated and scaled to calculate the output rms extrinsic jitter component[19].

[19] Note that while the scale factor in the integral is T_{ref}, due to the simplification proposed in Sect. 5.4.10, $D(f)$ is still periodic at the PLL output frequency, $1/T$.

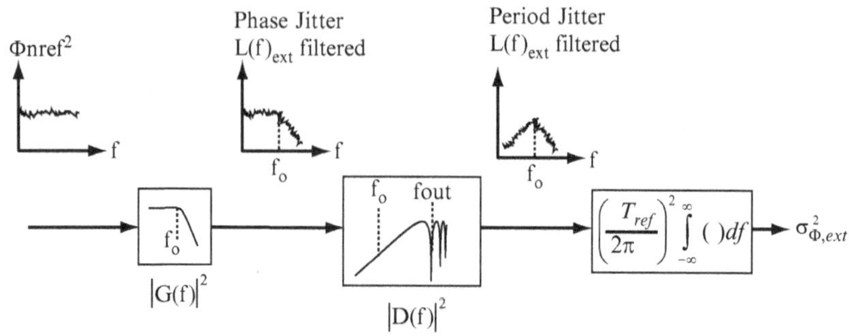

Fig. 5.18. Reference clock jitter to period jitter model

The output rms jitter due to reference clock, $\sigma_{\Phi,\text{ext}}$, is summed with the PLL intrinsic rms jitter according to Eq. (5.14) to determine total system jitter. The question then becomes: "Given an rms period jitter value, how many standard deviations should be used to determine a peak-to-peak value?" We propose that the metric to be employed to specify the number of standard deviations used to calculate peak-to-peak reference clock jitter for a microprocessor should be Mean Time Between Failures (MTBF).

MTBF in our case means how long the microprocessor can run continuously without exhibiting any timing failures, which is to say that the microprocessor does not experience any periods smaller than a critical value. If the MTBF for a 1 GHz processor is 1 year, we calculate the number of clock periods in 1 year as:

$$\text{num\_per} = (1\,\text{year})\left(365.25\frac{\text{days}}{\text{year}}\right)\left(86400\frac{\text{s}}{\text{day}}\right)\left(\frac{1\text{ period}}{1E-9\text{ s}}\right) = 3.16\text{E}+16\text{ periods} \tag{5.22}$$

We can think of $1/\text{num\_per} = 3.2\text{E}-17$ as equivalent to a bit error rate (BER), which we will use in Sect. 5.7 as the key component in determining how many σ to use for calculating peak-to-peak PLL jitter for a serial link. Simply stated, we are allowing 1 period out of $3.16\text{E}+16$ periods to be too small. This corresponds to an error rate of $3.2\text{E}-17$, which is also equivalent to a probability measure.

We utilize the fact that the jitter distribution is Gaussian (Normal) to determine how many σ are required to achieve a probability of error of $1/\text{num\_per}$ using an approximation to the *error function*, $Q(\sigma)$. For $\sigma > 3$, the error function can be approximated by [5]:

$$Q(\sigma) = \frac{1}{\sigma\sqrt{2\pi}}e^{\frac{-\sigma^2}{2}} \tag{5.23}$$

and the number of σ for a given peak-to-peak error of $1/\mathit{num\_per}$ can be read from the plot of $Q(\sigma)$ vs. σ.

Figure 5.19 shows the result of sweeping σ in $Q(\sigma)$. We can find the point on the chart that corresponds to the desired BER and determine how many σ should be used to determine the peak-to-peak value. In our example, we require a peak-to-peak value of 16.7σ, which corresponds to $\pm 8.35\sigma$ about the mean of zero.

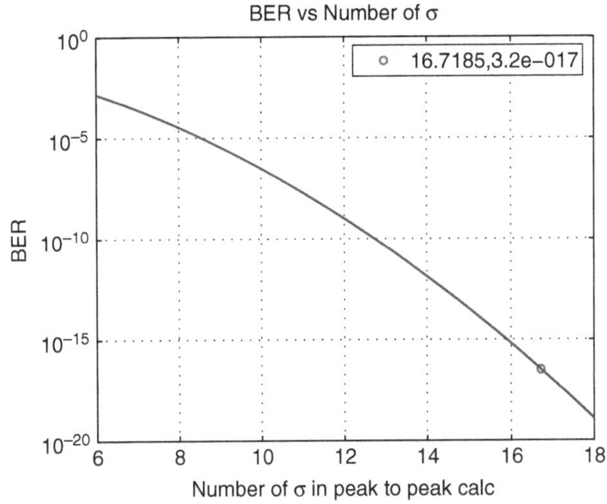

Fig. 5.19. Number of σ in peak-to-peak calculation for normal distribution

Jitter statistics reported on data sheets and in the literature typically provide rms and peak-to-peak values, where the peak-to-peak value is only seven to ten times the rms value. Part of the reason peak-to-peak values reported in data sheets vary in the number of standard deviations, σ, for which they are calculated, may be due to limits in the number of samples taken. This is an area that is often not well defined or explained in the average data sheet. We will proceed using the rms jitter value and MTBF calculation with the understanding that it may be on the conservative side.

In our example, we require $\pm 8.4\sigma$ to calculate peak-to-peak jitter. The problem is reduced to calculating rms period jitter at the PLL output. Figure 5.20 presents the filter functions that we will use in this example. The PLL closed loop transfer function, $G(f)$, is plotted along with the period jitter difference function, $D(f)$, and the cascaded transfer function, $G(f)D(f)$, which filters the reference clock phase noise. The PLL is modeled as a type II[20], second order system with a 10 MHz bandwidth and approximately 1.8 dB of peaking. $D(f)$ is the same as depicted in Fig. 5.14 for a 1 GHz PLL output clock.

Figure 5.20 shows that the shape of the filter cascade, $G(f)D(f)$, follows $D(f)$ at low frequencies and then flattens out when the falling slope of $G(f)$ equalizes the rising slope of $D(f)$. The cascade function then decreases at very high frequencies where $D(f)$ rolls off again.

To calculate the reference clock induced output jitter, we process the reference clock phase noise, Φn_{ref}^2, with the $G(f)D(f)$ cascade, as depicted in the top plot of Fig. 5.21. Φn_{ref}^2 has been set to have $1/f$ and $1/f^2$ regions at low frequencies, and a white thermal noise floor of -120 dBc/Hz at higher frequencies. This is consistent

[20] This means that there are two integrators inside the PLL loop. For example, the VCO is an ideal integrator, and the charge-pump approximates an ideal integrator very well.

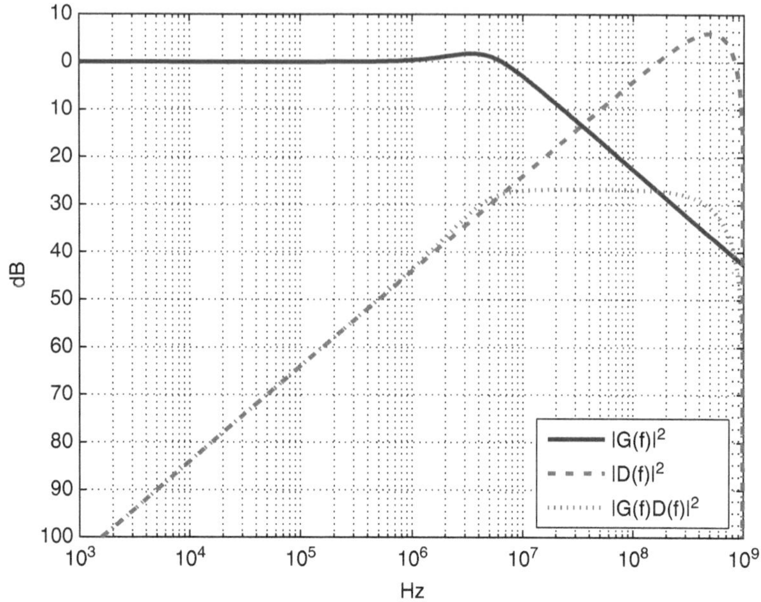

Fig. 5.20. Example transfer functions for period jitter calculation

with a crystal oscillator phase noise profile. Real crystals also have $1/f^3$ and $1/f^4$ regions at lower and lower frequencies, which we have left out for simplicity. Real crystals also typically do not report phase noise at frequencies above a few hundred MHz because of limitations in measurement equipment. We have extended the thermal noise floor up to 1 GHz for this analysis to demonstrate the impact of $D(f)$.

The filtered[21] Φn_{ref}^2 is then integrated and scaled in accordance with the model shown in Fig.5.18. The bottom plot presents the result of the integration. The period jitter value increases as the noise components are added up over frequency. We observe that the noise increases most at high frequencies where Φn_{ref}^2 flattens out. This is intuitive because Fig.5.21 has a logarithmic x-axis, so the flat region at very high frequencies contains more noise energy than the low frequency range.

Integration of the phase noise profile results in an rms period jitter value of $\sigma_{\Phi',\mathrm{ext}}$ =240 fs. Earlier,we calculated that we required approximately 17σ to determine the peak-to-peak value for a 1-year MTBF. The total result is $\sigma_{\Phi',\mathrm{pp,ext}}$ =4 ps, where $\sigma_{\Phi',\mathrm{pp,ext}}$ is the peak-to-peak period jitter at the PLL output due to the reference clock. This number is RSS (root sum of squares) summed with the PLL intrinsic period jitter to determine overall period jitter:

---

[21] We are being a little loose with the term "filter" in our discussions. $G(f)$ truly does filter the reference clock phase noise in the traditional sense. $D(f)$, by contrast, is a mathematical by-product of the definition of period jitter. So it is not a true, physical filter. It does have a frequency response that has an impact equivalent to a filter, so we use the term "filter" to describe the cascade of $G(f)$ and $D(f)$.

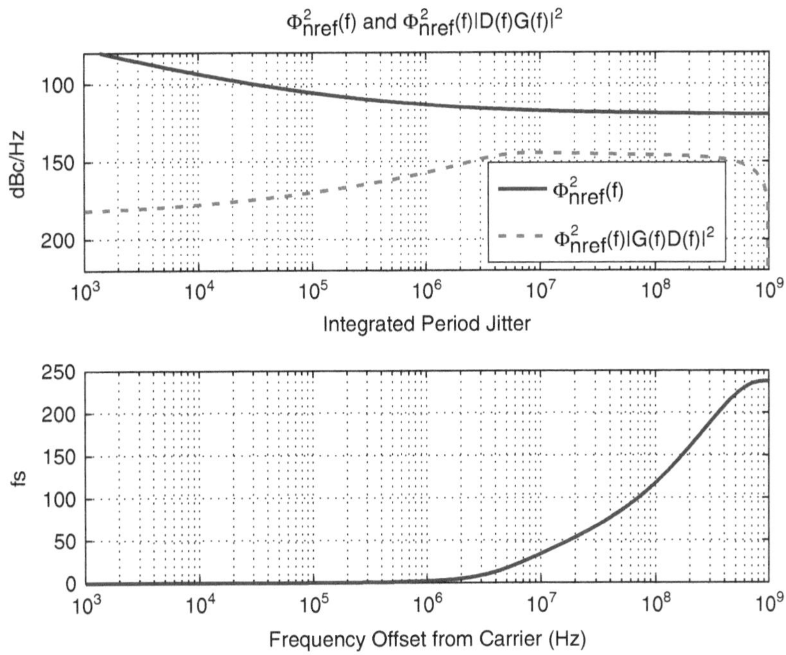

Fig. 5.21. Example reference clock Φn^2_{ref} and filtered Φn^2_{ref}

$$\sqrt{\sigma^2_{\Phi',\mathrm{pp,ext}} + \sigma^2_{\Phi',\mathrm{pp,int}}} \le t_{budget} \tag{5.24}$$

The total peak-to-peak period jitter number is then divided by two, and compared with the critical path budget, to ensure that timing margins are not violated. The reason for the division by two is that, under the assumption that the jitter is Gaussian with zero mean, half of the period jitter results in longer periods, which will not violate a critical path, while the other half of the period jitter is associated with shorter periods that will potentially violate the critical path specification.

For situations where the designer is attempting to define a reference clock phase noise profile, the analysis can be performed with a family of different phase noise responses. By taking the profile that results in the maximum PLL output period jitter, the designer can define the reference clock phase noise mask. Such a group of simulations is not trivial, so some assumptions need to be made. One reasonable assumption is that a crystal oscillator supplying the reference clock will generally have a broadband phase noise profile that is lower than the low frequency noise limit if the reference clock is supplied by a PLL.[22] Additionally, if a PLL is used to supply the reference clock, worst case output jitter will result if the bandwidth of the two

[22] A PLL that will generally have a crystal reference supplied to its reference clock input.

PLLs align, an effect that is amplified if both PLLs have peaking in their respective $G(f)$'s.[23]

As a final demonstration, we examine the impact of a high frequency pole in the PLL closed loop transfer function on total period jitter.

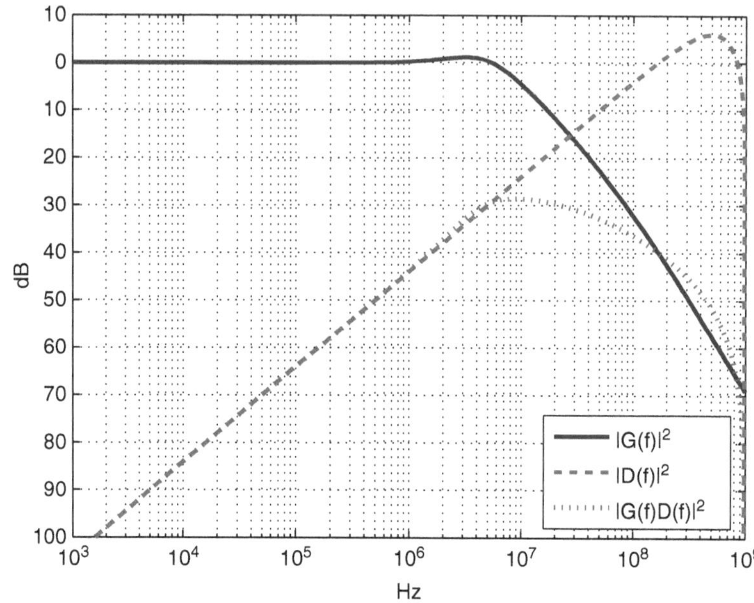

Fig. 5.22. Example transfer functions for period jitter calculation with added pole

Figures 5.22 and 5.23 present the results of the period jitter analysis with a 50 MHz pole added to the PLL closed loop transfer function. A high frequency pole is often added by the PLL designer to suppress high frequency noise caused by the phase comparison process occurring at the PFD. In some cases, several poles at high frequency are added to obtain additional suppression. As the figures show, the added pole helps greatly to reduce the high frequency phase noise present in the period jitter profile. The rms jitter value is reduced from 240 fs to 77 fs, and the peak-to-peak value from 4 ps to 1.3 ps. Clearly, high frequency poles in $G(f)$, whether explicit or parasitic, help reduce total period jitter.

5.6 Non-Random Jitter Distributions

Thus far the assumption of zero mean, Gaussian, random jitter distributions has been utilized when performing analysis. In practical systems, there will be non-random

[23] This is a phenomenon known as jitter peaking, which is of particular concern in long haul data links [5].

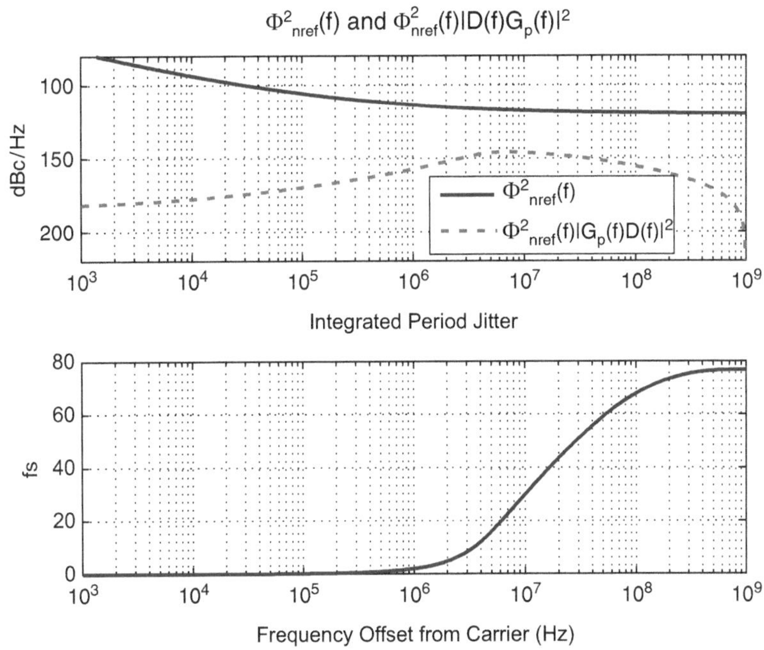

Fig. 5.23. Example reference clock Φ^2_{nref} and filtered Φ^2_{nref} with added pole

(deterministic) jitter sources that impact overall PLL jitter. Deterministic jitter is sometimes referred to as *static* jitter because unlike random jitter, which is defined by an rms value and whose peak-to-peak value depends on the application requirements, deterministic jitter can be defined by a single peak-to-peak value and manifests itself as a non-random modulation of clock phase. In this section, we present some sources of deterministic jitter and methods to include them in analysis.

5.6.1 Reference Spurs in PLLs

The feedback action inside a PLL compares the divider output to the reference input via a PFD. The output of the PFD is then filtered and used to control the VCO. Ideally, the only error components present in the signal at the VCO input relate to the phase error of the PLL or the noise sources inherent to the circuitry that comprises the PLL. In most practical PLLs, however, there is an additional error component due either to non-idealities in the PFD, or caused by design techniques aimed at improving the linearity of the PFD.

In Sect. 5.4.5, we discussed how an explicit offset is sometimes added in the reset path of a tri-state PFD to improve the PFD/charge-pump combination's linearity. We also discussed how the periodic modulation of the VCO control voltage introduced by the offset tri-state PFD appears in the PLL output spectrum as a spur. In the time domain, the reference spur appears as a deterministic jitter component.

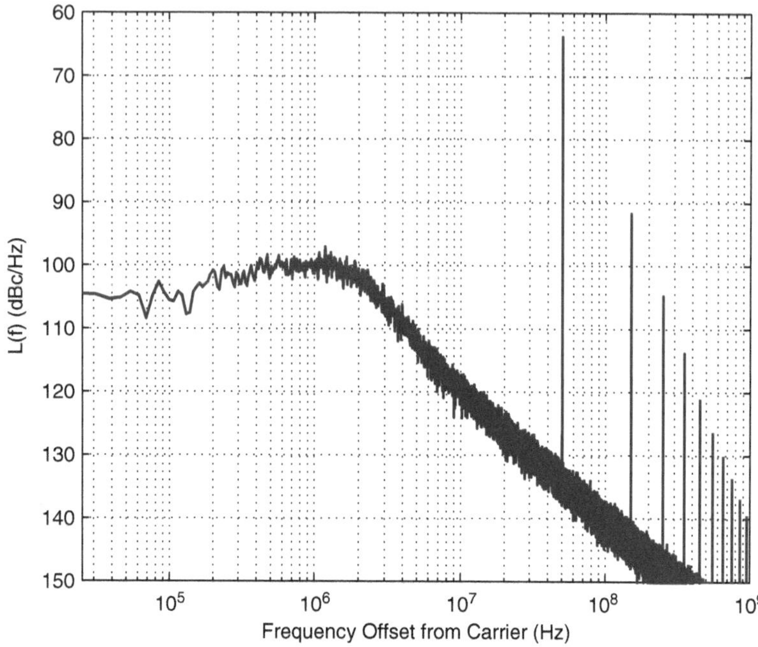

Fig. 5.24. PLL phase noise spectrum showing reference spurs

Figure 5.24 shows the phase noise profile of a PLL that exhibits reference spurs. For the PLL simulated, the reference clock frequency was set to 50 MHz and an offset tri-state PFD was used. For this PLL, the reference clock spur is significant, and we observe spurs at odd harmonics of the reference clock. The spurs are the result of the explicit offset introduced between the reference and divider flip-flop resets to improve PFD linearity. For a mathematical method to calculate the spur contribution to output jitter, see [18].

In the time domain, these spurs are observed as a deterministic component in the PLL output jitter, as shown in Fig. 5.25. We observe that the Gaussian jitter profile we expect from random noise sources is still present, but now there are two Gaussian profiles that overlap and whose peaks are separated by approximately 40 ps. The total jitter profile can, therefore, be thought of as a Gaussian jitter distribution convolved with two deterministic jitter impulses appearing 40 ps apart. For this example, the deterministic jitter component is expressed as a peak-to-peak value of approximately 40 ps.

Several techniques exist to reduce the reference spur. The simplest, and most commonly used technique, is to add high frequency poles to the PLL filter function, $G(f)$. Since the reference spur and its harmonics occur outside the PLL bandwidth, high order poles will suppress their impact while not significantly changing the PLL closed loop response. A second technique is to use a sample-and-hold loop filter [14, 17]. A sample-and-hold loop filter works by waiting until the PFD has completed its UP and DOWN pulses, and then sampling the loop filter cap. In this

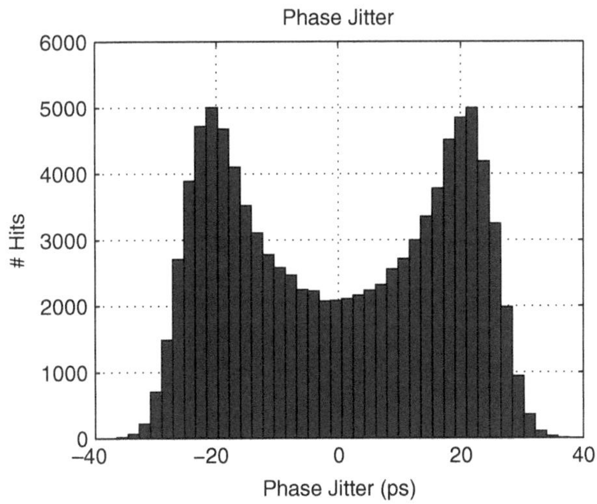

Fig. 5.25. Deterministic jitter histogram due to reference spurs

way, any periodic modulation of the VCO control voltage due to PFD operation is minimized.[24] Finally, the PFD itself can be architected to minimize the reference spur. In a tri-state PFD, this can be accomplished by minimizing the on-time of the UP and DOWN pulses when the PLL is in lock. The tradeoff with this approach is that any non-linearities present in the PFD and charge-pump combination will be exposed, as discussed in Sect. 5.4.4.

5.6.2 Duty Cycle Distortion (DCD)

In double data rate (DDR) systems, *both* edges of the clock are used as timing references. The clock frequency is, therefore, half the data rate, which eases bandwidth requirements for the clock distribution network, and, potentially, for a number of sub-systems that run at the clock rate.

The effective clock rate in a DDR system, which corresponds to the data rate, is, therefore, twice the clock frequency. However, if the clock signal exhibits any error in its duty cycle, the instantaneous data rate of the system varies from the nominal value according to the amount of duty cycle distortion present in the clock. Figure 5.26 depicts the impact of DCD in a DDR system.

*CLK*0 represents an ideal DDR clock. In this case, the time from rising edge to falling edge, t_{p0}, equals the time from falling edge to rising edge, t_{n0}, and the data rate, $1/T_d$, is equal to twice the clock frequency, $1/T$. To summarize:

$$t_{n0} = t_{p0} = T_d = \frac{1}{2}T \tag{5.25}$$

[24] Ideally, the periodic modulation of the VCO control voltage would be eliminated, but practical sample-and-hold loop filters will have some reference clock feed-through, and so a residual reference spur will remain.

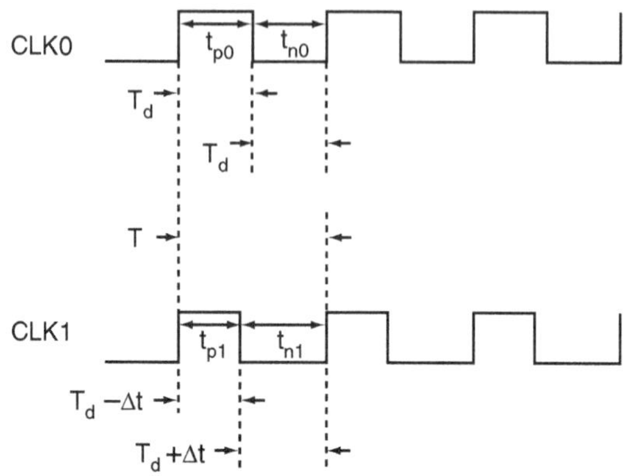

Fig. 5.26. Duty cycle distortion

*CLK*1, by contrast, exhibits DCD. While the frequency of *CLK*1 is equal to $1/T$ and is, therefore, the same as that of *CLK*0, the time from rising edge to falling edge, t_{p1}, does not equal the time from falling edge to rising edge, t_{n1}, and the data rate changes from bit to bit. For t_{p1}, the bit period is

$$t_{p1} = T_d - \Delta t \tag{5.26}$$

while for t_{n1} the bit period is

$$t_{n1} = T_d + \Delta t \tag{5.27}$$

The average bit period can be expressed as

$$< bit period >= \frac{t_{p1} + t_{n1}}{2} = T_d \tag{5.28}$$

The average bit period for the DCD case is, therefore, the same as that for the ideal case, but the instantaneous bit periods exhibit a deterministic jitter component with a peak-to-peak value $2\Delta t$.

5.6.3 Power Supply Noise

Power supply noise is another potential source of deterministic jitter. As mentioned in Sect. 5.4.8, power supply noise can couple directly into the VCO control voltage node and modulate the PLL output frequency. In this case, the impact of the supply noise lingers in the system because of the integration of phase that takes place in the VCO. Additionally, supply noise can modulate the delay through a buffer chain used to distribute the clock across chip (Sect. 6.8). In both the cases, if the supply noise is caused by a periodic signal (such as the clock itself, or sub-harmonics of the clock introduced by clock dividers or a piece of software running a loop on a microprocessor), the noise will be exhibited as a periodic modulation in the clock output phase, which appears as a deterministic jitter component.

5.6.4 Inter-Symbol Interference (ISI)

Yet another source of deterministic jitter in a communication system is inter-symbol interference (ISI). ISI is a phenomenon of particular concern in a serial data link, where the finite bandwidth of the channel filters the data being sent through it. To illustrate the impact of ISI, a CppSim behavioral simulation of a 2.5 Gb/s random data signal passing through a channel modeled as a simple low-pass filter with a −3 dB bandwidth of 1 GHz was performed. The results are presented in Figs. 5.27 and 5.28.

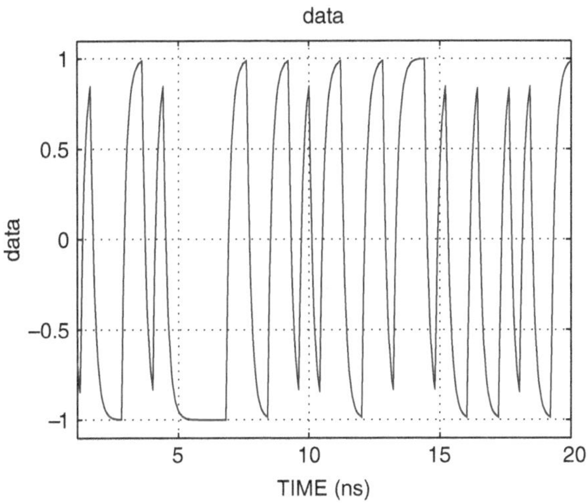

Fig. 5.27. Random data signal filtering by a channel

In Fig. 5.27, the filtered data signal is plotted. Note that sequences of multiple bits of the same polarity (sequences of multiple ones or multiple zeroes) have a larger final magnitude than one-zero-one or zero-one-zero sequences. This is because the finite bandwidth of the channel filters the high frequency data patterns more than the low frequency patterns.

An eye diagram of the filtered data signal is plotted in Fig. 5.28. An eye diagram is created by chopping the data signal in fixed increments of time, and then overlaying the sections. In Fig. 5.28, we plot the eye for two data periods. We observe that the finite bandwidth of the simple RC filter model for the channel results in incomplete settling for some bits and complete settling for others. Incomplete settling corresponds to the high frequency data sequences, while complete settling corresponds to the lower frequency data sequences.

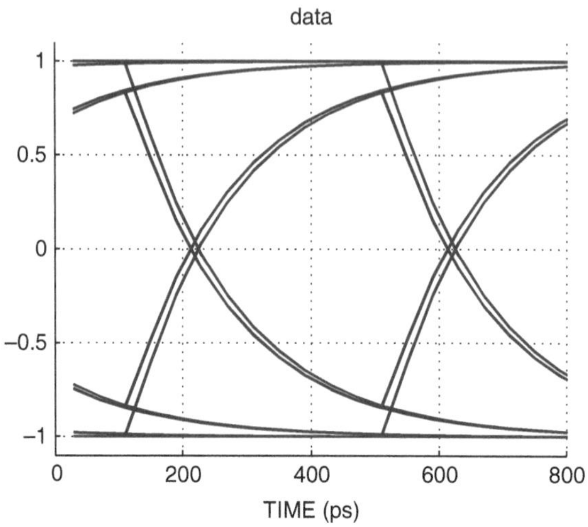

Fig. 5.28. Eye diagram of filtered data signal

Incomplete bit settling appears as a fixed component to the data jitter at the end of the channel, as observed in the two zero crossings of the data in the eye diagram. ISI can, therefore, be modeled as a deterministic jitter component.[25]

5.6.5 Including Deterministic Jitter in Analysis

The examples of non-random jitter mentioned above all have different origins and may introduce subtly different deterministic jitter components in the clock signal, but they can all be analyzed in a similar fashion.

As mentioned in Sect. 5.6.1, deterministic jitter can be modeled as impulses that occur at a finite peak-to-peak distance apart [19, 20]. These impulses are convolved with the random jitter distribution to calculate the overall jitter profile. For example, if the ISI induced deterministic jitter depicted in Fig.5.28 has a peak-to-peak value of 10 ps and system random jitter has a calculated peak-to-peak value of 50 ps, we calculate that the total peak-to-peak system jitter is 60 ps. If there are additional uncorrelated deterministic jitter sources, we add them linearly; that is, we add their peak-to-peak values. If there are additional random jitter sources, we add them in an RSS sense, under the assumption that they are statistically independent.

[25] In addition to ISI, cross-talk can introduce jitter components in a serial link. Cross-talk describes coupling between lanes within a link, or across links, in a serial data system. Coupling of an aggressor signal to the data signal can move the data signal's edge. Cross-talk is very specific to the exact board layout of a system and is, therefore, beyond the scope of this text and assumed to be minimal in our analyses.

5.7 Reference Clock Jitter Transfer Example: Serial Link

In this section, we present an analysis of the reference clock contribution to the output phase jitter of a PLL used in a serial link. The jitter analysis for a common reference clock architecture follows the example set forth in [4] and illustrates an important point about serial links, which is that if the reference clock is common to both ends of the link, the reference clock phase noise requirements may be greatly relaxed.

5.7.1 Serial Link Budgeting

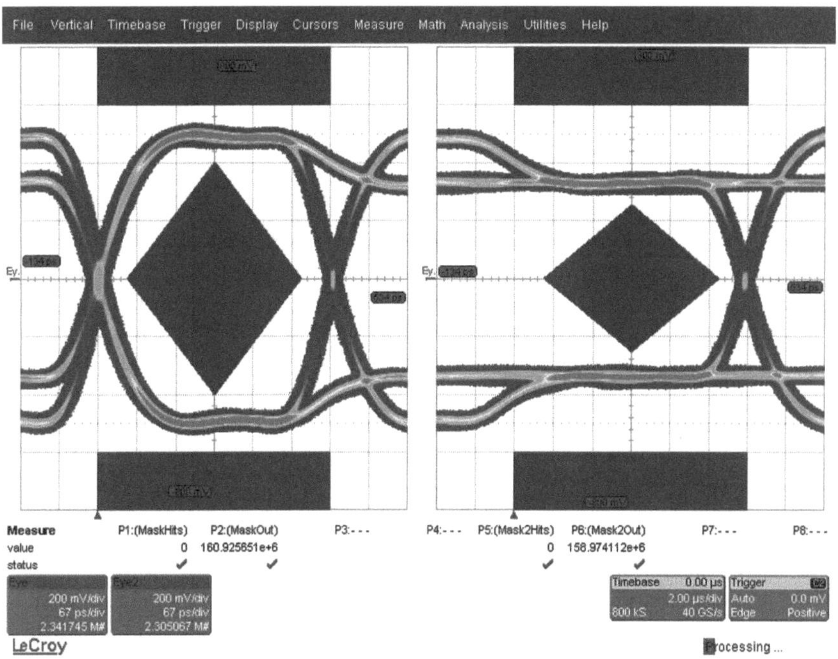

Fig. 5.29. Eye diagram

Serial link budgets are expressed in terms of an eye diagram. Figure 5.29 depicts the measured eye diagram of a 2.5 Gb/s transmitter (TX) with the PCI Express transmitter mask superimposed [21]. The eye mask has both amplitude requirements (the transmitter output must fall within the eye limits) and jitter requirements (the transmitter total output jitter must be less than the mask limits). We will focus on the jitter mask limit. In the case of the PCI Express specification, the transmitter applies some pre-filtering to overcome the finite bandwidth of the data channel. This technique, known as de-emphasis, results in two separate eye specifications. The first eye is for "transition" bits, which are bits that correspond to the data transitioning from zero to one to "don't care" or from one to zero to "don't care". The second eye is for "non-transition" bits, which are patterns of successive ones or zeros (one-one-"don't care", zero-zero-"don't care"). De-emphasis as a technique transmits larger amplitudes for transition bits and lower amplitudes for non-transition bits. One can think of this as placing emphasis on edge transitions in the data, which corresponds to a high-pass filter action. This high-pass filter action counters the low-pass filter nature of the channel.

The eye limits for jitter are expressed in UI, which represent a percentage of a "unit interval". A unit interval is simply the bit time. For a 2.5 Gb/s serial data link, 1 UI = 400 ps. The PCI Express mask depicted in Fig.5.29 requires that the total transmitter output signal not touch a mask that is 750 mUI wide. This means that total transmitter output jitter must be less than 1 UI–750 mUI = 250 mUI, which corresponds to 100 ps.

5.7.2 Bit Error Rate

In Sect. 5.2, we noted that phase jitter is the jitter type appropriate for serial links. In Sect. 5.3.1, we expressed the relationship between phase noise and phase jitter using Eq. (5.6). We also discussed, in Sect. 5.5.1, how bit error rate can be calculated using the Q function, and demonstrated an example in Fig.5.19. We can use the same approach for calculating transmitter output jitter in an example serial link. In the case of the PCI Express serial standard, the link bit error rate limit is $1\text{E}-12$, which corresponds to taking 14σ when calculating the peak-to-peak jitter value.[26]

5.7.3 Serial Link Block Diagram

For our example, we will make the assumption that the PLLs that control the transmitter and receiver output edges exhibit purely random jitter profiles. Figure 5.30 depicts a serial link where the two ends of the link have independent reference clocks. The TX clock is produced by a PLL internal to the TX IC. The TX PLL intrinsic jitter, $\sigma^2_{\Phi,int1}$ is summed with the extrinsic jitter introduced by the TX PLL reference clock, ref1. Ref1 phase noise is filtered by the TX PLL closed loop filter function, $G1(f)$. Given a phase noise profile for ref1, and given $G1(f)$, the total phase jitter at the output of the TX PLL can be calculated using the filter analysis techniques previously described in this chapter. The rms extrinsic jitter is then RSS summed with the intrinsic jitter as:

$$\sigma_{\Phi,TX} = \sqrt{\sigma^2_{\Phi,int1} + \sigma^2_{\Phi,ext1}} \tag{5.29}$$

where $\sigma^2_{\Phi,ext1}$ is the extrinsic jitter variance and the peak-to-peak output jitter for a $\text{BER} = 1\text{E}-12$ is $14\sigma_{\Phi,TX}$.

While our goal is to calculate the total TX output jitter, and we have just enumerated a method to do so, it is worth considering the receiver (RX) jitter as well. We see

[26] Actually, 14.07σ, but who's counting...

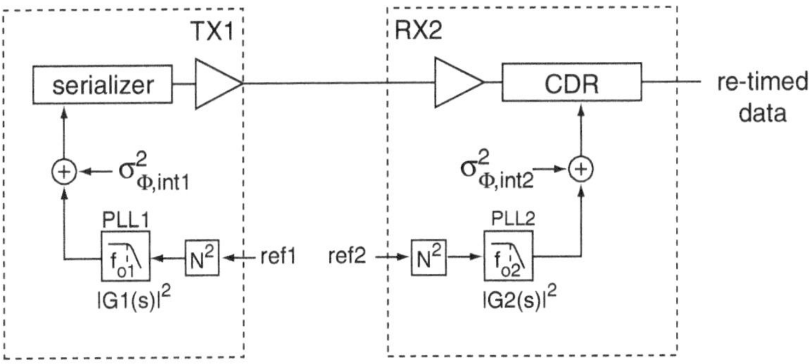

Fig. 5.30. Serial link with separate TX and RX reference clocks

that the receiver consists of a PLL and a clock and data recovery (CDR) block. This is a typical serial link design, where a single PLL is used to create a timing reference for the RX lane and then a CDR in the form of a DLL is used to recover the clock [5, 22]. This architecture allows a single RX PLL to be shared among several serial links, reducing overall power and area compared to an approach that uses one PLL per RX lane.

The RX jitter can be calculated in exactly the same manner as was the TX jitter. $G2(f)$, along with the phase noise profile of ref2 and $\sigma_{\Phi,int2}$, are used to calculate a total jitter for the RX link. If the sum of total TX and RX jitter components, along with any timing margin requirements in the RX circuitry and any distortion contributed by the channel (ISI, cross-talk, etc.) is less than 1 UI, the link operates successfully. For an accurate calculation, and under the assumption that the PLL jitters are totally random, we add the rms RX jitter and rms TX jitter in an RSS sense to get a total rms value for the link, multiply this value by 14 to get a peak-to-peak value, and then add this total to any deterministic components present elsewhere in the system.

In the architecture of Fig.5.30, the fact that the reference clocks at either end of the link are independent means that their jitter is uncorrelated and so each reference clock phase noise adds in an RSS sense to total link jitter. A different approach, that can relax reference clock phase noise requirements, is depicted in Fig.5.31.

In Fig.5.31, the reference clock is common to both the TX and RX PLLs, and so this link structure is called a "common clock" architecture [4]. On the basis of early discussions in this chapter, we know that any reference clock phase noise appearing at frequencies inside the PLL bandwidth are not filtered by the PLL. This means that the jitter transfer from reference clock to PLL output for in-band noise is unity, which has an interesting consequence in the link architecture presented in Fig.5.31. Since both PLL's are provided with the same reference clock, they both see the same in-band phase noise. To the extent that the PLL filter functions $G1(f)$ and $G2(f)$ match, both PLLs will track the in-band reference clock noise in the same way, and the net effect will be that the in-band portion of the reference clock phase noise will

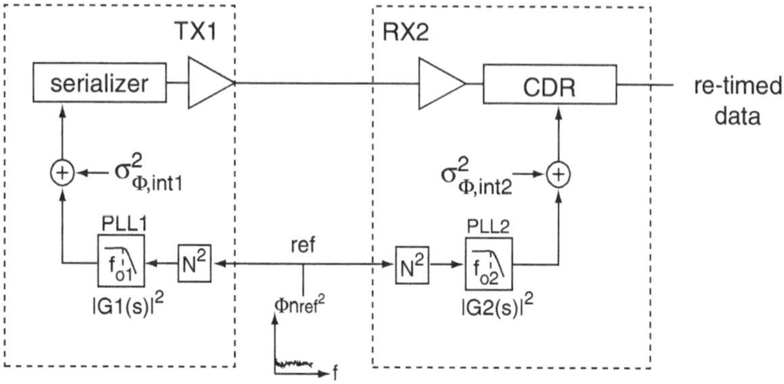

Fig. 5.31. Serial link with common reference clock

largely cancel and not contribute to the overall jitter budget! This behavior would, therefore, result in a relaxed specification for the reference clock phase noise.[27]

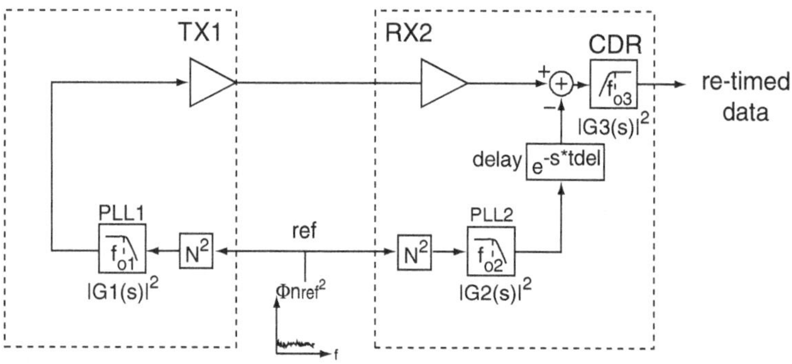

Fig. 5.32. Serial link model with common reference clock and skew

Figure 5.32 presents a modified noise model for the common clock serial link. The intrinsic jitter of each PLL has been removed, since we are now primarily interested in seeing how much reference clock phase noise impacts the jitter budget of the common reference clock link.[28] Additionally, we have introduced a delay term,

[27] This also explains why serial links that use spread spectrum clocking (SSC) to reduce electromagnetic interference (EMI) by modulating the reference clock frequency with a slow signal often require a "common clock" architecture. Since the reference clock spreading is in-band, if a common reference clock is used, both PLLs track the modulation identically, and there is minimal impact to system jitter performance.

[28] The intrinsic TX and RX jitter would, of course, be included in a total link jitter calculation.

$e^{-st_{del}}$, into the path for the RX PLL, where t_{del} represents any skew between the distribution of the reference clock to PLL1 (TX) and PLL2 (RX). A skew amounts to a phase difference, which impacts the degree to which the in-band reference clock induced jitter tracked by PLL1 matches that of PLL2, and therefore impacts the degree to which the in-band reference clock jitter cancels across the link.

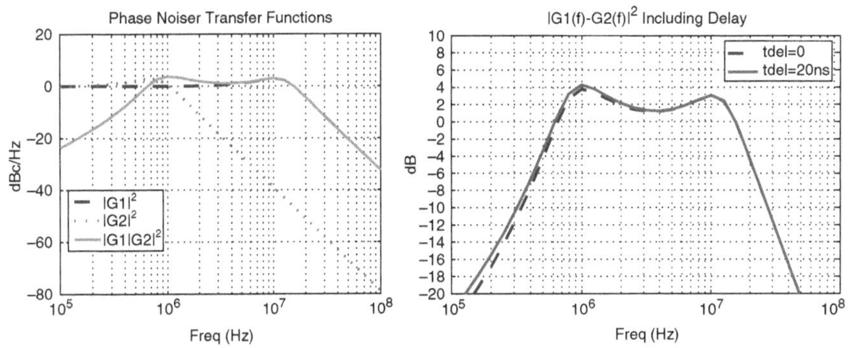

Fig. 5.33. Cancelation of reference clock noise in common clock architecture 1

Figure 5.33 shows results of jitter filter calculation for the model presented in Fig.5.32. $G1(f)$ has been set up as a second order PLL with a bandwidth of 22 MHz, while $G2(f)$ has been set up as a second order PLL with a bandwidth of 1.5 MHz. These values represent the extremes of the PLL bandwidth limits specified in the PCI Express 1.1 specification [21]. The PLL filter functions $G1(f)$ and $G2(f)$ are plotted, along with the difference between them, $|G1(f) - G2(f)|$. The skew between the two reference clock paths has been set to zero in the left plot. The right plot is a zoom-in of the difference function for two different values of skew. There are three points to note. First, low frequency noise will be tracked identically by both PLLs, and, therefore, cancels. Second, at higher frequencies, the difference function increases until it follows the shape of the higher of the two individual filter functions where the PLLs do not track one another. The net result is a band-pass filter function. Third, there is more suppression at low frequencies for a low value of skew than for a large value of skew. However, the impact of skew is relatively small, because, for in-band frequencies where the two individual PLL filter functions track one another, the skew value is a relatively small portion of the period of the frequencies in question. For example, the in-band cancelation has a low frequency -3 dB bandwidth of approximately 500 KHz, which corresponds to a noise period of 2μs. A 20 ns skew, therefore, represents 1% of the noise period at this frequency, which is not enough to result in significant cancelation error.

To calculate the reference clock impact on the jitter budget, we follow the same procedure as for an individual PLL, with the difference that we process the reference clock phase noise profile with the difference function, $G1(f) - G2(f)$, rather than a single PLL filter function, $G(f)$.

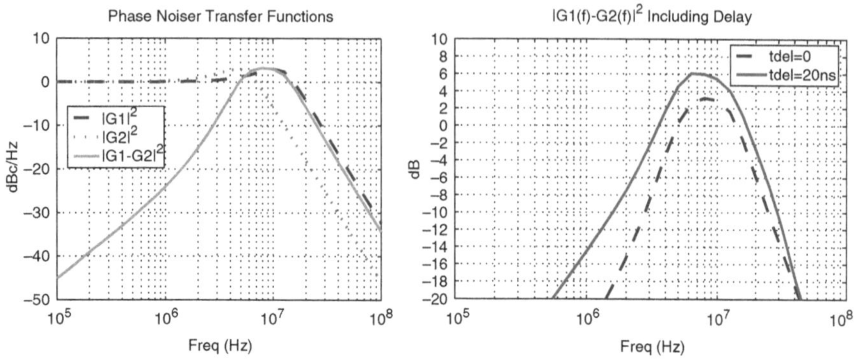

Fig. 5.34. Cancellation of reference clock noise in common clock architecture 2

The questions arises, "What happens if the individual PLL filter functions are closer together?" We examine the impact of such a situation in Fig.5.34. PLL1 bandwidth is kept at 22 MHz, while PLL2 bandwidth is changed to 10 MHz. In this case, there is more low frequency reference clock phase noise cancellation, since PLL2 bandwidth has been increased. We also see that the peak value of the difference function is higher (6 dB) for the 20 ns skew case than for the previous example (4 dB), but this is counteracted by the fact that the bandwidth of the band-pass difference function has decreased. Whether this is an overall advantage or not depends on the shape of the reference clock phase noise profile, and whether or not it has any peaking (if it is provided by a PLL), and where that peaking occurs relative to the difference function peaking. We also see that since, with more closely matched bandwidths, while the difference filter function suppresses noise up to a much higher frequency, it varies more with changes in the reference clock distribution skew. The difference function -3 dB bandwidth is now between 3 MHz and 5 MHz, and so a 20 ns skew results in a 6% to 10% error in cancelation, an amount which is significant.

For the zero skew case, there appears to be a clear advantage of having the PLL bandwidths more closely matched, as there is a narrower band-pass bandwidth in addition to a lower overall difference filter function peak. The skew case performing worse makes sense, since any skew will produce phase differences between the two paths that grow worse with increasing noise frequency.

Overall, use of a common clock architecture can potentially reduce the contribution of reference clock phase noise in a serial link. The exact extent to which the common clock architecture is advantageous depends on the difference filter function, $G1(f) - G2(f)$, the total skew in the reference clock distribution, and the shape of the reference clock phase noise profile.

5.8 Delay Locked Loops (DLLs) and Jitter

In this chapter, we make note of the fact that there are other systems in VLSI circuits that process timing signals and generate, and/or filter, jitter. One such system is the delay locked loop (DLL). Since DLLs are discussed in detail in Chap. 6, we will only mention one point about the jitter behavior of DLLs in this section.

There is no VCO in a DLL, but rather a delay line is used to align the phase of a reference with the output signal. As discussed earlier in this chapter, a delay element will exhibit jitter caused by supply noise only during the edges when the noise event occurs. Another way of stating this is that there is no memory in a delay element. This behavior is in contrast to a VCO, where noise introduced to the VCO control voltage node will remain in the system up to the point at which the closed loop can respond [9].[29] DLLs, therefore, have inherent noise advantages over PLLs [22]. The tradeoff, of course, is that frequency multiplication with DLLs is more difficult. Multiplying DLLs (MDLLs) attempt a compromise by having the equivalent of a gated ring VCO, where the phase of the oscillator is "resynchronized" to a reference every M cycles, where M is the multiplication factor [23, 24]. In this way, the memory impact associated with a VCO is reduced by the phase reset action. The most significant challenge with an MDLL is that the oscillator synchronization operation results in a potentially significant deterministic jitter component at the reference frequency, which must then be addressed by additional design techniques [25].

5.9 Conclusion

In this chapter, we have defined three primary types of jitter and related these jitter types to one another through simple discrete time filter functions. We have presented the relationship between phase noise and jitter to establish a frequency domain based approach for PLL jitter filtering analysis. Through the use of the behavioral model proposed in [1] and frequency domain filter analysis techniques, we have examined how the various intrinsic and extrinsic noise sources present in a PLL are filtered by the PLL closed loop transfer function, $G(f)$. In cases where it is desired to calculate period jitter or cycle-to-cycle jitter, simple discrete time filter functions may be included in the analysis. Examples for a core clock microprocessor period jitter calculation and a serial-link PLL phase jitter calculation were used to demonstrate the effectiveness of frequency domain filter based analysis in determining overall system jitter budgeting.

Acknowledgements

The author would like to thank Duke Xanthopoulos for providing the opportunity to contribute this chapter as well as thoughtful feedback. He would also like to thank his wife Kim for her encouragement and support throughout the writing process.

[29] In an open loop VCO, the impact of noise on the VCO control node stays around forever, since there is no feedback action to suppress it [9].

References

[1] M. Perrott, M. Trott, and C. Sodini, A modeling approach for $\Sigma - \Delta$ fractional-N frequency synthesizers allowing straightforward noise analysis. *IEEE J. Solid-State Circuits*, 37(8), 1028–1038, 2002.
[2] JEDEC Standard JESD65B, 2003.
[3] PCI Express Jitter Modeling, Revision 1.0RD, July 2004.
[4] PCI Express Jitter and BER, Revision 1.0, February 2005.
[5] B. Razavi, *Design of Integrated Circuits for Optical Communications*. McGraw-Hill, New York, 2002.
[6] D. P. Bertsekas and J. N. Tsitsiklis, *Introduction to Probability*. Athena Scientific, Belmont, MA, 2002.
[7] S. Meninger and M. Perrott, Bandwidth extension of low noise fractional-N synthesizers. In: *Proc. Digest of Papers Radio Frequency integrated Circuits (RFIC) Symposium 2005 IEEE*, 12–14 June 2005, pp. 211–214.
[8] T. Weigandt, B. Kim, and P. Gray, Analysis of timing jitter in CMOS ring oscillators. In: *Proc. IEEE International Symposium on Circuits and Systems ISCAS '94*, 4, 27–30, 1994.
[9] J. McNeill, Jitter in ring oscillators. *IEEE J. Solid-State Circuits*, 32(6), 870–879, 1997.
[10] A. Hajimiri and T. Lee, A general theory of phase noise in electrical oscillators. *IEEE J. Solid-State Circuits*, 33(2), 179–194, 1998.
[11] B. Razavi, (ed.) *Monolithic Phase-Locked Loops and Clock Recovery Circuits: Theory and Design*. Wiley-IEEE Press, New York, 1996.
[12] B. Razavi, (ed.) *Phase-Locking in High-Performance Systems: From Devices to Architectures*. Wiley-IEEE Press, New York, 2003.
[13] H.-M. Chien, T.-H. Lin, B. Ibrahim, L. Zhang, M. Rofougaran, A. Rofougaran, and W. Kaiser, A 4 GHz fractional-N synthesizer for IEEE 802.11a. In: *Proc. Digest of Technical Papers VLSI Circuits 2004 Symposium*, 17–19 June 2004, 46–49.
[14] S. Meninger, Low phase noise, high bandwidth frequency synthesis techniques. Ph.D. dissertation, Massachusetts Institute of Technology, 2005.
[15] W. F. Egan, *Frequency Synthesis by Phase Lock*, 2nd edn. Wiley-Interscience, New York, 1999.
[16] M. Perrott, CppSim behavioral simulation package. [Online]. Available: http: //www.cppsim.com
[17] M. Cassia, P. Shah, and E. Bruun, Analytical model and behavioral simulation approach for a $\Sigma\Delta$ fractional-N synthesizer employing a sample-hold element. *IEEE Trans. Circuits Syst. II*, 50(11), 850–859, 2003.
[18] Integrated phase noise. Silicon Laboratories AN256, 2006.
[19] Measuring Jitter in Digital Systems. Agilent Technologies Application Note 1448-1.
[20] M. Li, J. Wilstrup, R. Jessen, and D. Petrich, New jitter decomposition method and its applications. Wavecrest Corporation, MN, 1999.
[21] PCI Express Base Specification Revision 1.1, March 2005.

[22] S. Sidiropoulos and M. Horowitz, A semidigital dual delay-locked loop. *IEEE J. Solid-State Circuits*, 32(11), 1683–1692, 1997.
[23] R. Farjad-Rad, W. Dally, H.-T. Ng, R. Senthinathan, M.-J. Lee, R. Rathi, and J. Poulton, A low-power multiplying DLL for low-jitter multigigahertz clock generation in highly integrated digital chips. *IEEE J. Solid-State Circuits*, 37(12), 1804–1812, 2002.
[24] S. Ye, L. Jansson, and I. Galton, A multiple-crystal interface PLL with VCO realignment to reduce phase noise. *IEEE J. Solid-State Circuits*, 37(12), 1795–1803, 2002.
[25] B. Helal, M. Straayer, G.-Y. Wei, and M. Perrott, A highly digital MDLL-based clock multiplier that leverages a self-scrambling time-to-digital converter to achieve subpicosecond jitter performance. *IEEE J. Solid-State Circuits*, 43(4), 855–863, 2008.
[26] B. Razavi, RF Microelectronics. Prentice Hall, New Jersy, 1998.

6

Digital Delay Lock Techniques

Thucydides Xanthopoulos

Cavium Networks

6.1 Introduction

Digital delay locked loops are highly prevalent in integrated systems. They are essentially delay lines under feedback control that can generate derived clocks based on an input reference. Applications include clock distribution, I/O interfaces, clock generation, and frequency multiplication. Digital delay locked loops also have time-to-digital conversion properties and can be used in monitoring and sensing applications.

While DLLs can be designed with digital-only methods, their design involves direct manipulation of clock signals. Therefore, additional techniques are involved as opposed to standard custom digital datapath design. This chapter presents an identification of all essential digital delay locked loop components and addresses relevant design aspects for each part. It concludes with global design issues and an overview of advanced applications.

6.2 What Constitutes a Digital Delay Locked Loop?

The digital delay locked loop (DLL henceforth) is a simple closed loop system that is capable of generating a clock signal that has a precise phase relationship with an input reference clock. Because of feedback, this phase relationship tracks across input frequencies, process, voltage, and temperature. The accuracy of the phase relationship between input and output clocks depends on DLL design parameters, process mismatch characteristics, and on deterministic noise sources such as independent supply noise and forms of coupling.

Digital DLLs can easily be unconditionally stable and are analyzed in the time domain. Because of their all digital nature, they can be ported across process nodes, can be simulated using fast digital simulators, and can be easily monitored and characterized in silicon.

Figure 6.1 shows a simplified DLL block diagram, which identifies the three main system components: The phase detector, the control block, and the delay line.

T. Xanthopoulos (ed.), *Clocking in Modern VLSI Systems*, Integrated Circuits and Systems,
DOI 10.1007/978-1-4419-0261-0_6,

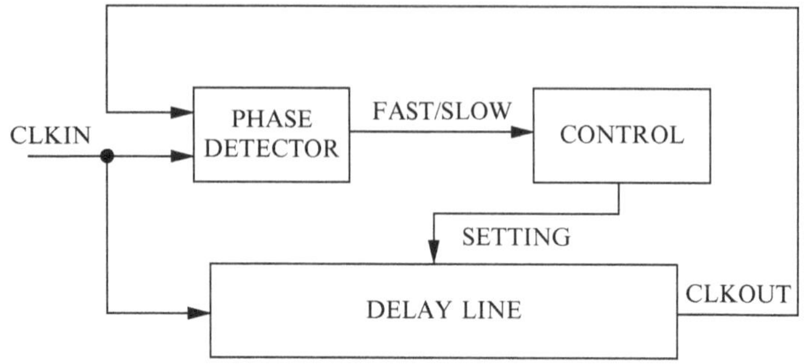

Fig. 6.1. Generalized DLL block diagram

The type of phase detector that will concern us in this chapter has a single bit output that only changes on the positive edge of the reference clock. It is commonly referred to as "bang–bang" in the literature. Such a phase detector can be thought of as a system that has the transfer function depicted in Fig.6.2. It compares the phase difference between two clocks and outputs a single logic value indicating which clock is ahead in time. It can be thought of as a flip-flop that has the reference clock (CLKIN) as its clock input and the controlled clock (CLKOUT) as the data input. When the flop evaluates to a logic 1, it means the controlled clock is faster than the reference. On the other hand, when the controlled clock is slower than the reference, the flop will evaluate to logic 0. This behavior can be guaranteed as long as the data input is faster than the clock input at least by the flop setup time (T_s in Fig.6.2) or slower at least by the flop hold time (T_h in Fig.6.2). If the timing between the two clocks falls within the gray area of Fig.6.2, the phase detector behavior is not defined and the output can be either a logic 0, a logic 1 or potentially a metastable value. The width of the gray rectangle ($T_s + T_h$) is the primary figure of merit of phase detectors and one of the main DLL design parameters. It is called the phase detector *sampling window* d_{sw} or *dead zone*. It affects the phase locking accuracy of the entire system in addition to other important specifications. Phase detectors will be discussed in Sect. 6.4.

The most design intensive component of the DLL is the Digitally Controlled Delay Line (DCDL). A sample transfer function is shown in Fig.6.3. A DCDL is a combinational circuit that delays its input by an open loop value that typically has a monotonic relationship with the digital setting input. Such delay value is not precisely defined and is subject to process, voltage, and temperature conditions. A DCDL is primarily characterized by three design parameters: its minimum delay D_{min} (delay value at setting 0), its maximum delay D_{max} (delay value at the maximum setting $N-1$), and its resolution d_r (incremental delay per setting). The dynamic range is defined as $D_{max} - D_{min}$ and is directly related to the capability of the overall DLL to track significant PVT variations or work with an extended range of input clock frequencies. The resolution d_r affects the DLL accuracy along with d_{sw}. DCDL design will be discussed in Sect. 6.5.

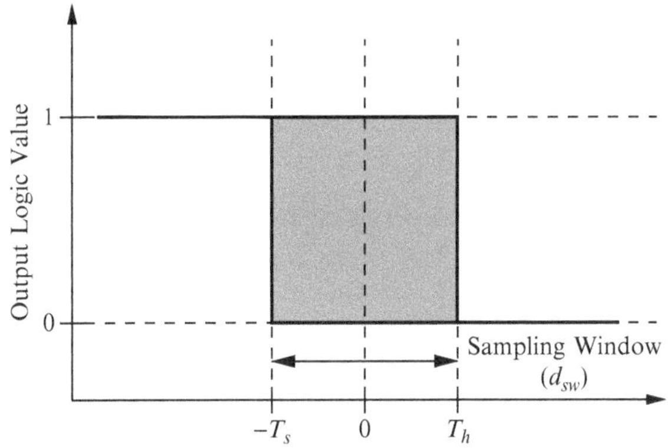

Fig. 6.2. Phase detector transfer function

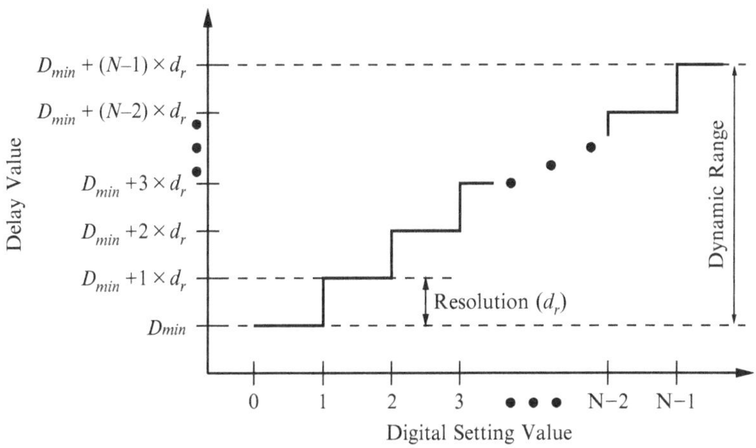

Fig. 6.3. Digitally controlled delay line transfer function

The final building block of Fig.6.1 is the control module. The control block increases and decreases the DCDL settings based on the output of the phase detector. In its simplest form, it is an up/down counter controlled by the phase detector. In its most general form, it is a finite state machine (FSM) that controls the DCDL settings based on the output of the phase detector and internal state. The inclusion of additional state information can support more complex behavior and extend the DLL capabilities. Control structures will be addressed in Sect. 6.6.

The combination of a phase detector, a delay line, and a control block produces a simple and useful feedback system that can find multiple applications in modern systems-on-a-chip. The DLL of Fig.6.1 will adjust its delay line until CLKIN and

CLKOUT are matched in phase. At this point, the DLL has *locked*, and the delay through its DCDL is one CLKIN period (or potentially an integral multiple of input clock periods).

6.3 An Overview of DLL Applications

The average modern microprocessor contains multiple digital delay locked loops embedded in various subsystems. Figure 6.4 demonstrates different uses of a basic DLL structure.

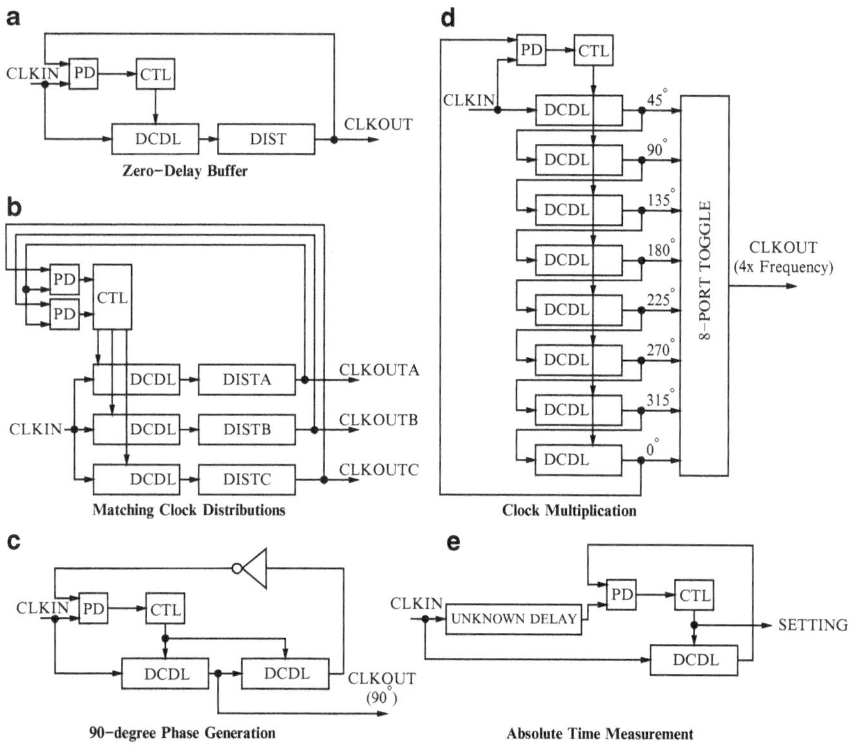

Fig. 6.4. DLL applications

The vast majority of DLL applications are related to clocking. Figure 6.4a demonstrates *zero-delay buffering*. Such a topology is well suited for synchronous I/O interfaces (e.g., PCI/PCI-X). A common clock (CLKIN) is being distributed to multiple bus end points. Each end point is buffering and distributing it to a number of flip-flops. A DLL in the loop ensures that the buffered clock version (CLKOUT) is phase-locked to the master interface clock (CLKIN).

Figure 6.4b shows an application where three separate clock domains are synchronized. Each domain has a separate clock distribution (DISTA, DISTB, DISTC) involving multiple buffer stages. Open loop matching is infeasible in the presence of PVT and random variations. A DLL-controlled delay line at the root of each distribution can guarantee phase matching among all three clocks. The DLL controller is more complex than the one shown in Fig.6.1, and it uses information from two separate phase detectors. First, CLKOUTA and CLKOUTB are phase locked. As soon as this happens, the controller locks CLKOUTC to CLKOUTA. This scheme is extensible to multiple clocks.

Quadrature clock generation is another application suitable for a DLL. Such a topology is shown in Fig.6.4c. When this DLL locks, the total delay through both DCDLs and the feedback path inverter is one half period of CLKIN (180°). Therefore, the delay through one DCDL is virtually a quarter period (90°) and CLKOUT is a quadrature clock.

The previous technique can be extended and used for constant factor clock multiplication. In Fig.6.4d, the delay through all eight DCDLs is one CLKIN period. The delay through a single DCDL is 45°. The outputs of the eight DCDLs are equally spaced phases spanning the entire CLKIN period. The toggle element can be thought of as a toggle flop with eight independent clock ports. Every positive edge of each of the eight phases can toggle the flop, thus, producing a CLKOUT that has a frequency equal to four times that of the input.

The final example (Fig.6.4e) is a deviation from strictly clocking applications. A DLL can be used for absolute measurements of unknown delays (time-to-digital conversion). First, a 2-point calibration is necessary. The DLL is placed in calibration mode (unknown delay is bypassed) and an input clock of a known period T_0 is fed into the CLKIN input. The DLL locks and the setting is recorded (s_0). The input clock period is changed to T_1, the DLL is allowed to lock and the setting is recorded again (s_1). We now have a system of two equations with two unknowns:

$$D_{\mathrm{min}} + s_0 \cdot d_{\mathrm{r}} = T_0, \tag{6.1}$$

$$D_{\mathrm{min}} + s_1 \cdot d_{\mathrm{r}} = T_1, \tag{6.2}$$

where D_{min} is the DCDL delay at the minimum delay setting and d_{r} is the DCDL resolution. We can solve the above system and obtain values for D_{min} and d_{r}. The DLL now is placed out of calibration mode, and the unknown delay is multiplexed into the system. The DLL locks and the setting (s_u) is recorded. The absolute delay is, therefore, $D_{\mathrm{min}} + s_u \cdot d_{\mathrm{r}}$.

6.4 Phase Detectors

In a digital delay locked loop, the output of the phase detector is typically processed by a digital circuit such as an up/down counter or in the most general case an FSM controller. The most common phase detector for such an application is a specially-designed flip-flop that has a single bit output indicating leading or lagging feedback clock as described in Sect. 6.2 and shown in Fig.6.2. Traditionally, such a structure with a single-bit digital output is called a "bang–bang" phase detector. The main design goals for a flip-flop used as a phase detector are:

1. It must be a fully static design containing cross-coupled nodes exhibiting exponential voltage development with time (Sect. 6.4.1) to minimize time spent in a potential metastable state and prevent system failure.
2. It must have a small sampling window d_{sw}, which is the sum of the underlying flop setup and hold times ($T_s + T_h$) to guarantee good phase matching between feedback clock and reference clock.
3. The setup and hold times must be well-balanced to avoid deterministic bias during phase detection and result in a significant systematic phase error between feedback clock and reference clock. A differential design can be desirable, but it is not a requirement.
4. The open loop gain of the cross coupled gates must be high to guarantee quick exit from a potential metastable condition (Sect. 6.4.1).
5. The capacitance of the cross-coupled nodes must be kept to a minimum given the other constraints in order to guarantee quick exit from metastability (Sect. 6.4.1). This will also help minimize setup requirements but may hurt hold time.
6. Unlike regular flop designs, a short clock-to-q delay is not a critical design requirement, since there is typically a full reference clock cycle available until the phase detector output needs to be setup and processed by the FSM controller. This path is unlikely to be critical. Moreover, unlike a proportional phase detector, balancing clock-to-q delays for a 0-to-1 vs. a 1-to-0 transition is not necessary. There is no phase information encoded in this delay for bang–bang operation.

The edge-triggered fully static flop designs of Chap. 3 such as the master-slave latch (flop) of Fig.3.4 or the sense-amp flip-flop of Fig.3.13 can be used as a bang–bang phase detector assuming that the design is tuned to meet the design goals outlined above. One design goal which won't be met is the balancing of the setup and hold times (and tracking across PVT) due to the assymetry in the clock and data path in any regular flip-flop. One way around this problem is illustrated in Fig.6.5 [1].

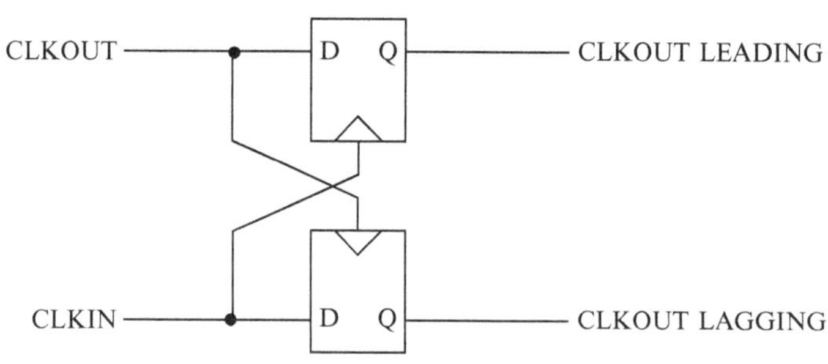

Fig. 6.5. Symmetric phase detector out of asymmetric flops

In a single-flop phase detector, the feedback clock leading decision depends on the setup time and the feedback clock lagging decision depends on the hold time. Asymmetry between these two properties can introduce a systematic phase error. In the coupled flops of Fig.6.5, both decisions depend on the setup time of the flop and assuming identical flops, the systematic phase error is removed. The coupled flops of Fig.6.5 are also a virtual ternary phase detector where a 11 or 00 state can be interpreted as a NOP (no operation) where no adjustment to the delay line takes place.

A flip-flop design that addresses all design goals outlined above is certainly possible. Figure 6.6 shows a non-traditional flop design used as a phase detector in [2]. CLKIN constitutes the clock input and CLKOUT is the data input. This edge-triggered flop is composed of three separate RS latches. Latch (A1,A2) is the master latch, latch (C1,C2) is the slave latch, and latch (B1,B2) is an auxiliary latch whose additional state is necessary for correct operation. Setup and hold timing behavior is entirely set by the master latch. The auxiliary latch is not in the signal path during sampling and the slave latch only affects clock-Q delay which is not relevant for bang–bang phase detector operation. Gates D1 and D2 are only present for load balancing along the CLKIN and CLKOUT paths.

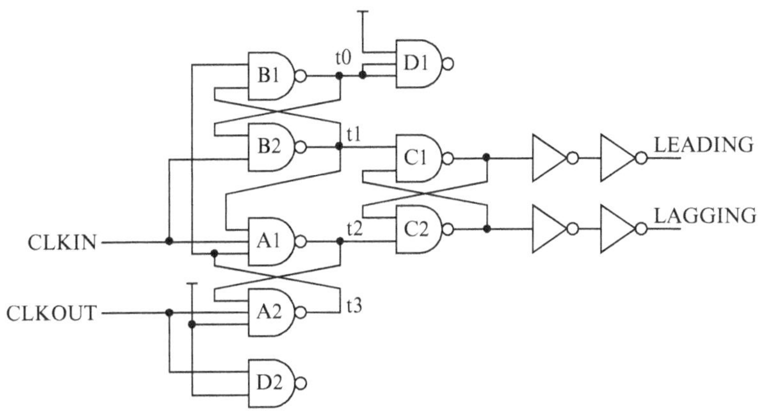

Fig. 6.6. Bang–bang phase detector. Reproduced with permission in a form similar to that in [2], ©1988 IEEE

Table 6.1 shows a truth table for the internal nodes $t0$–$t3$ of Fig.6.6. We observe that when CLKIN = 0, the inputs $t1$ and $t2$ of the slave latch are at logic 1 which means that the phase detector outputs are holding. On the positive edge of CLKIN, the internal nodes will assume a value conditional on the state of CLKOUT around the sampling edge, and the slave latch will be updated accordingly. The full operation of this structure is illustrated in the state sequence Tables 6.2 and 6.3 which show the sequence of internal node and output values for CLKOUT lagging and leading, respectively.

Table 6.1. Truth table for bang–bang phase detector internal nodes

CLKIN	CLKOUT	$t0$	$t1$	$t2$	$t3$
0	0	0	1	1	1
0	1	1	1	1	0
1	0	Hold	Hold	$\overline{t1}$	1
1	1	$\overline{t3}$	$t3$	Hold	Hold

Table 6.2. State sequence for CLKOUT lagging CLKIN

CLKIN	CLKOUT	$t0$	$t1$	$t2$	$t3$	*Leading*	*Lagging*
0	0	0	1	1	1	Hold	Hold
1	0	0	1	0	1	0	1
1	1	0	1	0	1	0	1
0	1	1	1	1	0	0 (Hold)	1 (Hold)
0	0	0	1	1	1	0 (Hold)	1 (Hold)

Table 6.3. State sequence for CLKOUT leading CLKIN

CLKIN	CLKOUT	$t0$	$t1$	$t2$	$t3$	*Leading*	*Lagging*
0	0	0	1	1	1	Hold	Hold
0	1	1	1	1	0	Hold	Hold
1	1	1	0	1	0	1	0
1	0	1	0	1	1	1	0
0	0	0	1	1	1	1 (Hold)	0 (Hold)

The sampling window of this phase detector is set entirely by the master latch (A1,A2). The setup and hold times of interest occur when the CLKIN and CLKOUT positive edges are virtually coincident. The setup time is set by the delay difference of NAND gate A2 on a 1-0 transition on node $t3$ and the delay of NAND gate A1 on a 1-0 transition on node $t2$:

$$T_s = t_{d(A2)} - t_{d(A1)}. \tag{6.3}$$

Similarly, the hold time is:

$$T_h = t_{d(A1)} - t_{d(A2)}. \tag{6.4}$$

Even though the setup and hold time of this phase detector can be minimized by removing all static bias from physical design, the true parameters should be determined statistically through Monte Carlo simulations. A positive setup and hold time must be determined that will guarantee correct decision with arbitrarily high probability in the presence of random process variations.

The phase detector of Fig.6.6 can satisfy all design requirements including being fully static and having nominally small and equal setup and hold times. Moreover, it is a true single phase design which requires no inversion on the clock or data inputs, thus, minimizing setup and hold requirements. The author has been hard pressed to find a better overall design with good portability across process nodes.

6.4.1 Metastability

A bang–bang phase detector will produce the correct lead/lag decision if the controlled clock input falls outside its sampling window (d_{SW}). A DLL in the locked state though will cause frequent controlled clock edges to fall within d_{SW}. A naive approach to this issue would be to assume that in this situation a wrong decision coupled with a small DCDL resolution d_r will not cause a DLL failure but rather a small increase in the DLL phase error which can be accommodated by the application.

A fourteenth century French philosopher Jean Buridan postulated that a donkey located at equal distances between two bales of hay should theoretically starve to death because it will be equally attracted to both [3, 4]. A more precise articulation of this principle in the words of Lamport [4] is as follows:

> *Buridan's Principle:* A discrete decision based upon an input having a continuous range of values cannot be made within a bounded length of time.

The continuous variable in the Buridan example is the position of the donkey along the axis connecting the two bales as a function of time and initial position. The discrete decision is which bale of hay to consume. In the case of the phase detector, the continuous variable is the voltage at the cross-coupled nodes of the flip-flop state element as a function of time and initial voltage. In this case, the initial voltage refers to the voltage established at the cross-coupled nodes right after sampling the flop data input. The discrete decision is whether a particular node will converge to logic 0 or logic 1.

This is of course the well-known synchronization failure problem [3]. No one can build a phase detector that can guarantee a valid logic output in bounded time if the controlled clock falls within the d_{SW} established by the sampling clock. In this case, the output can be at an undefined logic level (metastable state) for an arbitrarily long time, and eventually this undefined level may be interpreted by two separate logic receivers as different logic values. This can cause catastrophic DLL failure because it can drive the control automaton into a wrong or undefined state. Metastability in static flops has been extensively studied and observed in practice [5].

Although it is not possible to build a phase detector which will never experience metastability for unbounded time, it is possible to design a phase detector that minimizes the probability of system failure. We will study this problem by deriving an expression for the voltage in the cross coupled gates inside the phase detector as a function of time and initial conditions. Then, we will assign a probability measure to two events: Entering a metastable condition at time $t = 0$ and still being in a metastable condition at time $t = t_d$. Finally, we will derive an expression for the mean-time-between-failures (MTBF) which provides an indication of how often we can expect catastrophic synchronization failures in a DLL.

We develop a first order metastability model based on the analysis of Veendrick [6]. We begin by assuming a linear voltage transfer function for a CMOS inverter (for normalized supply voltage):

$$V_o = -A(V_i - V_{sw}) + 0.5, \tag{6.5}$$

where V_i and V_o are the voltages at the input and output of the inverter, respectively, V_{sw} is the switching threshold (defined as the value of V_i for which $V_o = 0.5$) and A is a large positive number denoting the inverter gain. There is an underlying assumption that V_o is further processed by a nonlinear limiter which clamps V_o to 1 for all $V_o > 1$ and also clamps V_o to 0 for all $V_o < 0$. All voltages are normalized (V_i, V_o, V_{sw}) and assume values between 0 and 1. Figure 6.7 plots Eq. (6.5) for $A = 10$, $V_{sw} = 0.5$. Figure 6.8 plots Eq. (6.5) for various A and V_{sw} parameter values.

Cross-coupling two inverters results in three equilibrium positions (Fig. 6.9): Two stable positions (a small voltage perturbation will result in the system canceling it out and remaining in the same state) that signify the two memory states, and an unstable equilibrium position (a small voltage perturbation will result in the system leaving this state and assuming one of the two stable states) signifying the metastable state. The metastable voltage V_m that the two system nodes converge is a function of the inverter gain A and switching threshold V_{sw} and can be computed easily by setting $V_i = V_o = V_m$ in Eq. (6.5) (Fig. 6.10):

$$V_m = \frac{2AV_{sw} + 1}{2(A+1)}. \tag{6.6}$$

We introduce transient behavior in the cross-coupled inverter model by adding an RC output stage to the ideal gain elements along the lines of [6] as shown in Fig. 6.11. We omit the limiters and make the assumption that this model is only valid for inverter input voltages that fall in the linear region of the transfer function $V_{sw} - 0.5/A \leq V_i \leq V_{sw} + 0.5/A$ (Fig. 6.7).

We assume that the sampling switch is ideal, and that clocking this latch consists of establishing the following initial conditions at $t = 0$ (please note the change of variable names $V_1(t)$ and $V_2(t)$ in Fig. 6.11):

$$V_2(0) = V_0, \tag{6.7}$$

$$V_1(0) = -A(V_0 - V_{sw}) + 0.5, \tag{6.8}$$

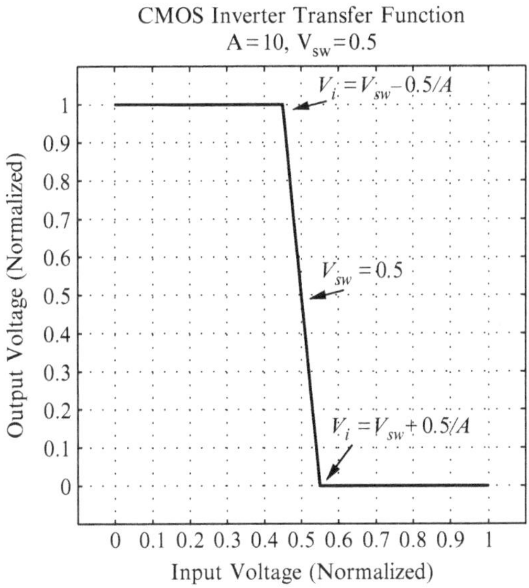

Fig. 6.7. CMOS inverter voltage transfer function

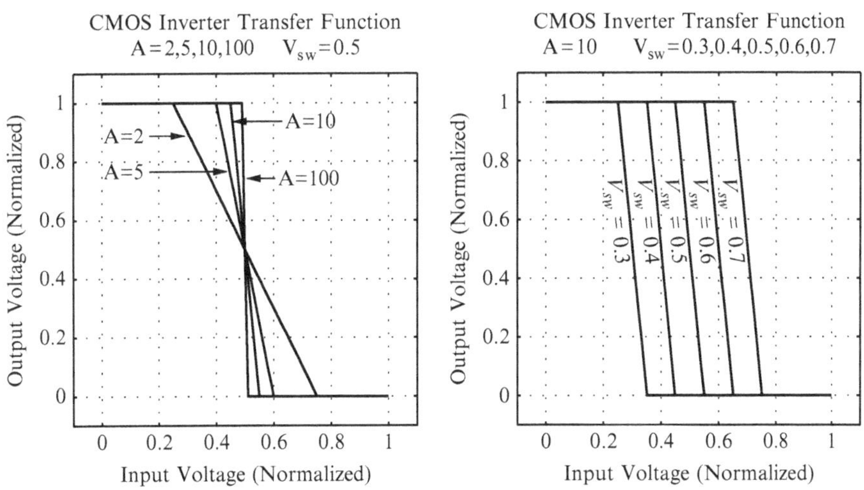

Fig. 6.8. CMOS inverter voltage transfer function parametrization

We want to derive an analytical expression for $V_1(t)$. We begin by enforcing KCL at $V_1(t)$ and $V_2(t)$, respectively:

$$\frac{\mathrm{d}V_1(t)}{\mathrm{d}t} + \frac{1}{RC}V_1(t) + \frac{A}{RC}V_2(t) - \frac{AV_{\mathrm{sw}} + 0.5}{RC} = 0, \tag{6.9}$$

$$\frac{\mathrm{d}V_2(t)}{\mathrm{d}t} + \frac{1}{RC}V_2(t) + \frac{A}{RC}V_1(t) - \frac{AV_{\mathrm{sw}} + 0.5}{RC} = 0. \tag{6.10}$$

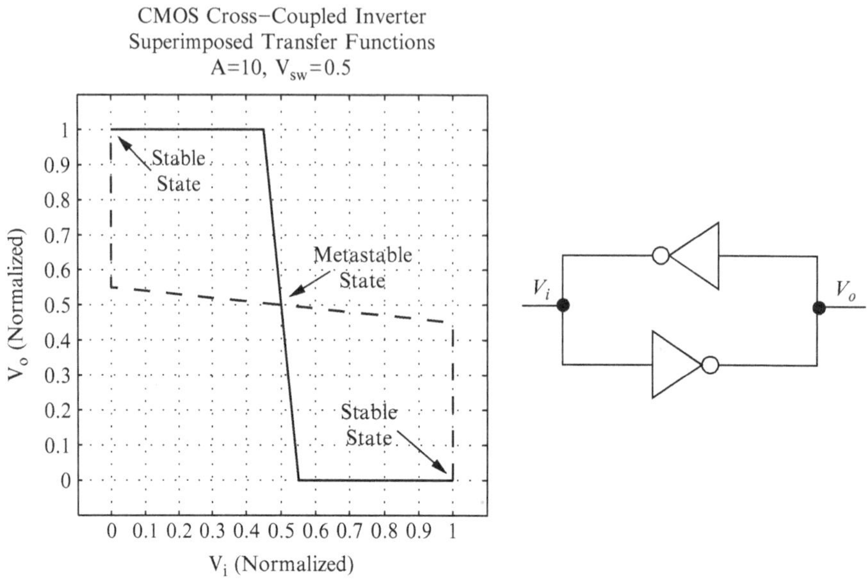

Fig. 6.9. Metastable state in CMOS cross-coupled inverters

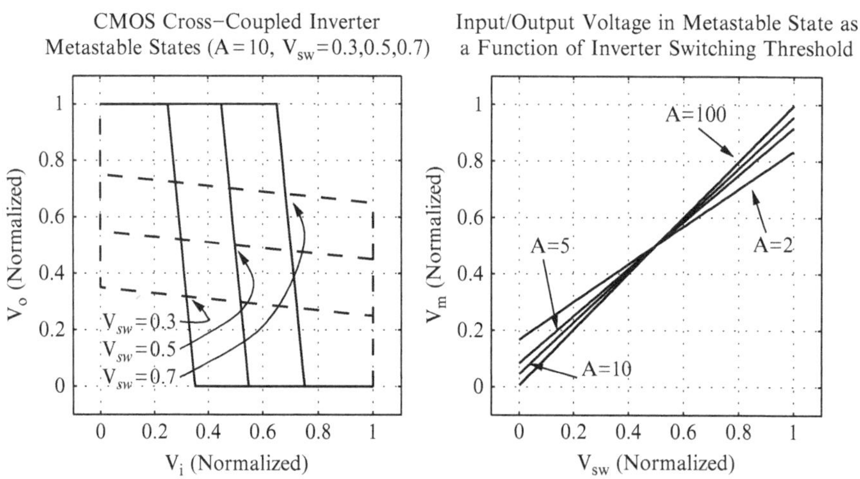

Fig. 6.10. Voltage in metastable state

In order to simplify the analysis, we introduce two new functions:

$$V_d(t) = V_1(t) - V_2(t),$$
$$V_s(t) = V_1(t) + V_2(t).$$

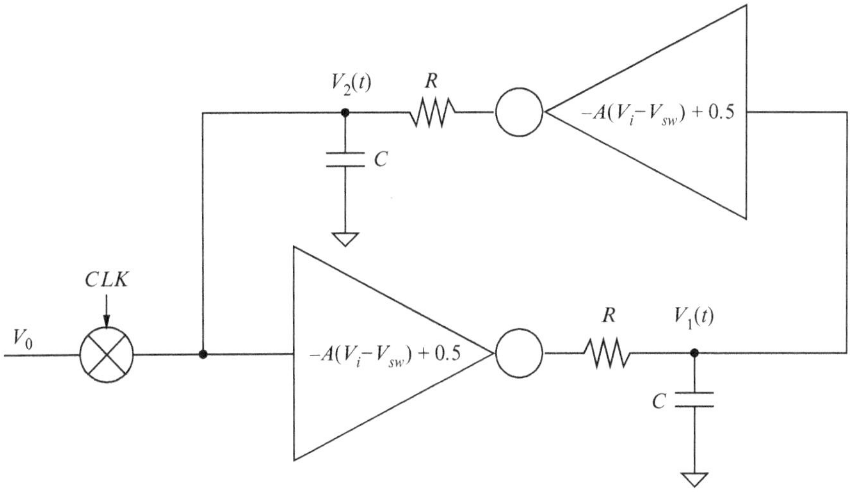

Fig. 6.11. Small signal model of cross-coupled inverters. Reproduced with permission in a form similar to that in [6], ©1980 IEEE

Subtracting (6.10) from (6.9) yields:

$$\frac{dV_d(t)}{dt} - \frac{A-1}{RC} V_d(t) = 0. \tag{6.11}$$

Adding (6.10) to (6.9) yields:

$$\frac{dV_s(t)}{dt} + \frac{A+1}{RC} V_s(t) - \frac{2AV_{\mathrm{sw}}+1}{RC} = 0. \tag{6.12}$$

Initial conditions for $V_{\mathrm{d}}(t)$ and $V_{\mathrm{s}}(t)$ have been established during sampling:

$$V_{\mathrm{d}}(0) = V_1(0) - V_2(0),$$

$$V_{\mathrm{s}}(0) = V_1(0) + V_2(0).$$

The solution of (6.11) is

$$V_d(t) = [V_1(0) - V_2(0)]\mathrm{e}^{\frac{A-1}{RC}t}. \tag{6.13}$$

The solution of (6.12) is

$$V_s(t) = \left(V_1(0) + V_2(0) - \frac{2AV_{\mathrm{sw}}+1}{A+1}\right)\mathrm{e}^{-\frac{A+1}{RC}t} + \frac{2AV_{\mathrm{sw}}+1}{A+1}. \tag{6.14}$$

Expression (6.14) can be simplified by substituting the expression for the metastable voltage (6.6):

$$V_s(t) = [V_1(0) + V_2(0) - 2V_\mathrm{m}]\,\mathrm{e}^{-\frac{A+1}{RC}t} + V_\mathrm{m}. \tag{6.15}$$

Reverting back to the original variable $V_1(t)$, we have

$$V_1(t) = \left(\frac{V_1(0)+V_2(0)}{2} - V_\mathrm{m}\right)\mathrm{e}^{-\frac{A+1}{RC}t} + \frac{V_1(0)-V_2(0)}{2}\mathrm{e}^{\frac{A-1}{RC}t} + V_\mathrm{m}. \tag{6.16}$$

Equation (6.16) can be simplified and become more meaningful if we make the following observation: The decaying exponential quickly vanishes with increasing t and the expression is dominated by the increasing exponential. We can approximate the expression above by only keeping the positive exponential with its coefficient adjusted for the initial condition:

$$\boxed{V_1(t) = (V_1(0) - V_\mathrm{m})\mathrm{e}^{\frac{A-1}{RC}t} + V_\mathrm{m}.} \tag{6.17}$$

Equation (6.17) is fundamental in the description of cross-coupled circuits and states that a perturbation from the unstable equilibrium position V_m results in a time-exponential path toward a stable state. We note that Eq. (6.17) has a consistent solution for $V_1(t=0)$ and also that for $V_1(0) = V_\mathrm{m}$, we have $V_1(t) = V_\mathrm{m}$ for all t which is another way of saying that a metastable state may persist indefinitely. We also note that Eq. (6.17) is identical to Eq. (1-18) in [3] although the derivation is very different. Mead and Conway derive the cross-coupled inverter node equation starting from basic circuit relationships in a depletion-loaded NMOS cross-coupled pair.

It is important to understand that Eq. (6.17) is only valid for node voltages in close proximity to the metastable voltage V_m. Both initial conditions $V_1(0)$ and $V_2(0)$ must lie in the shaded regions of Fig. 6.12 and be related through the gain curve. Unless this is true, the assumptions made during the derivation of (6.17) (both inverters are in their linear region) are no longer operative. We can still make the assertion that Eq. (6.17) fully describes the metastable state because reaching the voltage limit of the above equation signifies the exit from metastability: At that point, the node voltage has reached a value outside the linear gain limits and any receiver with a similar gain curve will interpret it as a well-defined logic state. Figure 6.13 plots (6.17) for various initial conditions of $V_1(0) - V_\mathrm{m}$ and exhibits the exponential nature of the trajectory leading the node voltage out of the metastable state.

Metastability is a stochastic phenomenon and must be studied with probabilistic tools. Equation (6.17) states that there exists an initial condition $V_1(0) = V_\mathrm{m}$ which will cause the phase detector to lie in the metastable state indefinitely. At the same time though, if $V_1(t)$ is modeled as a continuous random variable, then the probability of $V_1(t)$ assuming a discrete value V_m vanishes. It is not possible to design a perfect phase detector, one that is guaranteed not to lie in the metastable state indefinitely.

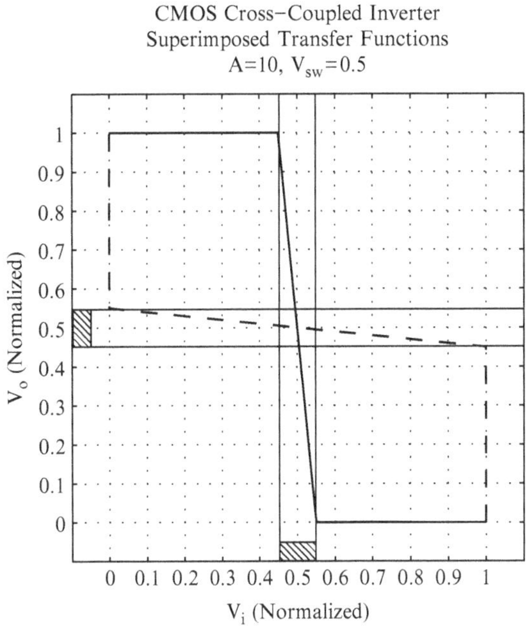

Fig. 6.12. Valid node voltage region for Eq. (6.17)

It is possible though to design a phase detector that has an arbitrarily small conditional probability of being in the metastable state at a given time t_d assuming it was in the metastable state at time $t = 0$.

Before proceeding to define the problem in stochastic terms, let us simplify Eq. (6.17) by referring all voltages to V_m instead of ground [6]:

$$V(t) = V_1 e^{\frac{t}{\tau_m}}, \tag{6.18}$$

$$V(t) = V_1(t) - V_m, \tag{6.19}$$

$$V_1 = V_1(0) - V_m, \tag{6.20}$$

$$\tau_m = \frac{RC}{A-1}. \tag{6.21}$$

Furthermore, we need to state two assumptions:

- We consider a phase detector to be metastable when its output voltage $V_1(t)$ is within ΔV of V_m:

$$|V(t)| < \Delta V, \tag{6.22}$$

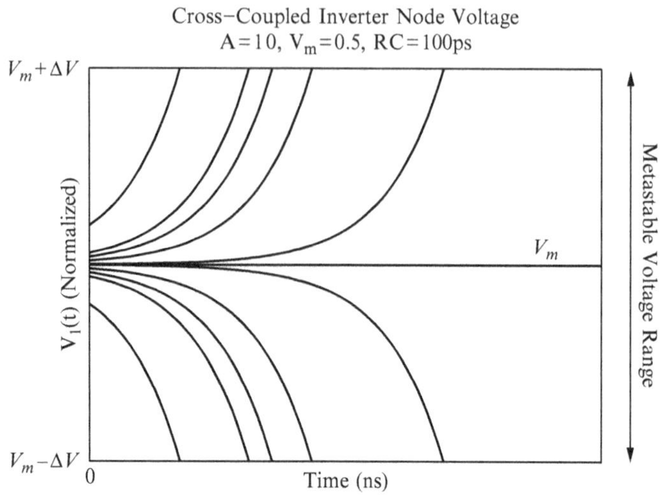

Fig. 6.13. Exponential trajectory towards a stable state

where ΔV is a phase detector/synchronizer parameter and is the maximum voltage deviation from V_m which will still guarantee that both cross-coupled nodes are in the linear gain region of the two cross-coupled CMOS gates.

- The initial condition V_1 is a random variable uniformly distributed between $-\Delta V$ and ΔV. By definition then, the system is metastable at $t = 0$.

We wish to quantify the probability that the phase detector is still metastable at $t = t_d$ given that it is metastable at $t = 0$:

$$\Pr(|V(t_d)| < \Delta V) = \Pr(\left|V_1 e^{\frac{t_d}{\tau_m}}\right| < \Delta V). \tag{6.23}$$

Multiplying both sides of the event inequality in (6.23) with $e^{-\frac{t_d}{\tau_m}}$ yields:

$$\Pr(|V_1| < \Delta V e^{-\frac{t_d}{\tau_m}}) = e^{-\frac{t_d}{\tau_m}}. \tag{6.24}$$

The last equation stems from the fact that V_1 is uniformly distributed between $-\Delta V$ and ΔV. We can arrive at the same result using a different approach [3]: Let us assume that the exit from the metastable state can be modeled as the first arrival of a Poisson process with rate λ. Modeling it as a Poisson arrival implies that the probability of leaving the metastable state within a very small time interval Δt is proportional to the duration of the interval and equal to $\lambda \Delta t$. This makes physical sense since in general, Poisson processes are heavily used to model simple stochastic phenomena in continuous time. The probability density function of the time until the first Poisson arrival (l_1) is given by the following Eq. ([7]):

$$f_{l_1}(t) = \lambda e^{-\lambda t}. \tag{6.25}$$

The probability that the phase detector is still metastable at time t_d is given by the following expression:

$$\Pr(l_1 > t_d) = \int_{t_d}^{\infty} \lambda e^{-\lambda t} dt = e^{-\lambda t_d}. \tag{6.26}$$

For $\lambda = 1/\tau_m$, expression (6.26) is in agreement with (6.24).

Equation (6.24) describes a conditional probability where the conditioning event is $|V_1| < \Delta V$ (phase detector is metastable at time $t = 0$). We will now quantify the probability of the conditioning event and determine the overall unconditional probability of phase detector failure. Let us assume that in the DLL locked state, the feedback clock edges are uniformly distributed between $-d_r$ and d_r with respect to the sampling reference clock edge (where d_r is the DCDL resolution). This is a reasonable assumption to make assuming that the digital control module will be designed with a small ± 1 LSB limit cycle and various noise processes will add uncertainty to the feedback clock edges. We further assume a feedback clock slew rate of L in V/s. Figure 6.14 shows graphically that such a distribution of feedback edges will cause a uniform distribution of sampled voltages with a range equal to $2Ld_r$. The probability of the conditioning event is, therefore, $\Delta V/(Ld_r)$, and the overall phase detector unconditional failure probability at t_d is,

$$\boxed{\Pr(\text{Metastable at } t_d) = \frac{\Delta V}{Ld_r} e^{-\frac{t_d}{\tau_m}}.} \tag{6.27}$$

So far, we have made enough assumptions and simplified our model sufficiently to be able to derive a simple formula for ΔV, the sampled voltage deviation from V_m which will place a cross-coupled structure in a metastable state. Without loss of generality, we can simplify the analysis by assuming $V_m = V_{sw} = 0.5$ in Eq. (6.5) (all voltages are normalized with respect to the nominal supply voltage). The range of input voltages which will cause a cross-coupled structure to become metastable should be determined by requiring that the result of applying the inverter transfer function (6.5) to such input voltage should yield a voltage which is still in the linear range of the same transfer function (shown in Fig.6.7). If this is not the case, then a logic gate with similar gain will interpret the cross-coupled output as a discrete logic value. Moreover, the second cross-coupled gate will also interpret such an output as

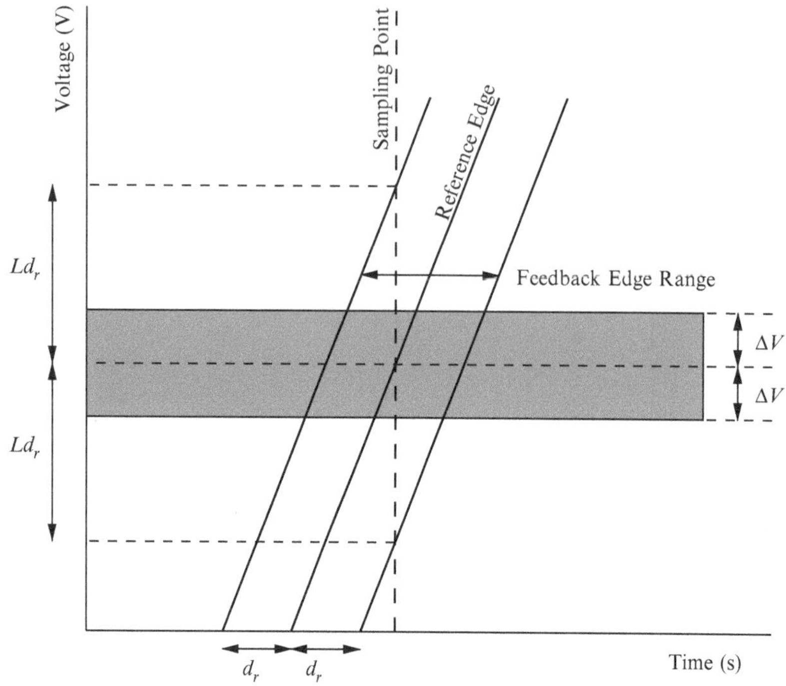

Fig. 6.14. Determining the probability of entering metastability

a discrete logic value, and the latch will converge to a non-metastable state as soon as the sample switch opens up. It is straightforward to calculate that the above condition will be satisfied for:

$$\Delta V = \frac{1}{2A^2}. \tag{6.28}$$

Let us now briefly focus on t_d, the allowed metastability resolution time. For a digital DLL with a bang–bang phase detector, t_d is at least equal to the reference clock period T. It is also common to add additional synchronization stages and effectively extend the resolution time to nT. Updating Eq. (6.27) with our latest observations and also expanding the expression for τ_m we have,

$$\boxed{\Pr(\text{PD Failure}) = \frac{1}{2nA^2Ld_\mathrm{r}} \mathrm{e}^{-\frac{A-1}{r_o C} nT},} \tag{6.29}$$

where

d_r DCDL resolution

L Feedback clock edge rate

r_o Small signal output resistance of gate around V_m.
Should be small signal saturation resistance of pullup device in parallel with pulldown device

A Small signal gate gain. Should be $g_{\mathrm{m}} r_{\mathrm{o}}$ where g_{m} is pullup transconductance in saturation in parallel with pulldown transconductance.
C Cross-coupled node capacitance
T Reference clock period
n Number of synchronization stages including the phase detector

An additional scaling factor of $1/n$ has been included to account for the fact that a digital controller typically implements a $1/n$ correction issue rate in order to minimize limit cycle amplitude and guarantee stability (Sect. 6.6). In such a case, the controller only looks at the output of the phase detector once every n cycles and the probability of entering a metastable state must be scaled accordingly. Equation (6.29) quantifies the risk of DLL failure for T seconds of operation. An alternative way of expressing this risk is the mean-time-between-failures (MTBF):

$$\mathrm{MTBF} = T \Pr(\mathrm{PD\ Failure}). \tag{6.30}$$

The MTBF figure of merit is typically expressed in years and signifies the average duration of error-free operation. In digital system applications, an MTBF in excess of 10–100 years is typically desired.

An Example of Phase Detector Failure Calculation

It is important to realize that Eq. (6.29) is based on a number of assumptions and approximations such as the inverter piecewise linear transfer function of Eq. (6.5). Nevertheless, it captures the right dependencies on design parameters and can be used as a phase detector design guideline. The MTBF value predicted by (6.30) should have substantial margin of a few orders of magnitude to guarantee correct operation even in the presence of modeling inaccuracies that can affect the final result in an exponential nature.

Although it is certainly possible to assign circuit parameters to all contributing factors of Eq. (6.29) [6], it is probably much easier to estimate MTBF using spice small signal analysis around V_{m}. Figure 6.15 shows the setup of a spice AC analysis for a 45 nm cross-coupled inverter using 45 nm bulk spice predictive models [8–10]. A self-biased inverter is used to compute the metastable voltage (V_{bias} in Fig. 6.15) which is then used as the bias point through an ideal buffer for a cross-coupled pair of identical sizing. An AC source is cascaded in series with the bias point to calculate gain, output resistance, and node capacitance.

For this example, Table 6.4 shows the MTBF calculation under certain assumptions for L, n, T, and d_{r}. The MTBF calculation is very sensitive to the metastability time constant (τ_m), and the parameters that affect it because of the exponential dependence. As an example, if the node capacitance gets doubled from the value in Table 6.4 due to the necessary addition of a receiving gate, MTBF becomes 4.7624×10^9 years. If it triples due to poor phase detector design, MTBF will become 24.58 years.

Table 6.4. Example MTBF calculation for the circuit of Fig. 6.15

Parameter	Value	Units
Clock period (T)	1×10^{-9}	s
Slew rate (L)	1×10^{10}	V/s
DLL subsampling factor (n)	1	–
DLL resolution (d_r)	50×10^{-12}	s
Gain (A)	4.5456	–
Output resistance (r_o)	9.3079×10^{3}	Ω
Node capacitance (C)	3.3271×10^{-15}	F
Probability of entering metastability: $\frac{1}{2nA^2Ld_r}$	0.0484	–
Metastability time constant: $\tau_m = \frac{r_oC}{A-1}$	8.7343×10^{-12}	s
Conditional probability of being metastable at nT: $e^{-\frac{nT}{\tau_m}}$	1.8928×10^{-50}	–
Unconditional synchronization failure probability: $\frac{1}{2nA^2Ld_r}e^{-\frac{nT}{\tau_m}}$	9.16×10^{-52}	–
MTBF	3.4615×10^{34}	years

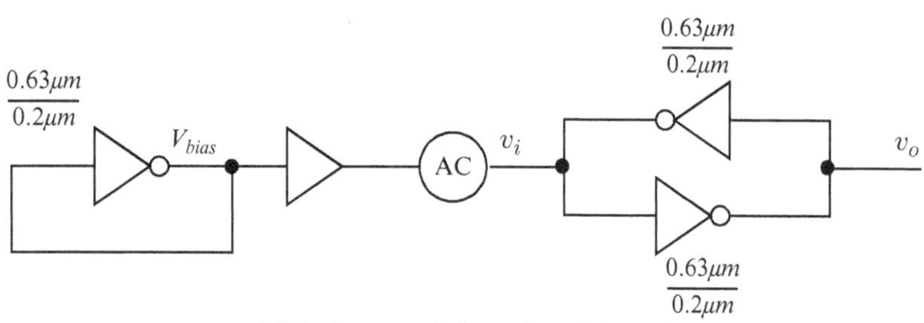

Fig. 6.15. Determining MTBF with a spice simulation

6.5 DCDL Design

There is a very broad range of delay line design options limited only by designer imagination. This section provides a structured overview of various DCDL design approaches. The structure based on DCDL characteristics adopted for this section does not follow a widely accepted classification in the field but is merely done for ease of presentation. Moreover, this section is by no means an all-inclusive exposition of all possible design options. It can be considered a stratified sampling of the design space that presents various alternatives based on the main DCDL characteristics of Sect. 6.2 ($D_{\min}, D_{\max}, d_r$). Additional characteristics (such as input capacitance, linearity, and potential for synchronous vs. asynchronous setting change) will also be identified.

For ease of presentation, we divide DCDLs into two main categories based on d_r: Gate-delay DCDLs (coarse) and Subgate-delay DCDLs (fine).

6.5.1 Gate-Delay DCDLs

Gate-delay DCDLs can be constructed by cascading standard CMOS logic gates to form a delay chain with intermediate outputs routed to a high-fanin multiplexing structure. Alternatively, the chain length can be modulated using low-fanin distributed multiplexing structures as part of the delay cell. They exhibit relatively small D_{min}, arbitrarily high D_{max} but relatively coarse d_r by design.

A very popular coarse delay line is shown in Fig.6.16 [11, 12]. This particular implementation has four hierarchical delay stages between input A and output Y. It is controlled by a bidirectional shift register with one-hot encoding ($Q[3:0]$). Resolution d_r is $2T_G$, where T_G is the average CMOS gate delay. NAND gates labeled with the letter B form a distributed multiplexer that controls the entry point of the clock (A input) in this cascaded structure. One important observation is that this design presents substantial load to the clock especially for a large number of delay stages since all delay element inputs are shorted. Clock buffering may be necessary which will increase D_{min} to a value larger than $2T_G$. An arbitrarily high number of stages can be cascaded to increase dynamic range provided that input A is sufficiently buffered. D_{max} is $2NT_G$, where N is the number of delay stages.

The folded design of Fig.6.17 eliminates the heavily loaded A input by modulating the delay line length in a telescopic fashion. This design is also controlled by a bidirectional one-hot shift register. For $Q[3:0] = 0001$ the signal path is Cell $0.A +$ Cell $0.B$. For $Q[3:0] = 0010$ the path is Cell $0.C +$ Cell $1.A +$ Cell $1.B +$ Cell $0.B$. The length modulation of this structure can be compared to the sliding action of a trombone. The wrap-around connection between $IN1$ and $OUT1$ of Delay Cell 3 is not part of the delayed signal path but is necessary to establish the propagating condition for the B NAND gate associated with the delay cell that constitutes the final telescope link (CTL $= 1$). If the delay cell had non inverting forward and reverse paths, then input $IN1$ of Delay Cell 3 could have been hard-wired to ground. The wrap-around connection makes it toggle between odd and even settings to ensure that the signal will propagate for all setting values. The characteristics of the telescopic line are identical to the previous one with the sole exception that D_{min} can be $2T_G$ since there is no need for input buffering.

A straightforward coarse DCDL implementation is shown in Fig.6.18. Two cascaded inverters constitute the delay element and a binary multiplexer selects the appropriate output tap. D_{min} is equal to $2T_G + \log_2 N \times T_G$ and d_r is still $2T_G$. This delay line can be controlled by a binary counter ($Q[1:0]$) requiring $\log_2 N$ storage elements as opposed to the N storage elements of the previous two designs. The control structure of this DCDL scales better with N as opposed to the NAND-based ones. D_{min} also increases though with increasing N and for certain applications this may be a concern.

DCDLs with a single T_G resolution are also possible. Figure 6.19 illustrates one based on a differential delay element. The relative sizing of the cross-coupled devices

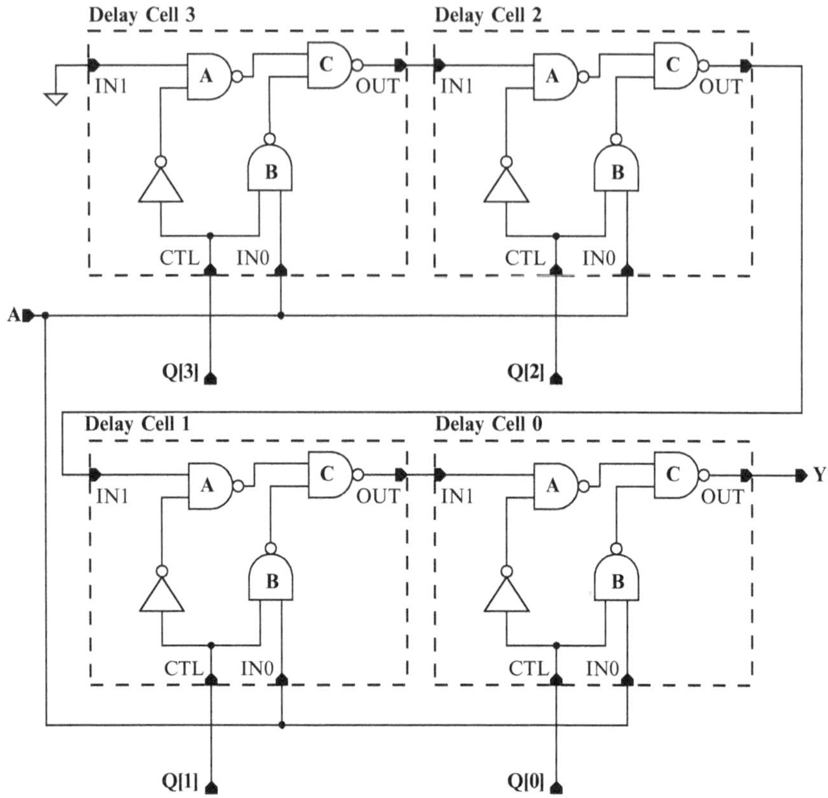

Fig. 6.16. NAND-based registered-controlled 4-stage delay line

vs. the forward devices can be used to fine tune d_r as long as writing the cross-coupled pair can be safely guaranteed across all PVT conditions. Figure 6.20 shows a single-ended single T_G DCDL with a conditionally inverting output stage. The output XNOR gate must be designed with equal delays from both input polarities to the output to ensure linearity.

Design options are certainly not limited to the five examples shown above. Not only different organizations are possible but also different logic families (i.e., dynamic or low swing current model logic) can be used as the base for the delay cell and the multiplexing structure. The characteristics of the examples presented so far are summarized in Table 6.5. Further assumptions are that the single-ended-to-differential converter of Fig.6.19 and the XNOR of Fig.6.20 can be implemented with two gates and, therefore, add $2T_G$ to D_{min} and D_{max} of the corresponding DCDLs.

Physical design is very important in precisely controlling DCDL specifications. All delay cells should be identical, and metal capacitance should be thoroughly characterized since it will probably be responsible for a substantial percentage of the cell delay. Post-layout simulation-based characterization across all PVT corners should

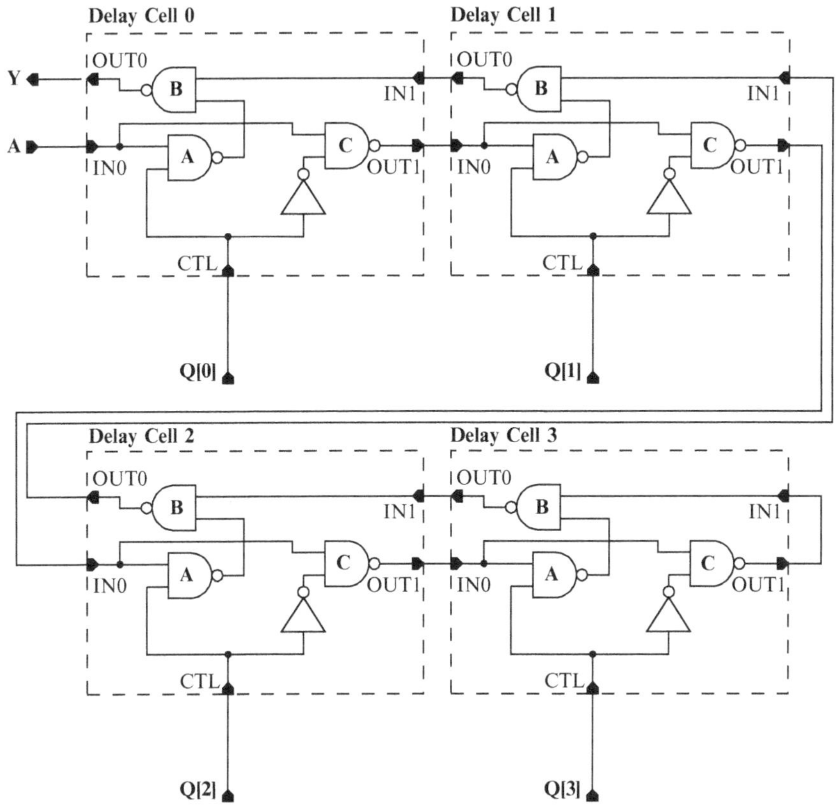

Fig. 6.17. NAND-based telescopic 4-stage delay line

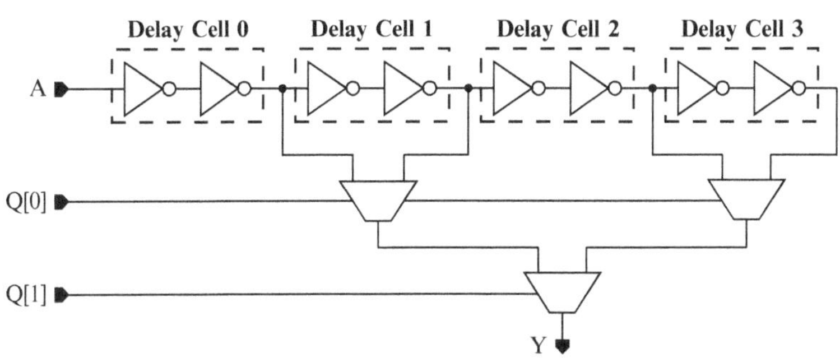

Fig. 6.18. Inverter-based logarithmic 4-stage delay line

be done in order to establish the ranges for D_{min}, D_{max}, and d_{r} and ensure that the application requirements are met. In most cases, coarse DCDLs will be operating on clock signals, where duty cycle fidelity is important. If there are significant

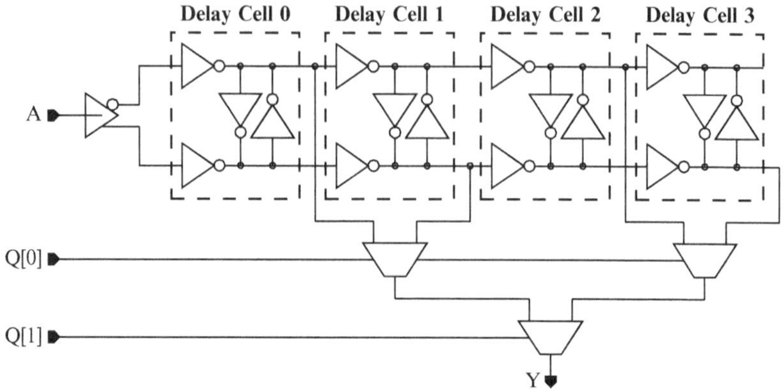

Fig. 6.19. Inverter-based differential 4-stage delay line

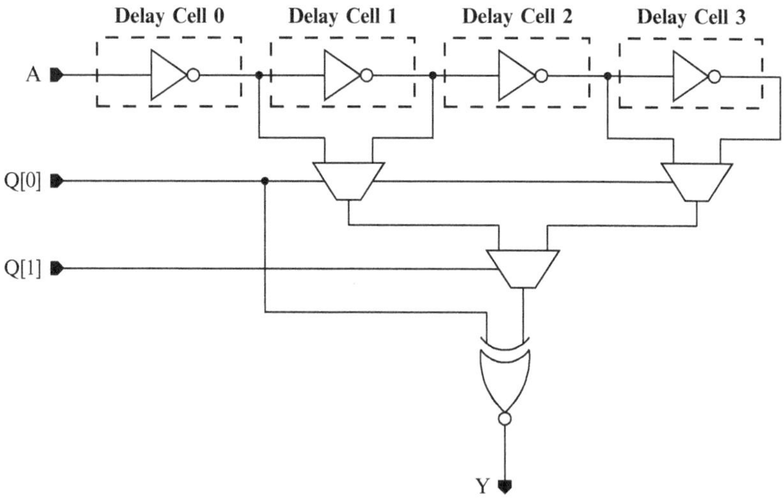

Fig. 6.20. Inverter-based conditional-output 4-stage delay line

Table 6.5. Characteristics of coarse DCDLs

	D_{min}	D_{max}	d_{r}	Dynamic range	Linearity	CTL flops
NAND (Fig. 6.16)	$> 2T_{\mathrm{G}}$	$> 2NT_{\mathrm{G}}$	$2T_{\mathrm{G}}$	$2(N-1)T_{\mathrm{G}}$	Good	N
Tel. (Fig. 6.17)	$2T_{\mathrm{G}}$	$2NT_{\mathrm{G}}$	$2T_{\mathrm{G}}$	$2(N-1)T_{\mathrm{G}}$	Good	N
Log. (Fig. 6.18)	$T_{\mathrm{G}}(2+\log_2 N)$	$T_{\mathrm{G}}(2N+\log_2 N)$	$2T_{\mathrm{G}}$	$2(N-1)T_{\mathrm{G}}$	Good	$\log_2 N$
Diff. (Fig. 6.19)	$T_{\mathrm{G}}(3+\log_2 N)$	$T_{\mathrm{G}}(2+N+\log_2 N)$	T_{G}	$(N-1)T_{\mathrm{G}}$	Good	$\log_2 N$
Cond. (Fig. 6.20)	$T_{\mathrm{G}}(3+\log_2 N)$	$T_{\mathrm{G}}(2+N+\log_2 N)$	T_{G}	$(N-1)T_{\mathrm{G}}$	Good	$\log_2 N$

imbalances between rise vs. fall delay in the delay cell and the multiplexing structure, severe duty cycle distortion may occur at the output signal especially for large delay settings. In extreme cases of high frequency clocks and long DCDLs, the square wave may completely disappear and a DC value may be observed at the output. NMOS vs. PMOS ratios for equal rise and fall times should be used and if possible inverting logic gates should be instantiated in cascaded pairs with similar fanouts to ensure equal treatment of positive vs. negative edges and good duty cycle PVT tracking.

Under certain circumstances, the NAND-based DCDLs of Figs. 6.16 and 6.17 can have a power advantage. Their switching activity is setting dependent and will exhibit reduced switching power at lower settings. A complete power comparison of the DCDLs presented so far is process and application dependent and will provide little if any additional insight.

Synchronous vs. Asynchronous Operation in Coarse DCDLs

DCDL outputs are typically clock signals and as such must always be well behaved and not undergo spurious transitions (glitches) that may cause erroneous circuit operation. In order to demonstrate an important differentiation among DCDLs, we will adopt the ad hoc definition that a DCDL capable of asynchronous operation is a DCDL that can undergo a valid setting change (± 1) at any time without the possibility of a glitch at the output. On the other hand, a DCDL capable of synchronous only operation is a DCDL that requires that valid setting changes be timed synchronously with respect to the input clock in order to guarantee glitch-free output behavior.

DCDLs are essentially combinational (stateless) circuits with clock being simply another input. As a result, spurious transitions at the output are certainly possible unless all inputs are guaranteed to transition at the same time and all inputs have equal delay paths to the output. Neither is true in the DCDL case. Clock is not guaranteed to transition at the same time as the control settings and clock has multiple paths to the DCDL output through the delay element chain. Under certain circumstances though, and using simple analysis, we can convince ourselves that a certain class of DCDLs can have a glitch-free output. Before proceeding with the analysis, we make two important assumptions: We are considering DCDLs where all the control setting inputs have equal delay paths to the output, and we only consider setting changes that increment or decrement the current setting by one position.

A simplified delay line that is consistent with these assumptions is shown in Fig. 6.21. It contains a single delay element of delay d_r and a 2-input multiplexer. For simplicity, we assume that the multiplexer input-to-output and select-to-output delays are both equal to t_m. The timing diagram on the right shows the conditions that can generate a spurious output transition. The select line needs to transition while the clock edge goes through the delay element. Furthermore, d_r needs to be large with respect to t_m for such a glitch to be formed. In reality, if d_r is reasonably close to t_m the glitch will never form because of low pass filtering. The right design approach is to set up the conditions of Fig. 6.21 and simulate across all PVT conditions to ensure that the glitch won't form.

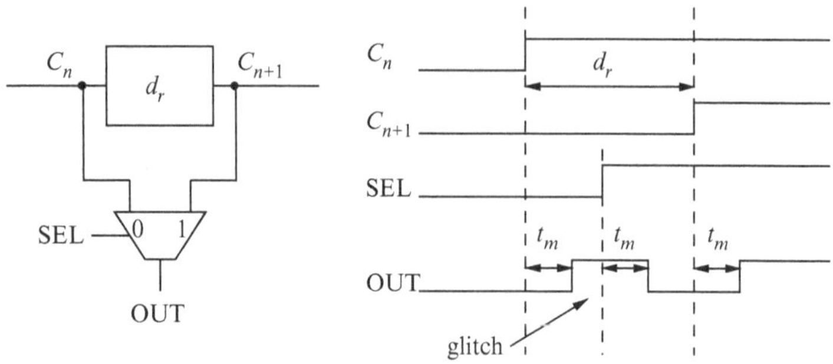

Fig. 6.21. DCDL spurious output transition

Interestingly enough, none of the DCDLs in Figs. 6.16 through 6.20 are glitch-free. Their setting vectors do not have equal delays to the output, and therefore the simple analysis of Fig.6.21 does not apply. They have a lot more exposure to spurious transitions, and their analysis is rather ad-hoc and can be quite cumbersome. It is safe to create a design restriction stating that *only coarse DCDLs with equal setting-to-output delays should be considered when glitch free operation is necessary*. Distributed and logarithmic multiplexer structures are not allowed.

The simple DCDL of Fig.6.18 can be a candidate for glitch-free operation if the output logarithmic multiplexer is replaced by a one-hot NAND-based (sum-of-products) multiplexing structure where each control line is at the same logic depth. Another potential solution is to pad the most significant bit with additional delay and the problem transforms to heterogeneous structure delay tracking across PVT and can be rather process-specific.

The coarse DCDLs of Figs. 6.16 through 6.20 are typically used in applications where the clock is not utilized while settings transition. Alternatively, they can be used in a synchronous fashion. The control inputs must be timed in a way that guarantees that when they change, the entire delay chain is at a constant value. This can be accomplished by latching the settings with an appropriate strobe which can either be the input clock or a delayed version of it (perhaps a particular delay chain tap) that meets this timing restriction. Extensive validation is required.

They can also be used asynchronously by adding significant hardware resources and control complexity. Such an example is shown in Fig.6.22. We have a pair of identical delay lines. Only one is on-line at any single time. When there is a settings change, it is performed on the off-line DCDL. When the change has stabilized and all spurious transitions are gone, the output multiplexer is flipped and the new settings change shows up at the output. At this point, the analysis of Fig.6.21 applies. In addition to consuming large hardware resources, this structure has a very subtle issue. It is based on the assumption that if we have two instantiations of a delay line on silicon and one has a setting equal to n and the other has a setting equal to $n+1$, then the one with the $n+1$ setting will have a longer delay. This may not be true in deep submicron processes due to large random V_t variation. Monte Carlo analysis can show a substantial probability of such a case. If this happens, then the composite delay line is non-monotonic. This will affect performance, and under certain circumstances, it may confuse the control automaton and cause catastrophic failure. A better approach is shown in Fig. 6.23 where a single dual output delay line is used. It uses less hardware and does not have potential monotonicity issues. It has the same control complexity though.

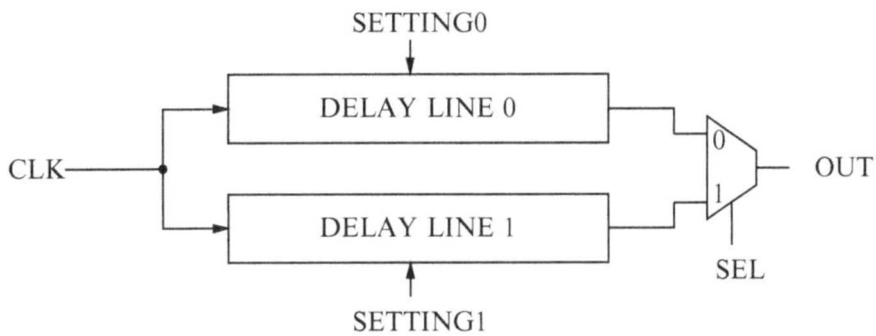

Fig. 6.22. Duplicating DCDLs for glitch suppression

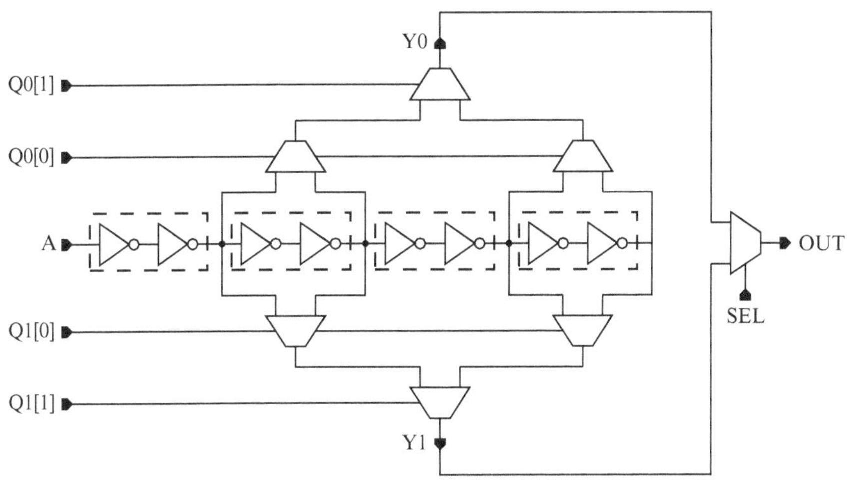

Fig. 6.23. Dual output DCDL for glitch suppression

6.5.2 Subgate-Delay DCDLs

Phase lock accuracy requirements can be tighter than a gate delay. In such applications, a subgate-delay (fine) DCDL must be employed.

A class of fine DCDLs relies on variable RC delays for delay generation. Two examples of fine DCDL stages are shown in Fig. 6.24. The delay stage on the left is based on variable cell driving resistance. Turning on more device branches

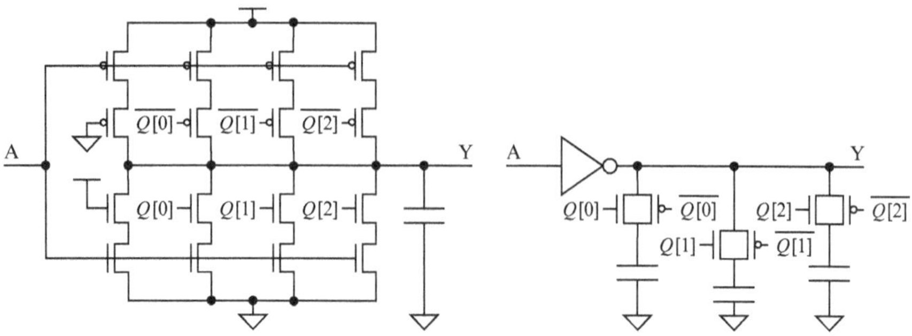

Fig. 6.24. RC-Based fine DCDL stages

through control vector $Q[2:0]$ reduces the overall RC delay through the stage. Transistors in the variable branches are sized for linear control vector encodings. Control bits $Q[2:0]$ are thermometer encoded and have a fixed switching order (i.e., $000 \rightarrow 001 \rightarrow 011 \rightarrow 111$). Logarithmic encoding is not possible due to the non-linearity of parallel resistance addition. Table 6.6 shows a normalized example of branch sizing for linearity. The example should be considered as a starting point only and process specific fine tuning will be required. The number of variable branches is limited by the extreme sizing that will be required to maintain linearity for an increasing number of control bits. The dynamic range can be extended by cascading multiple such stages at the expense of increasing D_{min}.

Table 6.6. Variable resistance DCDL branch sizing (thermometer-encoded control)

	Normalized width	Equivalent resistance	Capacitance	RC delay
Always-on branch (AO)	1	1	4	4
Branch 0 + AO	1/3 + 1	3/4	4	3
Branch 1 + 0 + AO	2/3 + 1/3 + 1	1/2	4	2
Branch 2 + 1 + 0 + AO	2 + 2/3 + 1/3 + 1	1/4	4	1

The stage on the right is based on variable output capacitance. Turning on more capacitive branches increases the RC delay through the cell. Both linear (thermometer) and logarithmic (binary) control encoding are possible due to the linearity of parallel capacitance addition. Output capacitors can be metal-based or device-based depending on accuracy and linearity requirements. Per-stage dynamic range is limited by the maximum edge rate tolerated by the design. The dynamic range can be extended by cascading multiple stages as in the previous case.

An alternative method of constructing fine delay lines is shown in Fig.6.25. It is based on the delay difference between the top vs. the bottom delay path. Control encoding is thermometer-based. This organization can achieve PVT-independent lin-

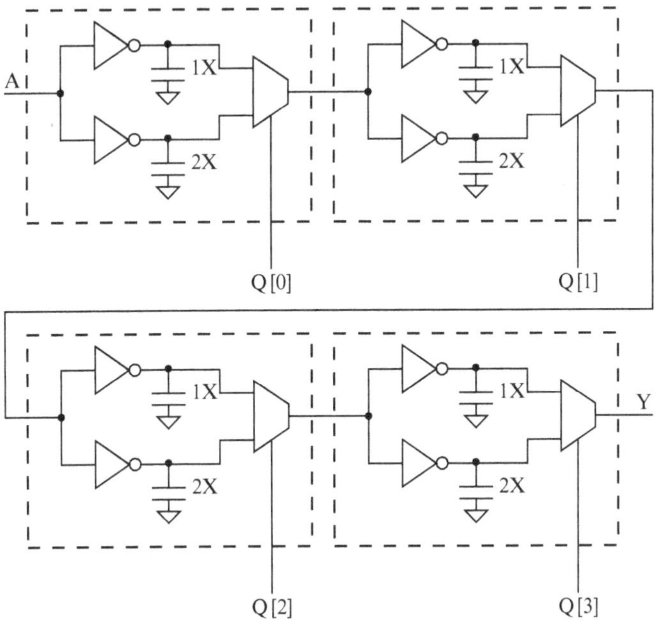

Fig. 6.25. Fine DCDL based on delay differences

earity subject only to random process variation. The main drawback is that both linearity and dynamic range are achieved through large $D_{\min}$ increases which can be a problem for certain applications.

An additional method of constructing subgate delays based on phase interpolation will be discussed in Sect. 6.5.3.

The delay lines of Figs. 6.24 and 6.25 can have resolution on the order of a few picoseconds but a very limited dynamic range. They can be operated asynchronously without the possibility of a glitch forming at the output. The DCDLs of Fig. 6.24 vary resistance and capacitance. Settings changes are not considered combinational logic switching events. On the other hand, the multiplexer-based DCDL of Fig. 6.25 can be analyzed using the method of Fig. 6.21 with a much smaller d_r.

6.5.3 Resolution vs. Dynamic Range in DCDLs

In typical applications, both fine resolution and large dynamic range are highly desirable. High resolution directly affects phase match accuracy. A large dynamic range ensures that the DLL won't be the limiting factor in selecting input clock frequencies or V_{MIN} for the overall design.

From the previous analysis on DCDLs, it would seem that designers must perform a tradeoff between resolution and range. This is rarely true in real applications since both are essential. Instead of using a single delay line that compromises d_r in favor of dynamic range for a given number of settings, multiple delay lines can be employed with different properties so that the overall design target is met [13]. Typically, a coarse high dynamic range DCDL is cascaded with a fine low dynamic range subgate DCDL to achieve both specifications. This design decision involves additional complexity in the control automaton which must first lock the coarse line to an appropriate setting (which tracks input clock frequency and static process and voltage variations) and then engage the fine delay line which tracks dynamic voltage and temperature variations. Depending on control implementation, the coarse DCDL can readily be of the synchronous type (as described in Sect. 6.5.1) if it is locked only once and settings never change while in mission mode. In this case, the designer must ensure that the fine delay line has enough dynamic range on either side of the lock point to guarantee phase lock maintenance in the presence of voltage and temperature ramps spanning the entire product voltage-temperature range in the worst possible scenario. If this is not possible, the control algorithm must ensure repeated engagement of the coarse DCDL to compensate for large V-T ramps without the possibility of a large phase change affecting system functionality. Such an example is a DDR memory incoming strobe phase lock system which engages the coarse DCDL only when there is no memory transaction present on the bus and, therefore, no one is looking at the data strobe. While memory reads or writes are in progress, only the fine DCDL is engaged which should guarantee DLL-induced deterministic jitter on the order of a few picoseconds and no spurious clock transitions. An additional concern with cascaded DCDLs is increased D_{min} which can be shown to be prohibitive for certain applications since it directly affects the overall minimum supply voltage allowed (V_{MIN}).

There exists a powerful method of achieving both low d_{r} (high resolution) and arbitrarily high dynamic range with low D_{min} and without cascading DCDLs. It is based on the "ping-pong" DCDL arrangement of Figs. 6.22 and 6.23 with a phase interpolator as the output stage instead of the 2-to-1 multiplexer. A phase interpolator receives two input clocks with a phase difference on the order of 1-2 gate delays and a digital setting value. For a setting equal to zero, the interpolator output simply delays the early input phase by a fixed amount. For a setting equal to the maximum possible value, the interpolator output delays the late input phase by the same amount. For any intermediate setting, the interpolator produces an output clock with a phase proportional to the applied setting as measured using the output at setting 0 as a reference.

A sample schematic is shown in Fig. 6.26. The two separate branches are implemented with tri-state inverters controlled by complementary settings. Typically, the output is restored with an output inverter (not shown) which also makes the overall structure non-inverting. For a first order analysis, we will use the methodology and terminology established in [14] for current-mode reduced-swing structures. The time delay between the two interpolated phases is Δt, and it will be used as a parameter throughout the analysis. We make the following assumptions:

1. The overall output resistance of a branch when fully enabled is R and has no voltage dependence.

2. The interpolation setting can be represented by a continuous variable w where $0 \leq w \leq 1$. When $w = 1$, the early phase is fully enabled, and the late phase does not affect the output. When $w = 0$, the early phase is disabled and only the late phase controls the output. For any other value, both the early and the late phase affect the output delay with weights equal to w and $1 - w$, respectively.

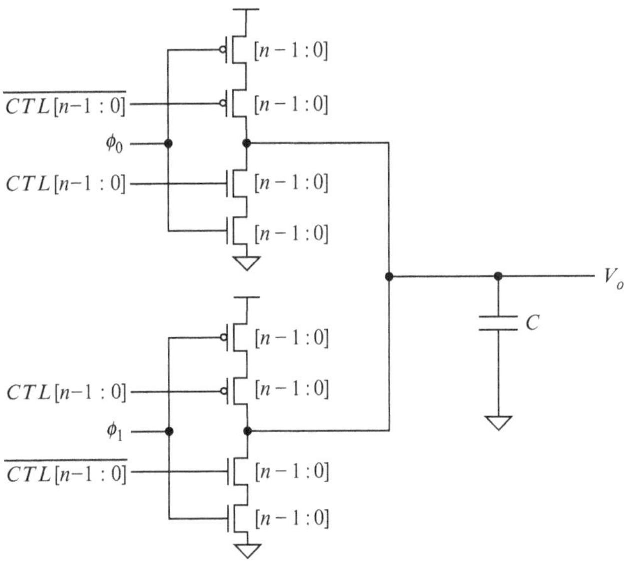

Fig. 6.26. Full swing phase interpolator ($\log_2 n$-bit control)

The interpolator time constant RC (where R is the on-resistance of a fully enabled branch and C is the output capacitance in Fig. 6.26) is a fundamental property which will affect transfer function linearity as will be shown. Without loss of generality, we assume that the interpolator is mixing two falling edges (ϕ_0 and ϕ_1) so that the output waveform is rising. Phase ϕ_0 occurs at time 0 and ϕ_1 occurs at time Δt. For $0 \leq t < \Delta t$, the equivalent circuit is shown in Fig. 6.27a. Writing KCL at V_o yields equation:

$$\frac{dV_o(t)}{dt} + \frac{1}{RC} V_o(t) - \frac{wV_{DD}}{RC} = 0, \tag{6.31}$$

which has the following solution assuming $V_o(0) = 0$:

$$V_o(t) = wV_{DD}(1 - e^{-\frac{t}{RC}}). \tag{6.32}$$

For $t \geq \Delta t$, both branches pull high and resistors R/w and $R/(1-w)$ are connected in parallel. The equivalent circuit is shown in Fig.6.27b. The governing equation is:

$$\frac{dV_o(t)}{dt} + \frac{1}{RC}V_o(t) - \frac{V_{DD}}{RC} = 0, \tag{6.33}$$

with the following well-known solution (assuming zero initial conditions):

$$V_o(t) = V_{DD}(1 - e^{-\frac{t}{RC}}). \tag{6.34}$$

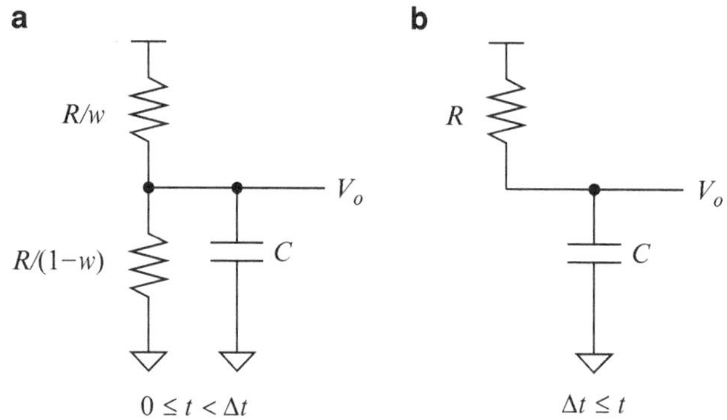

Fig. 6.27. Phase interpolator equivalent circuits

The overall solution is a composite waveform that consists of Eq. (6.32) for $0 \leq t < \Delta t$ and Eq. (6.34) for $t \geq \Delta t$ time-shifted by Δt and adjusted by the initial condition established by Eq. (6.32) at $t = \Delta t$:

$$V_o(t) = wV_{DD}(1 - e^{-\frac{t}{RC}})u(t)u(\Delta t - t) + V_{DD}[1 - (1 - w + we^{-\frac{\Delta t}{RC}})e^{-\frac{t-\Delta t}{RC}}]u(t - \Delta t), \tag{6.35}$$

where $u(t)$ is the unit step function.

Figure 6.28 plots Eq. (6.35) for various Δt using the interpolator weight w as a parameter. It is obvious from the $V_{DD}/2$ crossing point that linearity with respect to w is a strong function of Δt. The larger Δt is with respect to the interpolator time constant RC, the larger the deviation from linearity with respect to interpolation weight w. This is more clearly shown in Fig.6.29 that shows interpolator delay transfer function with respect to interpolation weight w using Δt (delay between two input phases) as a parameter. The top graph shows absolute inteprolator delay as a function of w adjusted so that the delay for $w = 1$ (early phase only affecting the output) is zero. The bottom graph shows the same data but normalized to the same $0-1$ range. Linearity is quite good for $\Delta t \ll RC$ but deteriorates quickly for increased spacing between input phases ϕ_0 and ϕ_1. The interpolator time constant RC is a fundamental property that has a strong effect on its linearity. Larger time constants are

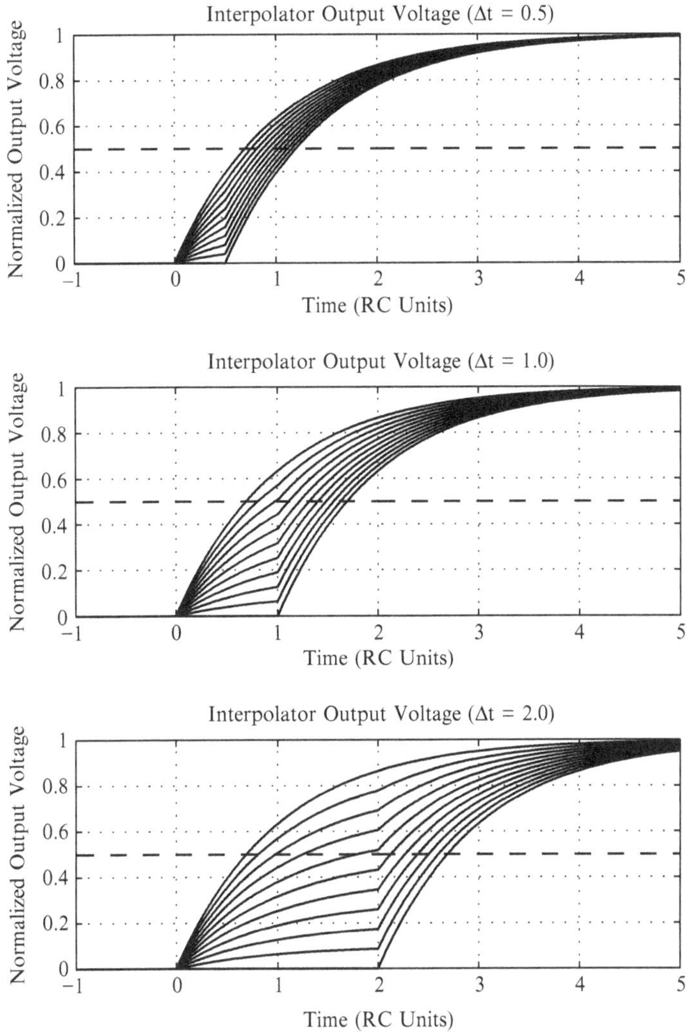

Fig. 6.28. Phase interpolator normalized voltage output for $\Delta t = 0.5, 1, 2$ (w is decreasing from 1 to 0 in steps of 0.1 from left to right in all three plots)

desired if linearity is important. An effect similar to an increased interpolator *RC* can be achieved by slowing the edge rates of input phases ϕ_0 and ϕ_1.

An interpolator based DCDL can achieve both high resolution and high dynamic range at the expense of increased control complexity. Such a DCDL can easily be designed to be free from spurious output transitions. When a coarse setting changes that affects one input of the interpolator, the interpolator weight should be such that only the other non-changing input affects the output.

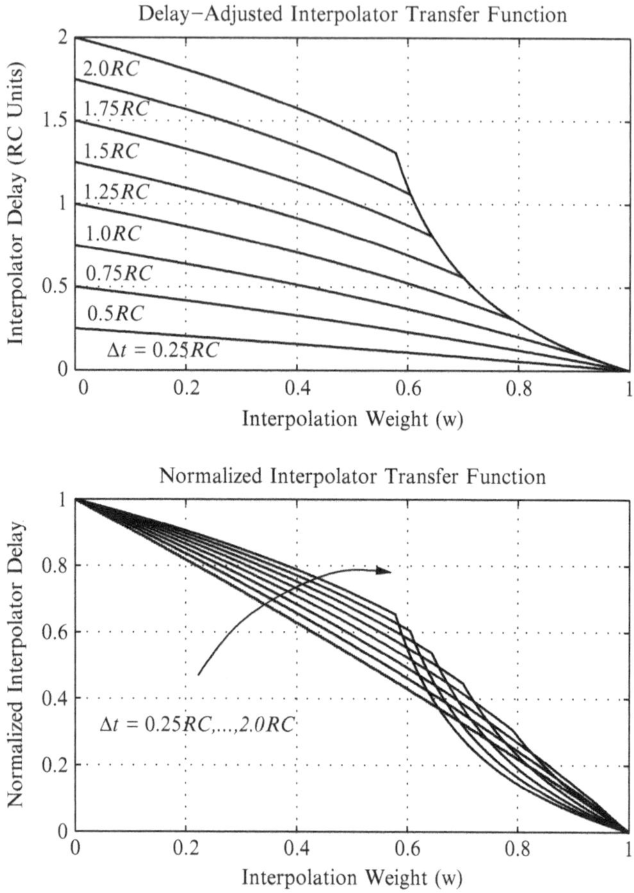

Fig. 6.29. Phase interpolator transfer function for varying Δt. Top graph reproduced in a form similar to that in [14], ©1998 Stefanos Sidiropoulos

6.6 Control

There is one major classification that needs to be made regarding digital DLL control: first order and higher order. A first order digital DLL contains a single DCDL state element (counter or shift register) that maintains the delay line setting and gets updated through some mechanism based on the phase difference between reference and controlled clock. In contrast, a second order DLL will have a second counter which maintains some representation of the *frequency* or rate of phase change between the two clocks. The DCDL is then adjusted by combining both such state variables. All DLL examples in this chapter assume that the reference clock at the phase detector and the source clock of the DCDL are the same. Therefore, there is no possibility of a frequency difference between the two inputs of the phase detector. An application where a second order control loop would make sense is clock-data recovery, which is briefly described in Sect. 6.9.4. The rest of this section is only concerned with first order control and identical frequencies of reference and feedback clocks.

A typical DLL control block contains a state structure that holds the DCDL settings and a mechanism to update the settings based on input from the phase detector. The state element can be a binary up/down counter if the DCDL has binary encoding. Different DCDL setting encodings can be easily handled with the addition of combinational decoding logic past the main counter. The designer should always keep in mind that post-flop combinational logic can generate spurious transitions and create more conditions for glitches at the output of the DCDL. For one-hot DCDL configurations, it is more common to use a bidirectional shift register which requires no decoding and will produce glitch free control outputs.

The simplest control arrangement is shown in Fig.6.30a. It involves a single up/down counter (or bidirectional shift register) under the direct control of the phase detector. When the phase detector evaluates to a 1 (controlled clock faster with respect to the reference), the counter is incremented and the DCDL has a larger (slower) setting. When the PD evaluates to a 0, the counter is decremented and the DCDL speeds up. A more general control arrangement is shown in Fig.6.30b. In this case, a Finite State Machine (FSM) is inserted between the phase detector output and the counter increment/decrement. The FSM will make setting change decisions based not only on the phase detector output but also based on internal state. In this fashion, more complex control algorithms can be implemented which can have a strong impact on DLL capabilities such as sensitivity to initial conditions, dynamic range, and stability.

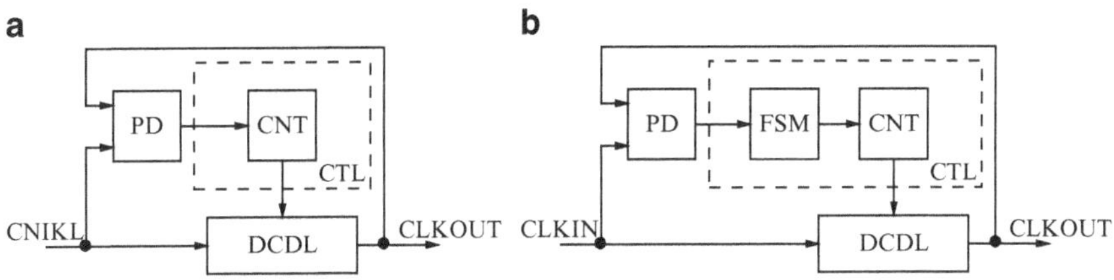

Fig. 6.30. DLL control options

6.6.1 Sensitivity to Initial Phase

Even though a DLL may contain a DCDL with a large dynamic range, a simplistic control design may result in an inability to achieve phase lock. An example is shown in Fig.6.31. Let us assume that the DLL configuration is the one in Fig.6.30a with $D_{\min} < 0.5T$ where T is the CLKIN period. The waveforms right after system reset (zero DCDL setting) are shown on the top part of Fig.6.31. The stateless control design of Fig.6.30a will attempt to push the DCDL toward the left and match edge 1 of CLKIN with edge 1 of CLKOUT. If the settings counter wraps around, a very large delay will be added to the DCDL and the DLL will either lock to an undesirable setting (not the minimum delay setting that will achieve phase lock) or will degenerate to an oscillation between D_{min} and D_{max} without ever achieving lock. If the settings counter is designed not to wrap-around but saturate at the minimum setting, the DCDL will stay fixed at D_{min} and again phase lock will never be achieved.

If on the other hand the FSM of Fig.6.31 (*bottom*) is inserted between the phase detector and the settings counter, lock can be easily achieved. The FSM will keep on incrementing the DCDL settings until a 1-to-0 transition is observed at the phase detector independent of the initial state of the phase detector output. Such a transition indicates a lock condition and the state machine will be bouncing back and forth between states INC and DEC. Edge 1 of CLKOUT will be aligned to edge 2 of CLKIN.

A word of caution is absolutely essential when a state machine similar to the one depicted in Fig.6.31 (*bottom*) is used. This particular FSM is designed to disregard a 0-to-1 transition from the phase detector and essentially lock the system when a 1-to-0 transition is observed. In reality, a 1-to-0 transition can easily be observed right after a 0-to-1 when the controlled clock (CLKOUT) phase traverses a negative edge of the reference clock (CLKIN). When the positive edge of CLKIN is aligned to the negative edge of CLKOUT the phase detector can undergo multiple transitions due to internal factors such as a large dead zone where its output is undefined or external factors such as supply noise and reference clock jitter. If state DEC is entered (Fig.6.31) while we have positive to negative edge alignment, the system will never

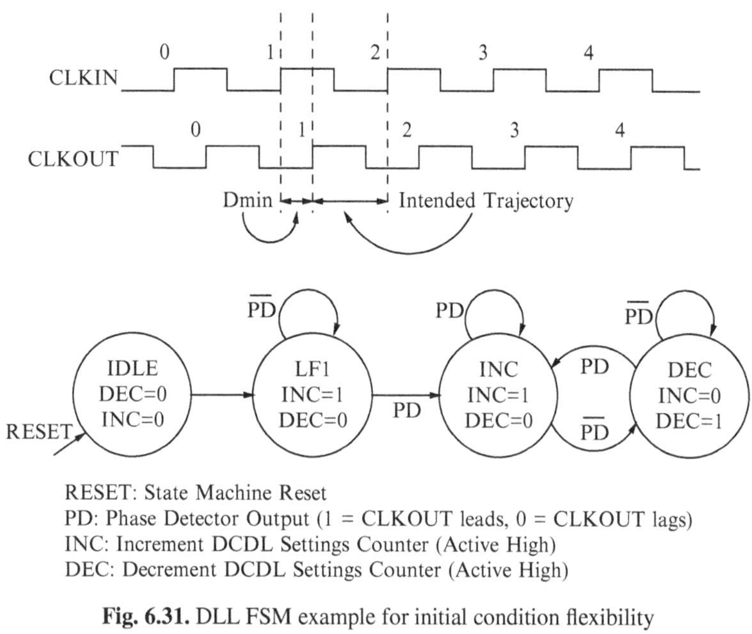

Fig. 6.31. DLL FSM example for initial condition flexibility

achieve lock but will likely be pushed back to its reset state. State DEC must only be entered when a positive CLKOUT edge has just started to lag a positive CLKIN edge. Otherwise, the phase detector decisions won't be interpreted correctly and the DCDL will be moved to the wrong direction.

There are multiple ways that this issue can be addressed. Digital low pass filtering of the phase detector output can be employed to remove spurious transitions. This method can reduce the probability of a false lock, but it is hard to show that it can totally eliminate it. A more robust method is to insert an intermediate state between LF1 and INC in which the controller will reside for n clock cycles. In this state, the DCDL setting is incremented n times independent of the phase detector output. In this way, we can guarantee that when we have negative edge traversal, the DCDL will be incremented enough to push CLKOUT out of the phase detector sampling window or dead zone (d_{SW}) given worst case supply noise and jitter conditions. The number n should be large enough to guarantee $n \cdot d_{\mathrm{r}} \gg d_{\mathrm{SW}}$ and at the same time small enough to guarantee $n \cdot d_{\mathrm{r}} \ll 0.5T$ (where T is the reference clock period) so that the next positive CLKIN edge is not traversed and the phase detector output does not change interpretation. This should be a rather loose constraint for most applications.

6.6.2 Dynamic Range Increase

It is possible to increase the dynamic range of a DLL with the addition of little extra hardware and incremental complexity in the control automaton. This method assumes that the DCDL output can be conditionally inverted (i.e., XOR output stage) and that the settings counter or bidirectional shift register can provide a carry out (CO) output to the FSM indicating that its maximum value has been reached. The concept is very simple. If the DLL reaches the maximum value of the settings counter without achieving lock (the phase detector output has not undergone a 1-to-0 transition), then the DCDL output is inverted, the counter is cleared and the state machine goes back to the initial state and attempts to lock the inverted DCDL. The inversion has the effect of adding 180° of phase and virtually doubles the DLL dynamic range. This is shown conceptually in Fig. 6.32 (*top*). An example FSM for this configuration is shown at the bottom of the figure.

6.6.3 Stability and Bandwidth

At the beginning of this chapter, a claim was made that a digital DLL can be unconditionally stable. In this section, we qualify this claim. The stability of feedback systems is typically analyzed in the frequency domain. A frequency domain model of a digital DLL is non-trivial and not particularly insightful due to the nonlinearity of a bang–bang phase detector and the sampled nature of the system. Instead, we choose to revisit the analysis of a linear (analog) delay locked loop, establish stability conditions and apply the intuition developed to the non-linear digital counterpart. We follow the methodology and notation of Maneatis in [15]. A very similar approach is also presented by Yang in [16].

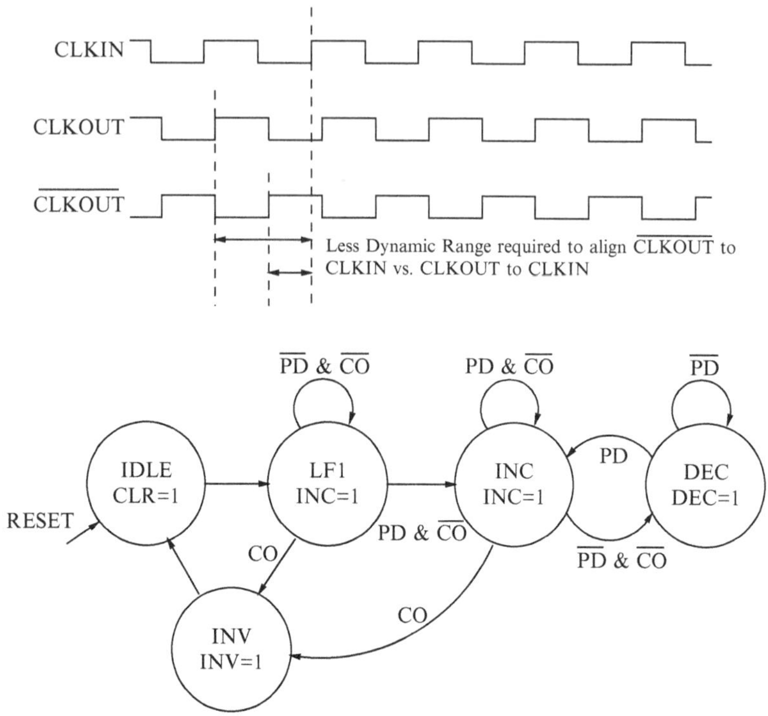

RESET: State Machine Reset
PD: Phase Detector Output (1 = CLKOUT leads, 0 = CLKOUT lags)
INC: Increment DCDL Settings Counter (Active High)
DEC: Decrement DCDL Settings Counter (Active High)
CLR: Clear DCDL Settings Counter (Active High)
INV: Invert DCDL Ouput (Active High, Sticky [sets RS Flop])
CO: DCDL Settings Counter Carry Out (Active High)

Fig. 6.32. Doubling DLL dynamic range using conditional inversion

Figure 6.33a shows a simple charge-pump based analog DLL. The phase detector is linear rather than bang–bang (Sect. 6.4). Its output is a pulse whose width is proportional to the difference between the CLKIN and CLKOUT phases. The proportional pulse controls a charge pump which integrates a constant current on capacitor C. The amount and sign of the integrated current (charge) depend on the amount and sign of the phase difference between the clocks. The capacitor voltage controls an analog delay line which in turn will adjust until CLKOUT matches CLKIN. There are many similarities between this analog DLL and the digital counterpart of Fig. 6.1. The settings counter/accumulator inside the control block performs the same action as the capacitor and essentially integrates the phase detector output. The main difference is that the digital DLL is a non-linear system because the correction in response to the phase error is not proportional but instead has a constant slew rate.

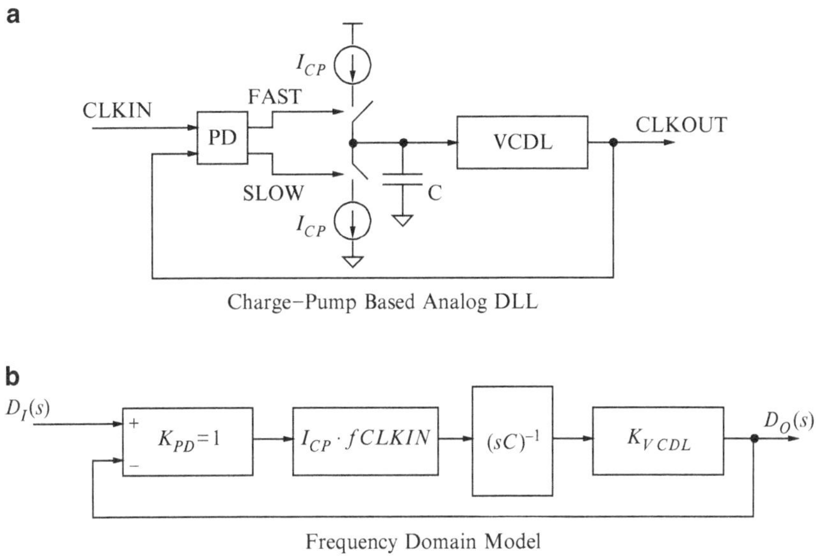

Fig. 6.33. Analog DLL frequency domain model

An additional important difference is that a digital DLL may have more delay around the loop due to inherent clock delays in the FSM controller and settings counter. This can be very important for stability and will be addressed in more detail later on.

A frequency domain model of the analog DLL is shown in Fig. 6.33b. The model terminal variables are $d_{\mathrm{I}}(t)$ (input delay), which is defined as the delay of input CLKIN with respect to an arbitrary point. The output delay of CLKOUT $d_{\mathrm{O}}(t)$ is also defined with respect to the same arbitrary point. The corresponding Laplace transforms are $D_{\mathrm{I}}(s)$ and $D_{\mathrm{O}}(s)$. The frequency domain representation of the phase detector is a simple subtractor with a constant gain of 1 since it generates a pulse with a width equal to the delay difference between CLKIN and CLKOUT. The units at the output of the phase detector are seconds (delay). The charge pump model is also a constant gain since it multiplies the proportional error pulse with a constant current after normalizing it to the CLKIN period ($I_{\mathrm{CP}} \cdot f_{\mathrm{CLKIN}}$, where f_{CLKIN} is the CLKIN frequency). The units at the output of the charge pump are Amperes (current). The charge pump capacitor C acts as an integrator with a transfer function $(sC)^{-1}$. The units at the integrator output are Volts. Finally, the Voltage-Controlled Delay Line (VCDL) is also modeled with a constant gain (K_{VCDL} s/V). The output variable $D_{\mathrm{O}}(s)$ has units of seconds (delay). Writing the loop equation yields:

$$D_{\mathrm{O}}(s) = [D_{\mathrm{I}}(s) - D_{\mathrm{O}}(s)] \cdot I_{\mathrm{CP}} \cdot f_{\mathrm{CLKIN}} \cdot \frac{K_{\mathrm{VCDL}}}{sC}. \tag{6.36}$$

Further manipulation yields the following transfer function [15]:

$$\frac{D_{\mathrm{O}}(s)}{D_{\mathrm{I}}(s)} = \frac{1}{1 + \frac{s}{\omega_N}}, \tag{6.37}$$

where

$$\omega_N = \frac{I_{\mathrm{CP}} \cdot K_{\mathrm{VCDL}} \cdot f_{\mathrm{CLKIN}}}{C}. \qquad (6.38)$$

The transfer function of Eq. (6.37) is of first order with a single pole at ω_N (loop bandwidth) and is unconditionally stable with a phase margin of 90°. This is true as long as the delay around the loop is virtually instantaneous and the phase margin is not reduced considerably. Even an analog DLL though is a sampled system and there is loop delay due to the sampled nature of the phase detector: A phase detector measurement won't be taken until the next positive clock edge, thus, reducing the phase margin at unity gain. Such margin reduction though is negligible if $\omega_N \ll 2\pi f_{\mathrm{CLKIN}}$. For all practical purposes, the loop delay can be considered virtually instantaneous if such delay is much lower than the DLL response time. Maneatis [15] sets the unconditional stability criterion at $\omega_N = 2\pi f_{\mathrm{CLKIN}}/10$.

Multiple authors [17, 18] have pointed out that the simple frequency domain model of Fig. 6.33 does not apply for the class of delay locked loops where the same reference clock is driving both the phase detector and the DCDL. Figure 6.33 models the delay line as a 2-terminal block (control input and clock output) and does not include a feed-forward path from the input reference clock to the output through the delay line. As a result, such a model cannot accurately model jitter transfer from input to output clock. The stability analysis changes due to the introduction of a zero in the first order model. However, unconditional or conditional stability can still be shown with root locus methods depending on model choice [18].

The main difficulty in applying a similar analysis to the digital DLL lies in the nonlinear transfer function of the phase detector (Fig. 6.2). The linear frequency domain analysis is not applicable and the steady state response of such a DLL will be oscillatory (limit cycle). Prior art [19, 20] has linearized the phase detector response around the dead zone, which is a valid approach for high CLKIN frequencies, high DCDL resolutions, and multiple averaged measurements since the expected phase error will be rather small and comparable to d_{sw}. For lower frequency applications though, the phase detector will mostly operate in its nonlinear regime.

Before proceeding, let us first define what we mean by stating that a non-linear DLL is stable. We know for a fact that a DLL with a non-linear (bang–bang) phase detector will exhibit limit cycle behavior due to the binary nature of the phase detector and its inability to encode a zero phase error. For the purposes of this discussion, let us define DLL stability to mean that the expected limit cycle should have by design the minimum possible amplitude of $\pm d_{\mathrm{r}}$. Frequency domain transformation is not really necessary to assess the stability of such a simple system. The time-domain behavior of the digital DLL (Fig. 6.1) in the locked state can be summarized by the following difference equations:

$$e[n] = \mathrm{sgn}(d_{\mathrm{O}}[n] - d_{\mathrm{I}}[n]), \qquad (6.39)$$

$$d_{\mathrm{O}}[n+1] = d_{\mathrm{O}}[n] + e[n]d_{\mathrm{r}}, \qquad (6.40)$$

where $e[n]$ is the error computed by the phase detector ($e[n] \in \{-1,1\}$), $d_{\rm I}[n]$ is the input delay at discrete time n (nth CLKIN positive edge), $d_{\rm O}[n]$ is the output delay at discrete time n and $d_{\rm r}$ is the DCDL resolution. We establish a time-domain stability criterion as follows: *The correction step in Eq. (6.40) at discrete time n should not produce an error ($d_{\rm O}[n+1]-d_{\rm I}[n+1]$) at $n+1$ of the same sign and greater magnitude than ($d_{\rm O}[n]-d_{\rm I}[n]$) at time n.* This criterion is easily satisfied if the correction term $e[n]d_{\rm r}$ in (6.40) does not result in crossing over two phase boundaries which will produce a phase error of the same sign ($mod T_{\rm CLKIN}$) and potentially larger magnitude. If $d_{\rm r} < 0.5T_{\rm CLKIN}$, this should never happen and this constitutes a constraint that is always met for all practical purposes. A DLL with such a large $d_{\rm r}$ would be highly impractical.

We now focus on the loop delay argument. Eqs. (6.39) and (6.40) do not capture the effect of excessive loop delay and the possibility of reduced "phase margin". A digital DLL has the potential of introducing substantial loop delay through its control mechanism and may turn negative feedback into positive. The reasons for the increased delay can be multiple:

- Additional flip-flop synchronization stages past the phase detector to ensure low probability of metastability.
- Low pass filtering of the phase detector output to remove spurious transitions and potentially emulate "ternary" error detection (fast, slow, NOP).
- Deserialization latency in a CDR (clock-data recovery) loop where phase detection occurs in the divided clock domain on the multiple deserialized bits.
- Pipeline stages in the FSM controller and settings counter.

Let us introduce two more general descriptive DLL design parameters (in addition to $d_{\rm r}$, $d_{\rm sw}$, $D_{\rm min}$ and $D_{\rm max}$) to help with the analysis of this issue:

$N_{\rm d}$ indicates the number of CLKIN delay cycles from the input of the phase detector to the setting input of the DCDL. This number is an integer with a minimum value of 1 (delay introduced by the bang–bang phase detector).

$N_{\rm bw}$ is the inverse of the rate at which the FSM controller issues corrections to the DCDL (once every $N_{\rm bw}$th cycle). One can think as $1/N_{\rm bw}$ as the "bandwidth" of this nonlinear system. The minimum value for $N_{\rm bw}$ is 1 (controller issues corrections on every cycle) and the maximum value can be arbitrarily high. $1/N_{\rm bw}$ is also referred to as the DLL sampling rate because it signifies the normalized frequency of looking at the phase detector output.

We introduce a second stability criterion which states that a digital DLL is stable if $N_{\rm d} < N_{\rm bw}$. This is the equivalent of stating that a *a digital delay locked loop is stable if it won't issue a DCDL correction before the outcome of the previous correction is fully evaluated by its control mechanism.*

We demonstrate this criterion with a cycle-accurate behavioral model of a DLL with the characteristics of Table 6.7 running at 250 MHz.

The controller is equivalent to the FSM of Fig.6.31 with the addition of wait states between the states that update the counter to implement $N_{\rm bw} \neq 1$. The modeled

Table 6.7. DLL behavioral model design parameters

D_{min}	1.525 ns	D_{max}	14.275 ns
d_r	0.050 ns	d_{sw}	0.0 ns
N_d	parameter	N_{bw}	4

DLL has a built-in N_d of 3 (phase detector plus 2 stages in the FSM/counter). We, therefore, make N_{bw} equal to 4 to guarantee stability when N_d is minimum. Fig.6.34 shows simulation results (DLL phase error over time) for various N_d values. Case $N_d = 3$ satisfies our stability criterion since the phase error changes sign every N_{bw} cycles. All other cases fail the criterion and the phase error is amplified. Loop delay causes the DLL to issue correction steps in the wrong direction. The number of wrong correction steps is $\lfloor N_d/N_{bw} \rfloor$ and the resulting phase error is bounded by $\pm(\lfloor N_d/N_{bw} \rfloor + 1)d_r$ as opposed to $\pm d_r$ in the stable case.

A digital DLL can be stabilized given a value for N_d by making $N_{bw} > N_d$. Such an example is shown in Fig.6.35 where a DLL with $N_d = 15$ is stabilized by increasing N_{bw} from 4 to 16. An expected increase in lock acquisition time is observed due to the reduced rate of issued DCDL corrections. Lock acquisition will be discussed in more detail in Sect. 6.6.4.

Increasing N_{bw} is analogous to reducing loop bandwidth ω_N in the analog domain. Higher N_{bw} increases loop response time and makes it more difficult to track a changing input. Spread spectrum clocking is one important application where tracking a changing input is necessary. In such a case, N_{bw} should be treated as an important design parameter and the designer must ensure that a changing frequency input won't generate a noticeable increase in phase error. A convenient way of describing the tracking ability of a nonlinear digital DLL is its delay slew rate r defined as follows:

$$r = \frac{d_r}{N_{bw} \cdot T_{CLKIN}}. \tag{6.41}$$

The slew rate is a unitless quantity that indicates the amount of delay correction that the DLL can produce per unit time. A simple way to determine whether a DLL can track a given spread spectrum (SS) clock is to calculate an equivalent slew rate r_s for the spread clock and compare it with the DLL rate r. SS clocks are typically defined with a spread factor A_s as a percentage which indicates modulation amplitude and the modulation frequency f_s on top of the clock period T_{CLKIN}. The equivalent slew rate of a spread clock indicates the period change per unit time and is given by the following formula:

$$r_s = 4A_s T_{CLKIN} f_s. \tag{6.42}$$

As an example, a 250 MHz SS clock with a 10% spread factor and a 500 KHz spread frequency will have an r_s of 0.8×10^{-3}. A DLL with d_r equal to 0.050 ns and a N_{bw} equal to 4 will have an r of 3.125×10^{-3}. Since $r > r_s$, we determine that the DLL can track the SS clock without change to its theoretical minimum phase error of $\pm d_r$. Phase error tracking is demonstrated in Fig.6.36 using the same behavioral

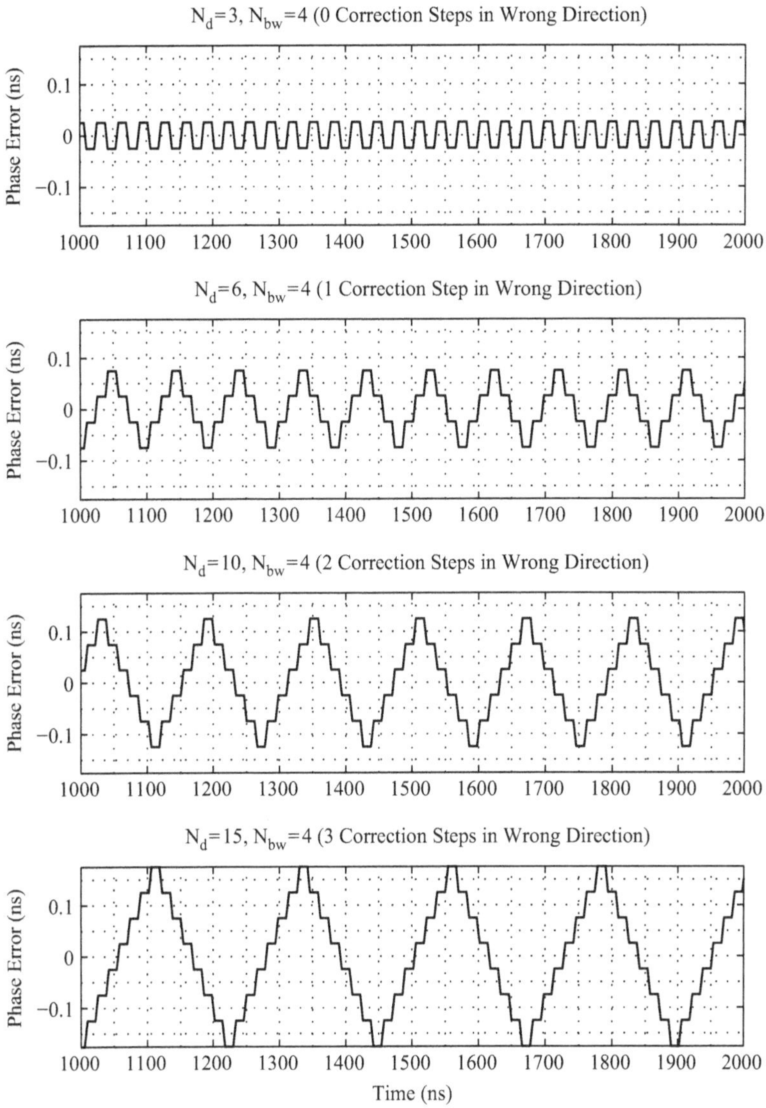

Fig. 6.34. DLL limit cycles as a function of N_d in the locked state

DLL model as past examples in this section. Clock is running at 250 MHz with 10%, 500 KHz spread, N_d is fixed at 3 and N_{bw} is parameterized to demonstrate the slew rate tracking criterion. For $N_{bw} = 4$, $r > r_s$ and phase error is close to the $\pm d_r$ structural value and spread spectrum modulation is tracked. For $N_{bw} = 16$, r is slightly less than r_s and phase error starts to get amplified. Finally, at $N_{bw} = 32$, the DLL has very small bandwidth and is incapable of tracking the SS input. Phase error

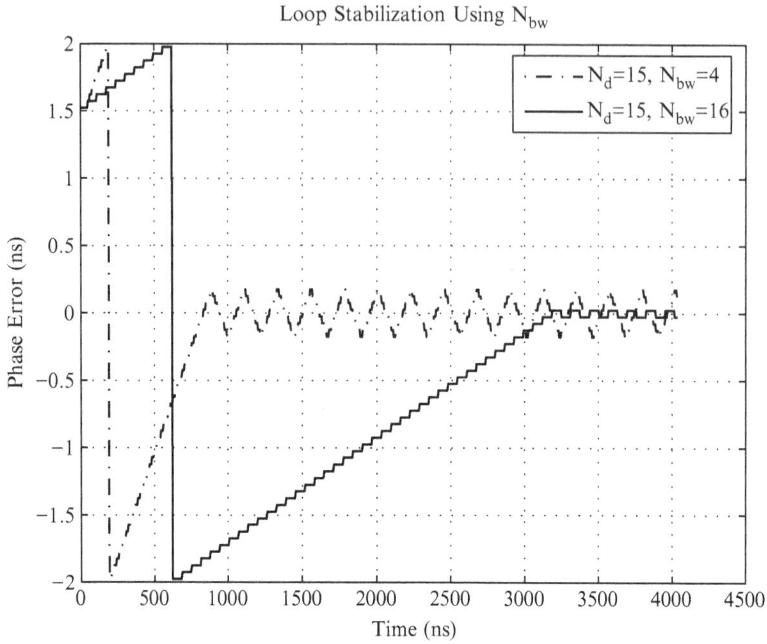

Fig. 6.35. DLL loop stabilization using N_{bw} (DLL is running at $T_{CLKIN} = 4$ ns)

is very similar to the SS period variation, and there is little difference between this DLL and an open loop delay line matched to the unmodulated input clock period.

It is worth mentioning that this tracking analysis is only relevant if the DCDL is matched to some function of T_{CLKIN}. If the application involves delay-only matching (Fig. 6.4b) then the phase error should not be affected by input clock spreading. All DLL outputs will vary in the same fashion and delay variation will be the same among all outputs resulting in zero additional phase errors.

This section has established two stability criteria and a tracking criterion for digital DLLs. It must be stressed that this analysis applies only to first order digital DLLs as defined in Sect. 6.6.

6.6.4 Lock Acquisition

The lock acquisition procedure is the most open-ended controller aspect from a design perspective. Multiple approaches are possible. A simple controller like the one depicted in Fig. 6.31 will produce lock acquisition profiles similar to the ones shown in Fig. 6.35, where lock is achieved with a constant slew rate equal to $d_r/(T_{CLKIN}N_{bw})$. Lock acquisition duration is, therefore, a strong function of N_{bw} selection.

Faster lock acquisition methods are possible such as variable slew rate (higher r during acquisition for lock time reduction and lower r during lock maintenance for stability), binary search [21] and open loop synchronous mirror delay locking

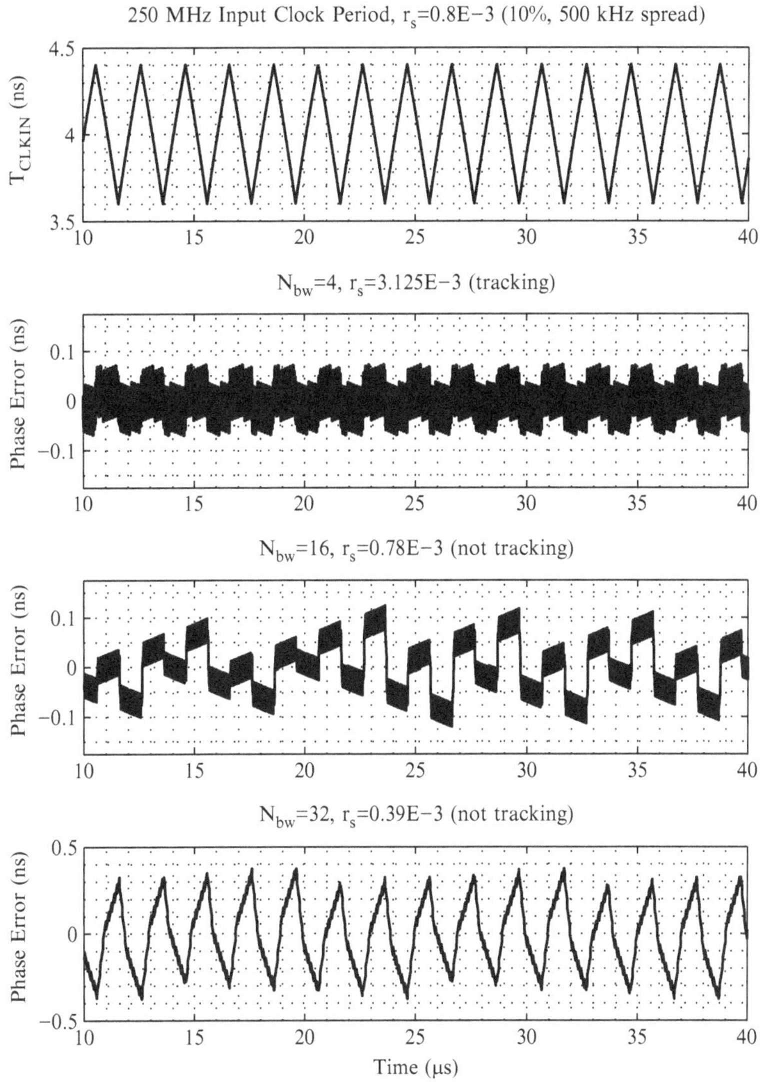

Fig. 6.36. DLL phase error tracking of SS clock as a function of N_{bw}

[22, 23]. The binary search method computes every bit of a logarithmic delay setting from MSB to LSB in a successive approximation fashion in $\log N$ steps where N is the total number of DCDL settings. The synchronous mirror delay method measures the difference between the unknown delay and a full clock period in a single step. This difference is then installed in the DCDL and the system can either revert to regular feedback DLL mode with standard control or remain open loop until the next time a measurement is made. This method is by far the fastest and can achieve delay lock within a few cycles independent of delay size. All of the above lock acquisition

methods are illustrated in Fig. 6.37 using our standard DLL behavioral model ($N_{bw} = 16$) for $T_{CLKIN} = 250$ MHz. The reduction in lock time is rather obvious going from top to bottom.

Fast lock acquisition can be essential for applications that involve clock gating and suspension modes such as memories and low-power/portable systems. Ability to relock quickly as opposed to saving the previous setting before entering the low power state and reinstating it is much more desirable due to voltage/temperature tracking and is mandatory for dynamic voltage/frequency applications.

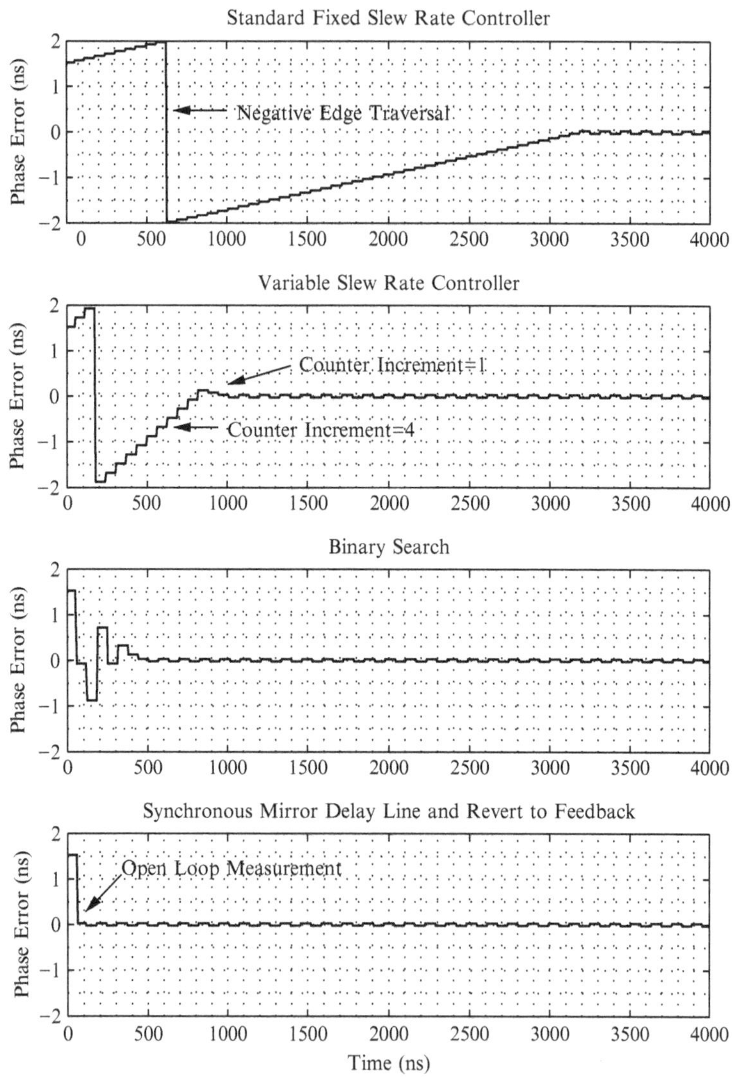

Fig. 6.37. DLL lock acquisition profiles ($T_{CLKIN} = 250$ MHz, $N_{bw} = 16$)

6.7 Putting it All Together

Sections 6.4 through 6.6 have described in considerable detail all DLL system components. In this section, we briefly review important design parameters and outline a basic design flow.

A digital DLL can be generally described by the following design parameters:

d_{sw} Phase detector sampling window
d_{r} Delay resolution
D_{min} Minimum DCDL delay
D_{max} Maximum DCDL delay
N_{d} Control mechanism delay
N_{bw} Inverse of DCDL correction issue rate

Phase detector design/selection should always minimize d_{sw}. There is really no tradeoff here. Typically, a designer develops a good phase detector with a small sampling window, and then this can be reused for the vast majority of DLL instances on the chip. Next, an appropriate d_{r} should be selected based on application requirements and desired DCDL complexity, area and power. D_{min} and D_{max} will be set by the entire range of input clock frequencies that need to be tracked and also by the entire process, voltage, and temperature operating space. Under no circumstances should a DLL prove to be the limiting factor in a chip design in terms of $(F_{\text{MAX}}, V_{\text{MAX}})$ or $(F_{\text{MIN}}, V_{\text{MIN}})$. Substantial margin should always be provided to allow for small changes in production specifications. The controller should be designed to perform successful lock acquisition for each PVT and for each input clock frequency. N_{d} should be minimized to the extent possible. In the lock state, the controller should maintain lock with the minimum theoretical phase error of $\pm d_{\text{r}}$ (wherever possible). N_{bw} should be greater than N_{d} but small enough to allow for input frequency variation tracking if the application demands it.

Other design aspects such as DCDL output glitching, controller robustness around non-locking edge traversal, etc., should be thoroughly investigated and analyzed because they can prove catastrophic.

6.8 Noise Considerations

Phase noise and jitter transfer have been traditionally analyzed in the frequency domain (Chap. 5.1). In this section, we focus on the time domain because of the difficulties introduced by the non-linear nature of the system. For a comprehensive analog DLL noise analysis in the frequency domain, the reader is referred to [17, 18]. Unlike a PLL which can filter reference clock jitter as described in Chap. 5.1, a digital DLL will add jitter to the reference clock through three primary mechanisms:

The first mechanism is the structural limit cycle which in a well designed DLL will add $\pm d_{\text{r}}$ of phase jitter to the output clock. From a frequency domain perspective, this noise will appear around the $f_{\text{CLKIN}}/N_{\text{bw}}$ frequency band. The second mechanism is the equivalent of jitter peaking in the analog domain: The digital DLL can amplify PLL jitter because of delayed response. This stems from the fact that the DLL is not capable of determining whether a disturbance in the reference vs. output clock must be tracked (i.e., spread spectrum or DCDL environmental change) or not tracked and potentially canceled out (reference clock jitter). Since all DLLs are designed to correct phase error, a jitter disturbance in the reference clock can be tracked before it shows up at the output of the delay line in a cycle later, and this can cause jitter amplification in the next cycle. In general though, this jitter amplification will be limited and on the order of $\pm d_{\mathrm{r}}$. Furthermore, it does not accumulate as in the PLL case (Chap. 5.1) and it will add a small cycle time penalty. It is very simple to alleviate both the structural and the jitter amplification components either by using a ternary phase detector with a NOP state [24] or by opening the loop and tracking the phase error with an auxiliary mechanism [25]. The final and most potent mechanism is supply noise induced jitter. Supply noise is particularly dangerous for DLLs because we have clock propagation through a long delay line which typically has significant delay sensitivity to power supply changes. The remainder of this section frames the problem and presents a simple analytical model.

For the analysis in this section, we will adopt the terminology of Sect. 5.2. We are trying to compute the maximum period jitter generated by a fixed delay D subject to known sinusoidal supply noise, as a function of noise amplitude, noise frequency, delay amount D, and clock period T. Before proceeding, we define the following symbols:

- $V(t)$ Power supply voltage as a function of time
- V_{DD} Nominal power supply voltage value
- A_{n} Supply voltage noise amplitude (normalized to V_{DD})
- f_n Supply voltage noise frequency
- ϕ Supply voltage noise phase at $t = 0$
- D DLL delay (also referred to as insertion delay)
- T Clock period
- t_i Time of ith clock positive edge at the input of insertion delay D with respect to an arbitrary reference

We define the kth order period jitter introduced by a DLL with a constant insertion delay D as:

$$\Phi_k' = \max_i |t_{i+k} - t_i - kT| \,. \tag{6.43}$$

Equation (6.43) is interpreted as the maximum amount that the sum of k consecutive cycles of a given clock of period T can differ from the nominal value. We have introduced an additional subscript k (order) with respect to the jitter definitions of Sect. 5.2 to allow one more parameter in this analysis and have the capability to address multi-cycle paths and long term jitter requirements imposed by certain applications. The subscript k is only relevant for the relative jitter definitions of Sect. 5.2 such as period jitter ($\Phi'[n]$) and cycle-to-cycle jitter ($\Phi''[n]$). The relationship between Eq. (6.43) and the period jitter definition of Sect. 5.2 is

$$\Phi_1' = \max_n \left|\Phi'[n]\right| \,. \tag{6.44}$$

In order to derive an analytical expression for Φ_k' which will help us gain some insight regarding supply noise induced jitter, we start by assuming sinusoidal supply noise:

$$V(t) = V_{\mathrm{DD}}[1 + A_n \sin(2\pi f_n t + \phi)]. \tag{6.45}$$

Period jitter is generated by modulating the insertion delay D through supply noise. We make a second assumption, that the actual value $D_m(t)$ of the insertion delay D at time t (where t is the time that an ideal positive edge arrives at the input of the delay D) is given by the following formula:

$$D_m(t) = D \times \left(2 - \frac{\overline{V}(t,D)}{V_{\mathrm{DD}}}\right), \tag{6.46}$$

where $\overline{V}(t,D)$ is simply the forward moving average of $V(t)$ over a period of time equal to insertion delay D:

$$\overline{V}(t,D) = \frac{1}{D} \times \int_t^{t+D} V(\tau)d\tau. \tag{6.47}$$

Equation (6.46) simply states that the DLL insertion delay is linearly dependent on its supply voltage averaged over its flight time (i.e., an $x\%$ increase in average supply voltage results in an $x\%$ decrease in delay). This is an acceptable approximation for values of $\overline{V}(t,D)$ reasonably close to nominal V_{DD}. Obviously, this is not true for much larger or smaller values of $\overline{V}(t,D)$ and the axes crossing points implied by Eq. (6.46) are meaningless. There is an additional approximation in Eq. (6.47) since the moving average is computed over a constant time interval D whereas in reality the time interval should vary due to supply noise modulation. This approximation should introduce little error yet make an analytical approach tractable.

Now that we have explicitly defined $D_m(t)$, we can write an expression for t_i (time of ith positive edge at the output of delay D):

$$t_i = iT + D_m(iT). \tag{6.48}$$

Substituting (6.48) in (6.43) yields

$$\Phi_k' = \max_i |D_m((i+k)T) - D_m(iT)|. \tag{6.49}$$

After substituting (6.46) in (6.49) and with minimal manipulation we have

$$\Phi_k' = \frac{D}{V_{\mathrm{DD}}} \times \max_i \left|\overline{V}(iT,D) - \overline{V}((i+k)T,D)\right|. \tag{6.50}$$

Since we have an expression for supply voltage, we can obtain a closed form expression for the moving average by substituting Eq. (6.45) in (6.47).

$$\overline{V}(t,D) = V_{\mathrm{DD}} + \frac{A_n V_{\mathrm{DD}}}{D} \int_t^{t+D} \sin(2\pi f_n \tau + \phi)d\tau. \tag{6.51}$$

Performing the simple integration yields

$$\overline{V}(t,D) = V_{\mathrm{DD}} + \frac{A_n V_{\mathrm{DD}}}{2\pi f_n D}[\cos(2\pi f_n t + \phi) - \cos(2\pi f_n (t+D) + \phi)]. \tag{6.52}$$

Substituting (6.52) in (6.50) yields

$$\Phi'_k = \frac{A_n}{2\pi f_n} \max_i \left| \begin{array}{l} \cos(2\pi f_n iT + \phi) - \cos(2\pi f_n (iT + D) + \phi) - \\ \cos(2\pi f_n (i+k)T + \phi) + \cos(2\pi f_n ((i+k)T + D) + \phi) \end{array} \right|. \tag{6.53}$$

We now recall the following well-known trigonometric identity:

$$\cos u + \cos v = 2\cos\left(\frac{u+v}{2}\right)\cos\left(\frac{u-v}{2}\right). \tag{6.54}$$

Applying (6.54) on (6.53) yields

$$\Phi'_k = \frac{A_n}{\pi f_n} \max_i \left| \cos(\pi f_n (2i+k)T + \phi)[\cos(\pi f_n (kT + D)) - \cos(\pi f_n (kT - D))] \right|. \tag{6.55}$$

Equation (6.55) is expressed in a very convenient form which will enable us to dispense with the maximization operation by inspection: Only the first cosine term $(\cos(\pi f_n (2i+k)T + \phi))$ is a function of i, and its maximum value for every possible i is 1. Therefore, Eq. (6.55) becomes

$$\Phi'_k = \frac{A_n}{\pi f_n} \left| \cos(\pi f_n (kT + D)) - \cos(\pi f_n (kT - D)) \right|. \tag{6.56}$$

Application of one more trigonometric identity results in the following expression which is more desirable since it separates the effects of variables T and D:

$$\boxed{\Phi'_k = \frac{2A_n}{\pi f_n} \left| \sin(\pi f_n kT)\sin(\pi f_n D) \right|.} \tag{6.57}$$

Equation (6.57) constitutes a closed form analytical expression of worst case period jitter as a function of supply noise frequency and amplitude (f_n,A_n), clock period (T), and DLL insertion delay (D). Parameter k denotes whether we are interested in single period jitter ($k = 1$) or longer term (multi-cycle) jitter ($k > 1$). In order to gain some insight into this expression and determine whether it makes physical sense, let us focus on the two plots of Fig. 6.38. Both diagrams plot Eq. (6.57) under certain conditions. The top graph shows Φ'_1 as a function of noise frequency f_n for 2 different insertion delays. We make the following observations: First, we note that for all insertion delay values we have zero jitter at $f_n = 0$ (trivial) and at noise frequencies that are integral multiples of the clock frequency (500 MHz and 1 GHz). This is entirely expected because when this occurs, consecutive clock edges will be delay-modulated identically going through any insertion delay D resulting in zero jitter. Consecutive clock edges will experience the same noise. A second observation is that we have zero jitter at noise frequencies that are integral multiples of the inverse insertion delay (333 MHz and 667 MHz in the top graph for the case where $D = 3$ ns). This is also expected because if the integration interval in the moving average Eq. (6.51) is an integral multiple of the period of the sinusoid that is being integrated, then the integration result is zero. Integrating a sinusoid over an integral multiple of periods always results in zero. This is also clearly observed on the bottom graph of Fig.6.38 where Φ_1' is plotted as a function of insertion delay for 3 different noise frequencies. All zeroes on the graph are at insertion delays equal to an integral multiple of the noise period. A final observation is that with increasing noise frequency, period jitter tends to decrease due to the $1/f_n$ term in Eq. (6.57). This happens because as f_n increases, the amount of noise averaging in delay line D also increases and the supply moving average will approach the nominal value.

Figure 6.39 shows the jitter dependence on noise frequency and insertion delay simultaneously in 2 dimensions for 1 GHz and 500 MHz clocks respectively. Such contour plots for each relevant clock frequency are very easy to construct using Eq. (6.57) and can provide a broad perspective early in the design phase in order to help determine important implementation aspects such as:

1. Identification of worst case period jitter and particularly undesired noise frequencies.
2. Derivation of detailed specifications of supply voltage frequency and amplitude if external to the IC.
3. Identification of need for internal voltage regulation to minimize A_n if external supply specifications turn out to be too stringent.

Application of Eq. (6.57) is not restricted to delay locked loops but can be applied to any open-loop clock buffer in order to determine supply induced period jitter. DLLs are very convenient because insertion delay D is well defined and tracked with feedback. In an open-loop buffer, D will vary with PVT and can only be estimated to a limited degree of accuracy. Yet, the analysis above reveals an interesting aspect of supply induced jitter: In an ideal world, it can be canceled out by selecting an appropriate clock buffer delay D as Figs. 6.38 and 6.39 indicate. This is not really possible because the supply noise frequency is never known a priori and typically does not consist of a single sinusoid. A supply voltage waveform will typically contain multiple natural frequencies introduced by the package, board, and external decoupling capacitance network. In addition, it will contain particular solutions related to the system clock frequency. Natural frequencies are unknown at IC design time, since they are system dependent. Particular solutions are also hard to estimate especially in programmable ICs. One could imagine the possibility of constructing a feedback system which measures supply noise frequency and then adjusts clock buffer delay accordingly to minimize jitter. Feedback must be close to instantaneous in order to avoid the worst case, and this makes a digital controller very hard to implement. Assuming that supply noise frequency can actually be measured, such a system could potentially track an undesired natural frequency that should not have been there in the first place. Controlling noise amplitude, undesired natural frequencies and clock

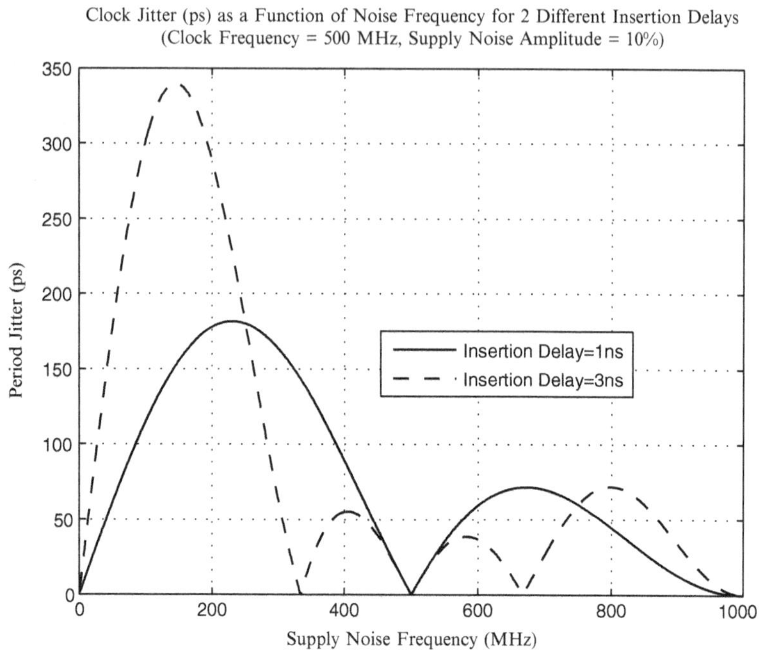

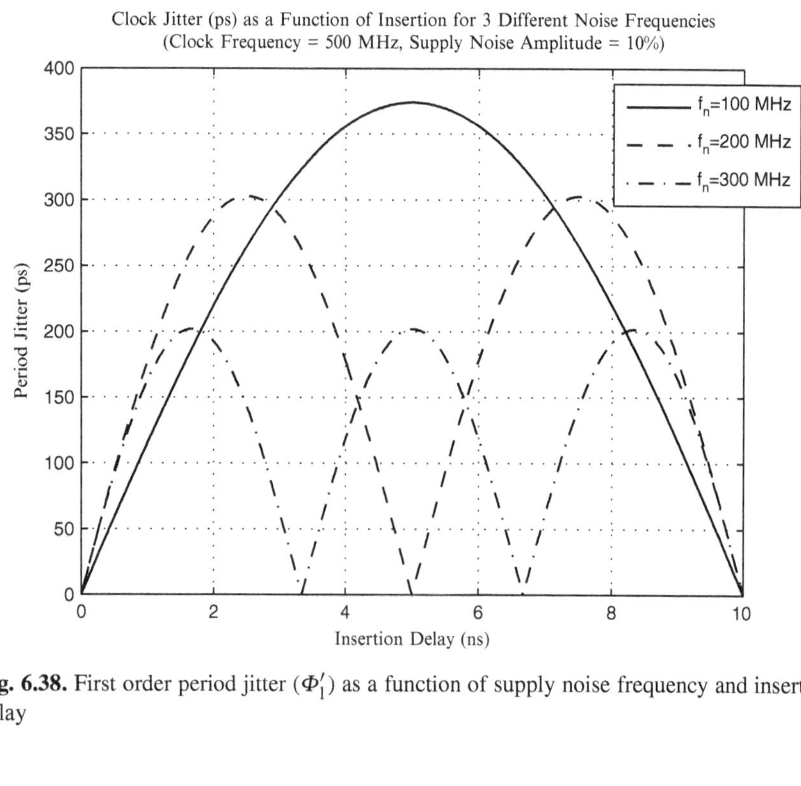

Fig. 6.38. First order period jitter (Φ_1') as a function of supply noise frequency and insertion delay

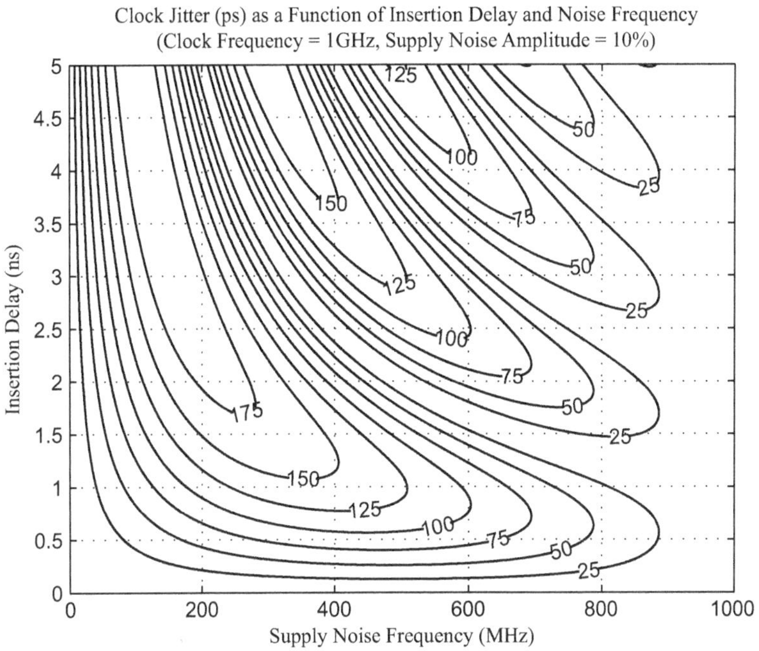

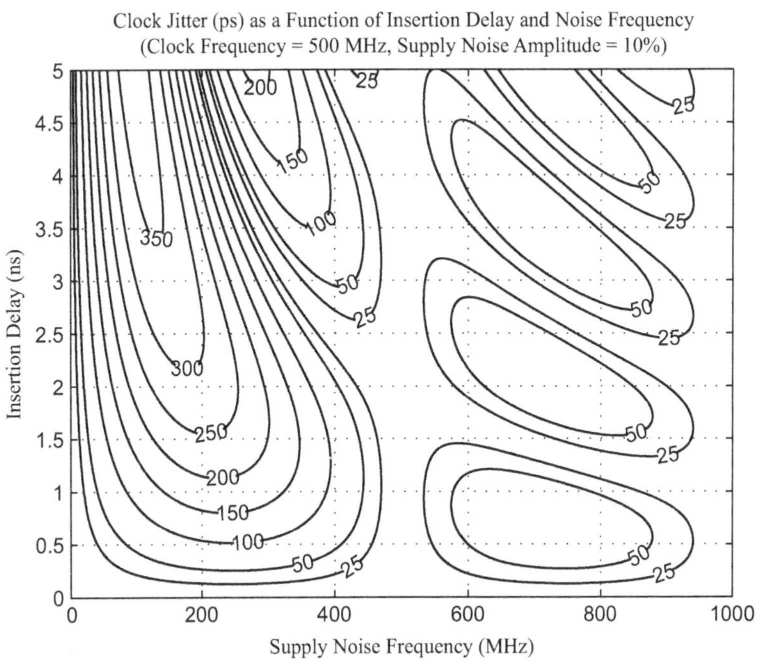

Fig. 6.39. First order period jitter (Φ_1') as a function of insertion delay and supply noise frequency in 2 dimensions

buffer delays are the only effective methods for reducing supply induced jitter that the author is aware of.

6.9 Advanced Applications

In this section, we review a number of non-standard DLL applications and demonstrate the versatility of delay lock techniques.

6.9.1 Duty Cycle Correction

DLLs are heavily used in I/O applications such as DRAM interfaces. In such applications, 50% duty cycle is very important because the interfaces are typically double-data-rate (DDR) and both clock phases are used for signaling. Duty cycle distortion is a source of deterministic jitter and must be reduced.

It is fairly common for DLLs to include a separate stage of duty cycle correction as an output stage [26] or even distributed duty cycle control at various points along the signal path [27]. These are typically analog methods that involve current-steering based structures. An alternative method of duty cycle correction implemented in DRAMs [28, 29] is shown in Fig. 6.40. It uses two independent DLLs. The first one (PD1, DCDL1, CTL1) is the main loop and is responsible for locking the output to the desired CLKIN phase. The second one (PD2, DCDL2, CTL2) has an inverting delay line and locks the positive edge of CLK2 (which has been generated by a negative edge of CLKIN) to the positive edge of CLK1 (which has been generated by a positive edge of CLKIN).

When the second loop locks, CLK1 and CLK2 will have the phase relationship shown in Fig. 6.40. A phase interpolator with equal weights (similar to the one in Fig. 6.26) is used as the final clock output stage. The positive edge of output clock CLKOUT will be generated by the positive edge of CLK1 since both CLK1 and CLK2 have phase locked positive edges. The negative edge of CLKOUT will be generated by the average of the CLK1 and CLK2 negative edges, resulting in duty cycle correction. Figure 6.40 assumes for simplicity that the phase interpolator is a zero-delay, thus non-causal circuit. In reality, there will be delay involved but it has no effect in the overall correction capability of this method. The amount of duty cycle correction possible depends on the *RC* of the interpolator as described in Sect. 6.5.3.

6.9.2 Clock Multiplication

In Sect. 6.3, the basic idea of DLL-based clock multiplication has been presented (Fig. 6.4d). This method of phase-locking a multi-tap DCDL to a reference clock period and then using the intermediate phases to generate a frequency multiple through a clock-multiplying structure such as a multi-port toggle flop has an obvious frequency limitation: it is limited by the minimum delay $D_{\min}$ through each DCDL segment. As an example, the highest output frequency that can be generated by the structure of Fig. 6.4d is subject to the following constraint:

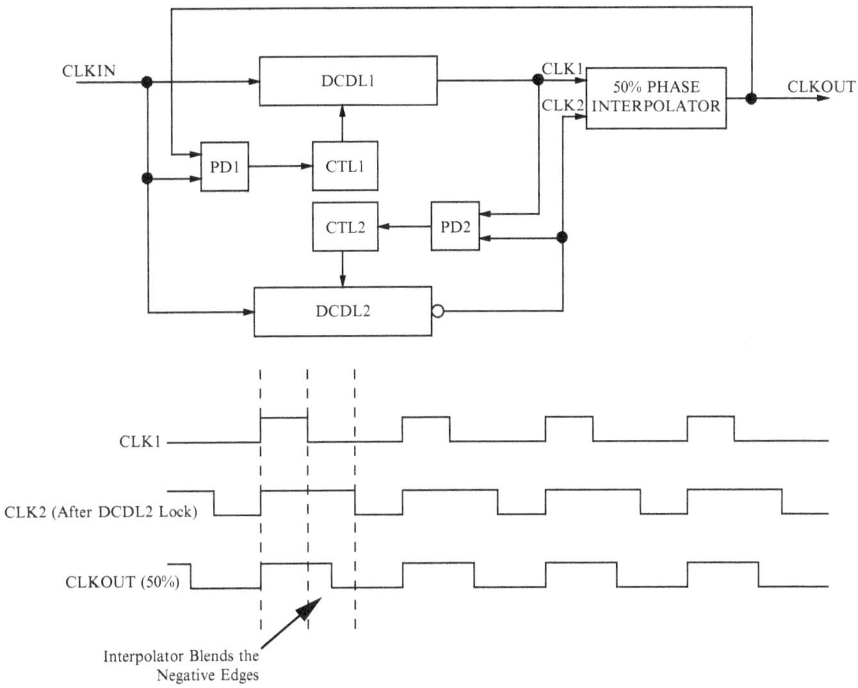

Fig. 6.40. Duty cycle correction in a DLL environment

$$f_{\text{out}} \leq \frac{1}{2D_{\text{min}}}, \tag{6.58}$$

where D_{min} is the minimum possible delay through each of the 8 DCDL segments.

Reference [30] proposes an interesting way to remove this upper bound. The key observation is that the sum of the N DCDLs (where $N/2$ is the intended clock multipler) can be phase-locked to multiple reference clock periods (M) instead of 1 and the resulting phases can be re-sorted before driving the final clock multipler stage. For this method to work, M must not divide N exactly but have a non-zero modulus. Figure 6.41 illustrates this concept. We wish to perform clock multiplication by 4 ($N = 8$), but we want to generate an output clock frequency that is faster than the constraint in (6.58) would allow for. Instead of phase-locking all 8 DCDL segments to one reference clock period T_{ref}, we choose to phase-lock them to $3T_{\text{ref}}$ ($M = 3$) which evidently allows for a much larger D_{min} per DCDL segment. Phases are sorted modulus T_{ref} before being routed to the final clock multiplication stage. The constraint in (6.58) now becomes

$$f_{\text{out}} \leq \frac{3}{2D_{\text{min}}}, \tag{6.59}$$

and the upper bound for the final output frequency has become more loose by a factor of 3. In [30], a 40 GHz output clock is generated by a 4.4445 GHz reference ($M = 2$,

$N = 9$) by employing an LC-based clock multiplier which can produce a full 360° oscillation per stimulating phase.

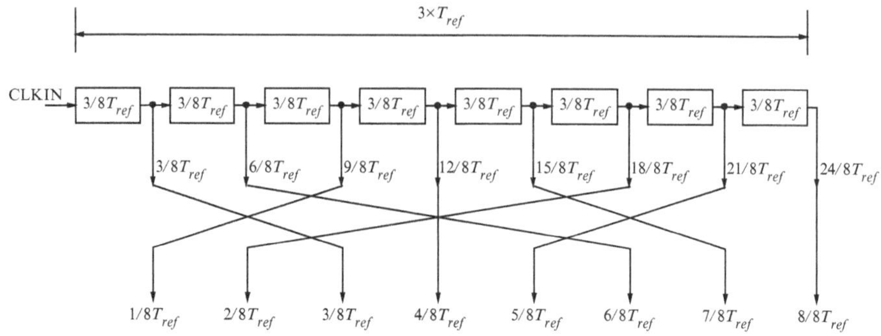

Fig. 6.41. Multiperiod DCDL locking for high output frequency clock multiplication

Clock multiplying DLLs with alternative organizations have been proposed in the past years [31, 32] and are actively pursued as a credible alternative to PLLs for clock generation due to superior jitter performance.

6.9.3 Infinite Dynamic Range

Lock acquisition and phase capture can be limited in standard DLLs with a reduced DCDL dynamic range as explained in Sects. 6.5 and 6.6. One way to address this issue is to extend the delay line dynamic range through additional DCDL stages or conditional inversion (Sect. 6.6.2) coupled with increased complexity in the FSM controller. An alternative method is to employ a dual DLL architecture where the core loop generates phases that span the input clock period range, and the peripheral loop blends a selected pair of consecutive phases to match the reference clock [1, 33]. This architecture is shown in Fig.6.42. The core DLL phase-locks at 180° and generates 6 phases that are 30° apart. The block labeled "Phase Selection" contains two 3×1 multiplexers that select a pair of phases ϕ and ψ to be interpolated. Before reaching the interpolator, these phases are conditionally inverted inside the "Selective Phase Inversion" block so that the cascade of these three structures can generate all phases at 30° increments that span the entire 0–360° phase interval. Finally, conditionally inverted phases ϕ' and ψ' are interpolated. Although the phase interpolator structure in [33] is CML-based, the analysis is very similar to the full swing interpolator of Sect. 6.5.3. The output of the interpolator Θ is phase-locked to the reference clock ("ref CLK" in Fig.6.42). The feedback of the peripheral DLL closes with a bang–bang phase detector and an FSM that controls the multiplexers, conditional inversion block and phase interpolator. The infinite dynamic range stems from the fact that the FSM controller can keep on selecting phase pairs to be interpolated around the entire 360° phase interval and eventually the selected pair will encompass the phase of the reference clock. Interpolation will then guarantee phase lock within the accuracy of the interpolator. In this architecture, the input core DLL clock ("in CLK") and the reference clock ("ref CLK") can be the same clock at 0-phase or identical frequency clocks at any phase relationship. They can even be plesiochronous clocks (of similar but not equal frequency), and this case will be addressed in Sect. 6.9.4.

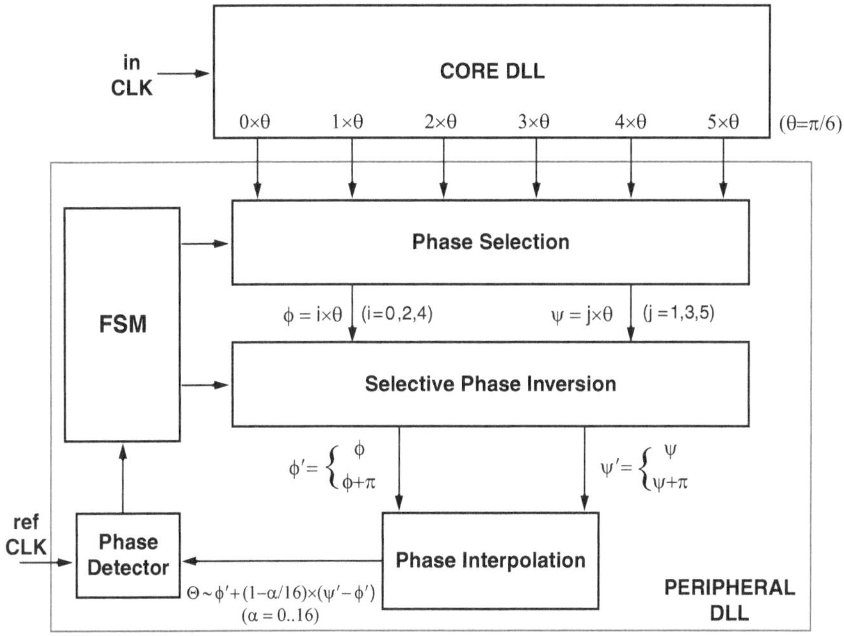

Fig. 6.42. Dual DLL architecture for virtually infinite phase capture range. Reproduced with permission from [33], ©1997 IEEE

DLLs with quadrature phase mixing had been proposed before [26] and they also exhibit infinite dynamic range. The main difference between the dual DLL architecture and quadrature mixing is that the phase interpolation is not limited to 90° clocks. Interpolation can blend phases at smaller intervals (30° in this case) which will have fewer slew rate limitations and better noise performance [33].

6.9.4 Clock-Data Recovery

The full power of the method of Fig. 6.42 cannot be fully appreciated unless we consider the case where "in CLK" and "ref CLK" are plesiochronous and vary slightly in frequency. This architecture can achieve lock in the plesiochronous case because the FSM controller can change the weight of the phase mixing in each cycle in such a way so as to generate a clock that is effectively faster or slower than "in CLK".

When the interpolator reaches the end of its dynamic range, then the phase selection multiplexer chooses a different pair of phases for mixing that are either leading or lagging the previous pair depending on whether "ref CLK" is faster or slower than "in CLK", respectively. This property makes this method ideally suited for clock-data recovery (CDR) applications where a sampling clock must be generated at the receiver and used to sample the incoming data. The generated clock must match the frequency of the received data using multiple free-running locally-generated phases.

This concept is illustrated in Fig.6.43. This structure forms the basis of a large class of clock-data recovery architectures implemented as part of various high-speed signaling standards such as PCI Express, XAUI, and FBD (Fully-Buffered-DIMM) [34]. The core DLL of Fig.6.42 has been replaced with a multi-phase PLL that generates a local high frequency clock which can be plesiochronous to the incoming data. In order to collect enough information for clock recovery, a CDR needs some form of data oversampling. In this case, oversampling is accomplished by generating both an in-phase clock (CLKI) and a quadrature clock (CLKQ). High-speed incoming data are sampled with both clocks using two separate slicers (I-SLICER, Q-SLICER). The data are then deserialized typically by a factor of 10 and retimed in a divided clock domain which is usually a divided-down version of CLKQ. The block labeled "EDGE DETECTION/PHASE DETECTION" examines both I and Q deserialized data and determines whether the current sampling clocks CLKI and CLKQ are fast or slow with respect to the incoming data. This decision can be made deterministically due to oversampling. This decision is then forwarded to the CDR FSM which implements a control algorithm and can change the settings of the phase selectors and interpolators to advance or delay the sampling clocks CLKI and CLKQ. The FSM algorithm will determine important capabilities such as phase capture profile, frequency difference between local PLL and incoming data that can be safely absorbed and limit cycle behavior which will cause deterministic jitter in the system. The output of the Q-DESERIALIZER represents parallel data sampled safely with maximum margin in the middle of the eye diagram and is forwarded to the rest of the system for higher layer processing.

All blocks of Fig.6.43 with the sole exception of the multi-phase PLL constitute essentially a DLL. The role of the delay line is fulfilled by the combination of phase selectors and interpolators that can effectively delay or advance an input clock and the phase detection is carried out by the slicers, deserializers, and decision block. In a typical SERDES application [34], this DLL structure is present on all receiver bits (*lanes* in SERDES terminology) whereas the multi-phase PLL is amortized over a number of receive and transmit lanes. In this application, it is desirable to implement a second order control structure in order to track more effectively the frequency differences between the incoming data and the locally generated clock. Second order control design has not been addressed in this chapter and is beyond our scope.

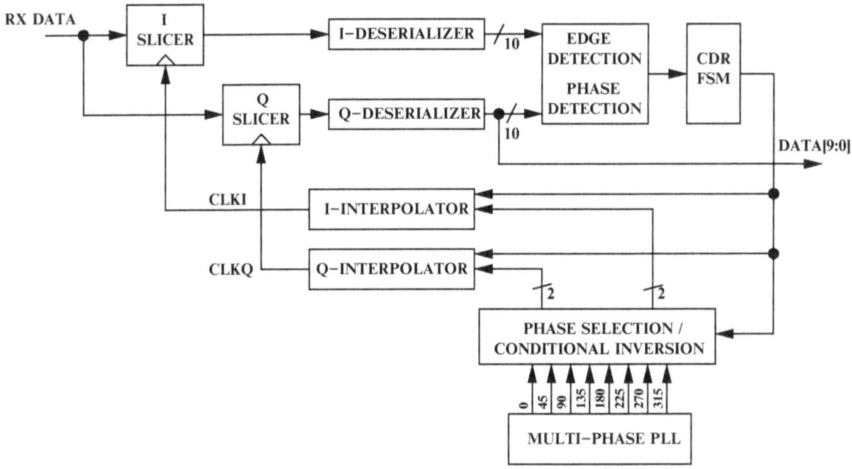

Fig. 6.43. DLL-based clock-data recovery loop

6.9.5 On-Chip Temperature Sensing

An appropriately calibrated and accurately measured delay line can be used as a temperature sensor since temperature affects digital gate delay. It has been shown [35] that when a delay line is trimmed to a fixed delay at a certain temperature in a production environment across multiple chips, it can act as an accurate temperature sensor. Trimming to a fixed delay at production has the effect of suppressing sensitivity to process and voltage. A trimmed delay line will vary only due to temperature across all fused chips and can have an accuracy of $\pm 2^{\circ}$C across the 0–100°C temperature range [35].

Two DLLs constitute this system: A reference DLL is phase locked to a known reference clock which is distributed to the sensor. Tapping at intermediate delay line elements can generate fixed reference delays that are fractions of the original reference clock period. A secondary DLL phase locks the sensing delay line to a fixed delay generated by tapping an intermediate delay output of the reference DLL. This step must be performed at a controlled junction temperature on all chips during manufacturing. The setting in the locked state is then fused and this step performs the trimming (normalization) of the sensing delay line.

During measurement, the temperature can vary and the trimmed sensing line will deviate from its fixed delay value. The secondary DLL then treats it as an unknown delay and measures it using intermediate delay outputs of the reference DLL as the measuring delay line. The digital setting output at the locked state will be an accurate representation of temperature. This dual DLL scheme can provide good accuracy and a high measurement bandwidth (5 ksamples/s) and is appropriate for microprocessor thermal monitoring.

6.10 Conclusion

This chapter has presented an overview of the basic digital DLL structure and its components in addition to a survey of a range of applications. DLLs are an essential component of modern VLSI systems. Increased clocking system complexity, process variation, and wider supply voltage ranges will make their use even more prevalent than it is today. Furthermore, their time-to-digital conversion capabilities make them an attractive candidate for measuring analog on-chip quantities (i.e., voltage drop, temperature, process variation) either for valuable data collection or system normalization to dynamically remove variation.

Acknowledgments

A preliminary version of the power supply noise analytical model in Sect. 6.8 was developed jointly with John Eble when we were colleagues at the Alpha Development Group, Compaq Computer.

References

[1] W. Dally and J. Poulton, *Digital Systems Engineering*. Cambridge University Press, New York, NY, 1998.
[2] M. Johnson and E. Hudson, A variable delay line PLL for CPU-coprocessor synchronization. *IEEE J. Solid-State Circuits*, 23(5), 1218–1223, 1988.
[3] C. Mead and L. Conway, *Introduction to VLSI Systems*. Addison Wesley, Reading, MA, 1980.
[4] L. Lamport, Buridan's principle, October 1984. [Online]. Available: http://research. microsoft.com/users/lamport/pubs/buridan.pdf
[5] T. Chaney and C. Molnar, Anomalous behavior of synchronizer and arbiter circuits. *IEEE Trans. Comput.*, C-22(4), 421–422, 1973.
[6] H. Veendrick, The behaviour of flip-flops used as synchronizers and prediction of their failure rate. *IEEE J. Solid-State Circuits*, 15(2), 169–176, 1980.
[7] A. Drake, *Fundamentals of Applied Probability Theory*. McGraw-Hill, New York, 1967.
[8] Nanoscale Integration and Modeling Group, Arizona State University, Predictive technology model. [Online]. Available: http://www.eas.asu.edu/ptm.
[9] Y. Cao, T. Sato, M. Orshansky, D. Sylvester, and C. Hu, New paradigm of predictive MOSFET and interconnect modeling for early circuit simulation. In: *Proc. CICC Custom Integrated Circuits Conference*, 2000, pp. 201–204.
[10] W. Zhao and Y. Cao, New generation of predictive technology model for sub-45 nm early design exploration. *IEEE Trans. Electron Devices*, 53(11), 2816–2823, 2006.
[11] A. Hatakeyama, H. Mochizuki, T. Aikawa, M. Takita, Y. Ishii, H. Tsuboi, S. Fujioka, S. Yamaguchi, M. Koga, Y. Serizawa, K. Nishimura, K. Kawabata, Y. Okajima, M. Kawano, H. Kojima, K. Mizutani, T. Anezaki, M. Hasegawa,

and M. Taguchi, A 256-Mb SDRAM using a register-controlled digital DLL. *IEEE J. Solid-State Circuits*, 32(11), 1728–1734, 1997.

[12] F. Lin, J. Miller, A. Schoenfeld, M. Ma, and R. Baker, A register-controlled symmetrical DLL for double-data-rate DRAM. *IEEE J. Solid-State Circuits*, 34(4), 565–568, 1999.

[13] T. Xanthopoulos, D. Bailey, A. Gangwar, M. Gowan, A. Jain, and B. Prewitt, The design and analysis of the clock distribution network for a 1.2 GHz Alpha microprocessor. In: *Digest of Technical Papers IEEE International Solid-State Circuits Conference (ISSCC 2001)*, 2001, pp. 402–403.

[14] S. Sidiropoulos, High performance inter-chip signalling, Ph.D. dissertation, Stanford University, 1998.

[15] J. Maneatis, Design of high-speed CMOS PLLs and DLLs. In: A. Chandrakasan, W. Bowhill, and F. Fox (eds.) *Design of High Performance Microprocessor Circuits*, IEEE Press, New York, 2001, pp. 235–260.

[16] C.-K. K. Yang, Delay-locked loops – an overview. In: B. Razavi, (ed.) *Phase-Locking in High-Peformance Systems*, Wiley-IEEE Press, New York, NY, 2003, pp. 13–22.

[17] M.-J. Lee, W. Dally, T. Greer, H.-T. Ng, R. Farjad-Rad, J. Poulton, and R. Senthinathan, Jitter transfer characteristics of delay-locked loops – theories and design techniques. *IEEE J. Solid-State Circuits*, 38(4), 614–621, 2003.

[18] J. Burnham, G.-Y. Wei, C.-K. K. Yang, and H. Hindi, A comprehensive phase-transfer model for delay-locked loops. In: *Proc. IEEE Custom Integrated Circuits Conference CICC '07*, 2007, pp. 627–630.

[19] J. Lee, K. Kundert, and B. Razavi, Analysis and modeling of bang-bang clock and data recovery circuits. *IEEE J. Solid-State Circuits*, 39(9), 1571–1580, 2004.

[20] J. Sonntag and J. Stonick, A digital clock and data recovery architecture for multi-gigabit/s binary links. *IEEE J. Solid-State Circuits*, 41(8), 1867–1875, 2006.

[21] G.-K. Dehng, J.-M. Hsu, C.-Y. Yang, and S.-I. Liu, Clock-deskew buffer using a SAR-controlled delay-locked loop. *IEEE J. Solid-State Circuits*, 35(8), 1128–1136, 2000.

[22] T. Saeki, Y. Nakaoka, M. Fujita, A. Tanaka, K. Nagata, K. Sakakibara, T. Matano, Y. Hoshino, K. Miyano, S. Isa, S. Nakazawa, E. Kakehashi, J. Drynan, M. Komuro, T. Fukase, H. Iwasaki, M. Takenaka, J. Sekine, M. Igeta, N. Nakanishi, T. Itani, I. Yoshida, K. Yoshino, S. Hashimoto, T. Yoshii, M. Ichinose, T. Imura, M. Uziie, S. Kikuchi, K. Koyama, Y. Fukuzo, and T. Okuda, A 2.5-ns clock access, 250-MHz, 256-Mb SDRAM with synchronous mirror delay. *IEEE J. Solid-State Circuits*, 31(11), 1656–1668, 1996.

[23] K. Sung and L.-S. Kim, A high-resolution synchronous mirror delay using successive approximation register. *IEEE J. Solid-State Circuits*, 39(11), 1997–2004, 2004.

[24] A. Efendovich, Y. Afek, C. Sella, and Z. Bikowsky, Multifrequency zero-jitter delay-locked loop. *IEEE J. Solid-State Circuits*, 29(1), 67–70, 1994.

[25] B. Mesgarzadeh, Low-power low-jitter clock generation and distribution, Ph.D. dissertation, Linkoping University, 2008.

[26] T. Lee, K. Donnelly, J. Ho, J. Zerbe, M. Johnson, and T. Ishikawa, A 2.5 V CMOS delay-locked loop for 18 Mbit, 500 megabyte/s DRAM. *IEEE J. Solid-State Circuits*, 29(12), 1491–1496, 1994.

[27] B. Garlepp, K. Donnelly, J. Kim, P. Chau, J. Zerbe, C. Huang, C. Tran, C. Portmann, D. Stark, Y.-F. Chan, T. Lee, and M. Horowitz, A portable digital DLL for high-speed CMOS interface circuits. *IEEE J. Solid-State Circuits*, 34(5), 632–644, 1999.

[28] J.-B. Lee, K.-H. Kim, C. Yoo, S. Lee, O.-G. Na, C.-Y. Lee, H.-Y. Song, J.-S. Lee, Z.-H. Lee, K.-W. Yeom, H.-J. Chung, I.-W. Seo, M.-S. Chae, Y.-H. Choi, and S.-I. Cho, Digitally-controlled DLL and I/O circuits for 500 Mb/s/pin x16 DDR SDRAM. In: *Digest of Technical Papers IEEE International Solid-State Circuits Conference (ISSCC 2001)*, 2001, pp. 68–69, 431.

[29] W.-J. Yun, H. W. Lee, D. Shin, S. D. Kang, J. Y. Yang, H. O. Lee, D. U. Lee, S. Sim, Y. J. Kim, W. J. Choi, K. S. Song, S. H. Shin, H. H. Choi, H. W. Moon, S. W. Kwack, J. W. Lee, Y. K. Choi, N. K. Park, K. W. Kim, Y. J. Choi, J.-H. Ahn, and Y. S. Yang, A 0.1-to-1.5GHz 4.2mW all-digital DLL with dual duty-cycle correction circuit and update gear circuit for DRAM in 66nm CMOS technology. In: *Digest of Technical Papers IEEE International Solid-State Circuits Conference (ISSCC 2008)*, 2008, pp. 282–283, 613.

[30] C.-N. Chuang and S. luan Liu, A 40GHz DLL-based clock generator in 90nm CMOS technology. In: *Digest of Technical Papers IEEE International Solid-State Circuits Conference (ISSCC 2007)*, 2007, pp. 178–179, 595.

[31] R. Farjad-Rad, W. Dally, H.-T. Ng, R. Senthinathan, M.-J. Lee, R. Rathi, and J. Poulton, A low-power multiplying DLL for low-jitter multigigahertz clock generation in highly integrated digital chips. *IEEE J. Solid-State Circuits*, 37(12), 1804–1812, 2002.

[32] B. Helal, M. Straayer, G.-Y. Wei, and M. Perrott, A highly digital MDLL-based clock multiplier that leverages a self-scrambling time-to-digital converter to achieve subpicosecond jitter performance. *IEEE J. Solid-State Circuits*, 43(4), 855–863, 2008.

[33] S. Sidiropoulos and M. Horowitz, A semidigital dual delay-locked loop. *IEEE J. Solid-State Circuits*, 32(11), 1683–1692, 1997.

[34] D. Stauffer et al., *High Speed Serdes Devices and Applications*. Springer Science and Business Media, New York, 2008.

[35] K. Woo, S. Meninger, T. Xanthopoulos, E. Crain, and D. Ham, Dual-DLL-based CMOS all-digital temperature sensor for microprocessor thermal monitoring. In: *Digest of Technical Papers IEEE International Solid-State Circuits Conference (ISSCC 2009)*, 2009, pp. 68–69.

7

Clocking and Variation

James Tschanz

Intel Corporation

7.1 Introduction

As process technology continues to scale, increasing numbers of transistors are integrated onto a single processor die, providing higher levels of performance and additional features (Fig. 1.1). However, the unwanted side-effect of this increased integration is the worsening of variations of all types: static process variations resulting from reduced device dimensions and dynamic voltage and temperature variations. At the same time, transistor degradation and early-life failure are always concerns, but becoming more critical as the number of devices increases.

The clock network on a microprocessor plays a special role in how the processor tolerates variations of all types. On one hand, the clock is very sensitive to variations, and any fluctuations in the clock due to variations can directly impact the frequency of the processor. Thus, it is absolutely critical to design the clock network in such a way that the impact of variations is reduced. On the other hand, variations also present an opportunity in clock network design. Because of the global nature of the clock, it can be used as a "knob" for tolerating variations or reducing their impacts. In this chapter, we will discuss both challenges and opportunities that variations present to clock network design.

7.2 Variation Reduction Through Design

The first step in reducing the impacts of variations on the clock network is to design the clocking system in such a way that it is tolerant of variations. This entails properly selecting the clock network topology, driving the clock with a high-quality clock source, and designing and sizing the clock wires and buffers to reduce skew and jitter. These techniques and tradeoffs have already been discussed in Chaps. 2 and 5 and will also be addressed from a physical design perspective in Chap. 8. However, variation-tolerant techniques can also be employed in the datapath to reduce the impacts of variations in both the clock network and the datapath. In this

T. Xanthopoulos (ed.), *Clocking in Modern VLSI Systems*, Integrated Circuits and Systems, DOI 10.1007/978-1-4419-0261-0_7, © Springer Science+Business Media, LLC 2009 chapter, we describe two techniques which employ latches and time-borrowing flip-flops as a method of hiding some of the impact of clock uncertainty and random device variation.

7.2.1 Skew and Jitter-Tolerant Design

One method for reducing the impact of clock skew and jitter on the maximum frequency (F_{MAX}) of a microprocessor is to employ two-phase latches rather than flip-flops in the critical path [1–3]. This technique has also been discussed in Sect. 3.3.2. Alternately, for speed-critical dynamic circuits, skew-tolerant domino gates can be used with overlapping clocks such that the clock edge never gates the latest-arriving data input [4]. Using latches and dynamic gates in this way provides several important benefits which translate to higher achievable clock frequency. First, this scheme allows time-borrowing: additional slack in one pipeline stage can be taken advantage of in the next pipeline stage so that ideally all stages become equally critical. This is an important feature for designs in which it is not naturally easy to divide the logic into equal-delay stages: rather than one stage becoming the limiter, the logic can be averaged over multiple stages. Secondly, when the data arrive during the transparency window of a latch or dynamic gate, fluctuations in clock edge timing due to clock skew or jitter do not affect the path delay. Therefore, if the transparency window is sufficient, clock skew and jitter do not impact the cycle time of the processor. Of course, proper selection of the transparency window is necessary in order to balance the amount of time-borrowing desired, the variation-tolerance needed, and the power and min-delay overhead for this technique.

Because many designs are flip-flop based and it is sometimes difficult to break up circuits into two phases for a latch-based design, time-borrowing flip-flops are also used [1, 3]. Time-borrowing flip-flops (Fig. 7.1) include a transparency window after the rising clock edge such that data arriving in this window are able to propagate through the flip-flop, similar to a latch. Time-borrowing flip-flops require additional clock buffers inside the sequential in order to generate the transparency window; therefore, they have increased power consumption in comparison to a standard non-time-borrowing design; however, this transparency window allows the usage of time-borrowing and tolerance to clock skew and jitter. Additional detailed discussion on time-borrowing flops has been presented in Sect. 3.3.1. Two-phase latches, skew-tolerant domino, and time-borrowing flip-flops are all used commonly in microprocessor circuits to reduce the impacts of clock skew and jitter.

7.2.2 Time Borrowing for Datapath Variation Reduction

The previous section described the usage of transparent latches and time-borrowing flip-flops in reducing the effects of clock skew and jitter, providing tolerance to variations in the clocking network of a microprocessor. In this section, we describe how time-borrowing flip-flops may also be used to provide tolerance to datapath variations. While this technique may be applied for any type of datapath block, we focus in this section on long-distance on-chip interconnects, since the differences in

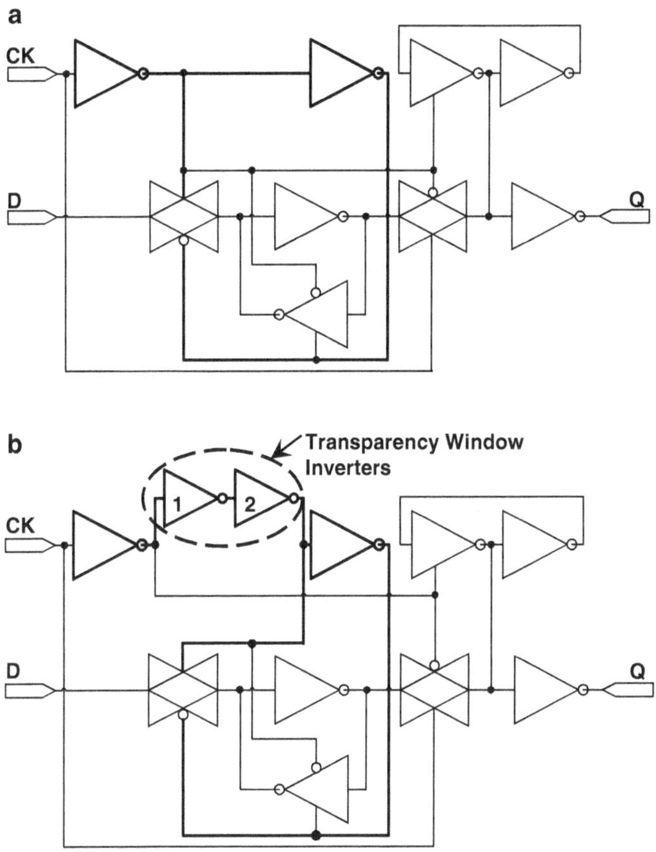

Fig. 7.1. (**a**) Standard master-slave flip-flop. (**b**) Creation of a time-borrowing flip-flop by inserting transparency window inverters

max-delay and min-delay for these interconnects are typically much smaller than those in logic blocks, allowing a larger transparency window in the flip-flop without causing min-delay failures. In microprocessor designs, many signals are routed over long repeater-based interconnects, requiring that these signals are propagated across multiple clock cycles. By replacing the flip-flops in these interconnects with time-borrowing flip-flops, delay variations across multiple segments can be averaged, mitigating the impact of systematic and random within-die (WID) variations on mean F_{MAX}. Moreover, the absorption of clock skew and jitter by the transparency windows provides additional slack in the nominal cycle time in comparison with a multi-cycle interconnect with conventional master-slave flip-flops.

Figure 7.2 shows a schematic for a multi-cycle, buffered interconnect using standard non-time-borrowing (NTB) and time-borrowing (TB) flip-flops [5]. The N-cycle interconnect consists of N repeater-based interconnect segments separated by flip-flops, where $t_{DATA(i)}$ denotes the delay of the ith interconnect segment, and

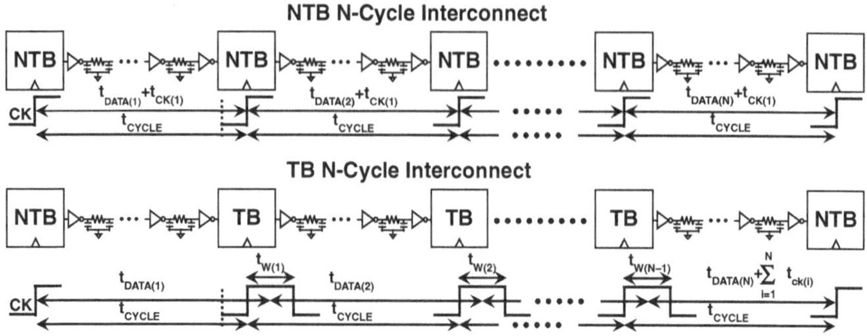

Fig. 7.2. NTB and TB N-cycle interconnects, where t_{CYCLE} is the cycle time, $t_{DATA(i)}$ is the ith interconnect segment delay, $t_{W(i)}$ is i^{th} transparency window, $t_{CK(1)}$ is single-cycle clock skew and jitter, and $t_{CK(i)}$ is additional long-term jitter component for i^{th} cycle ($i > 1$). Reproduced with permission from [5], ©2006 IEEE

$t_{CK(1)}$ represents the single-cycle clock skew and jitter. As shown in Fig. 7.2, an NTB flip-flop enforces a hard clock edge between the interconnect segments, where data are allowed to pass through the flip-flop only when the data arrive before the setup time. For an NTB interconnect segment, $t_{DATA(i)}$ equals the sum of the flip-flop clock-to-output delay, the delay through the repeater segments and the protection inverter, and the setup time for the receiving flip-flop. The cycle time of one NTB interconnect segment is the sum of $t_{DATA(i)}$ and $t_{CK(1)}$. For an N-cycle interconnect, the slowest interconnect segment limits the cycle time, resulting in more severe degradation in mean F_{MAX} as N increases.

The TB multi-cycle interconnect is constructed by replacing the flip-flops of the intermediate repeater segments with time-borrowing flip-flops. If the bus segments can be optimized such that the data nominally arrive within the time-borrowing window of the flip-flops, variations in datapath delay arising from process, voltage, or temperature variations along the interconnect can be averaged across multiple segments, and their impact on F_{MAX} will be reduced. As shown in Fig. 7.2, the cycle time for a TB multi-cycle interconnect, assuming all signals arrive within the time-borrowing window, is based on the average of the segment delays, while total clock skew and jitter (consisting of single-cycle skew and jitter $t_{CK(1)}$ plus the additional long-term jitter component $t_{CK(i)}$ in each additional cycle) can be amortized over multiple cycles. Figure 7.3a shows an example nominal timing diagram for a TB 2-cycle interconnect with zero clock skew and jitter. Ideal data delay averaging occurs in a multi-cycle interconnect where t_{CYCLE} equals the average of the interconnect segment delays, which is the best-case scenario for delay variation tolerance. The amount of delay variation averaging which occurs, then, depends on the delay of the two interconnect segments as well as the size of the time-borrowing window chosen. If $\delta_{DATA(i)}$ denotes the delay deviation from nominal conditions for the ith interconnect segment, the "delay variation mismatch" between the first and second interconnect segment is $\delta_{DATA(1)} - \delta_{DATA(2)}$. For ideal data delay averaging, the

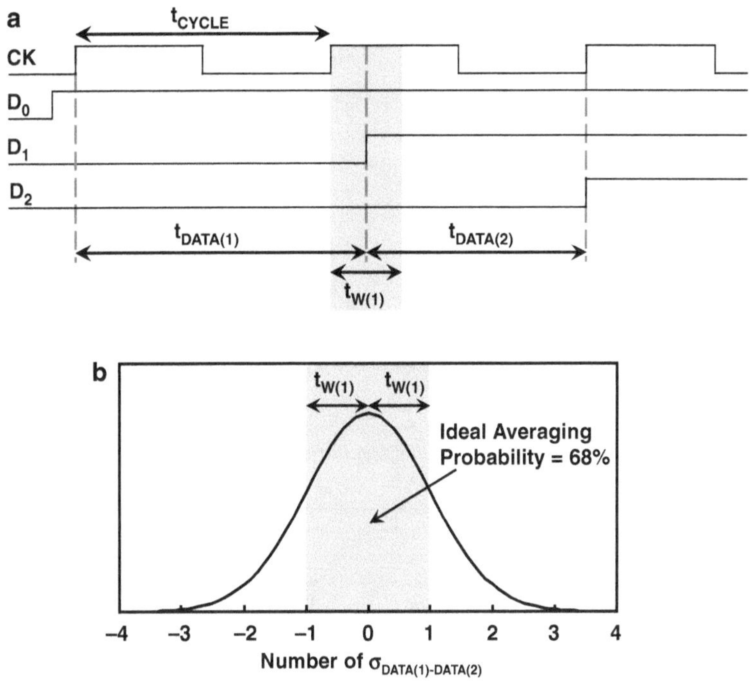

Fig. 7.3. (**a**) Nominal timing for a TB 2-cycle interconnect with zero clock skew and jitter, where D_0, D_1, and D_2 are input data for driving, intermediate, and receiving flip-flops, respectively. (**b**) Probability density versus number of standard deviations in $\delta_{DATA(1)} - \delta_{DATA(2)}$ ($\sigma_{DATA(1)-DATA(2)}$), where $\delta_{DATA(i)}$ is the 0-mean delay deviation for the ith interconnect segment. Reproduced with permission from [5], ©2006 IEEE

time-borrowing window must be at least as large as the difference in segment delay variation mismatches, $\delta_{DATA(1)} - \delta_{DATA(2)}$. As shown in Fig. 7.3b, the time-borrowing window size ($T_{W(1)}$) can be expressed in terms of the standard deviation of the difference between the two segment delays: $\sigma_{DATA(1)-DATA(2)}$. Choosing the TB window as 1σ ensures that 68% of the time, ideal delay averaging occurs. Choosing a larger TB window provides increased variation tolerance for the TB interconnect, at the cost of increased flip-flop clocking energy.

Simulation results for this technique are described in Fig. 7.4. The design of an NTB interconnect segment in M4 is optimized to provide the minimum energy for a nominal cycle time of 250 ps in a 65 nm CMOS technology [6] with a V_{DD} of 1.2V and a temperature of 110C, resulting in 6 repeaters, a protection inverter, and a flop-to-flop interconnect segment length of 1,500 μm. Starting with the optimal NTB interconnect segment, NTB flip-flops are replaced with TB flip-flops for the intermediate segments of a multi-cycle TB interconnect. Since clock skew and jitter are absorbed by the transparency window, additional slack is available, and TB flip-flop and repeater transistors are downsized until the nominal segment delay

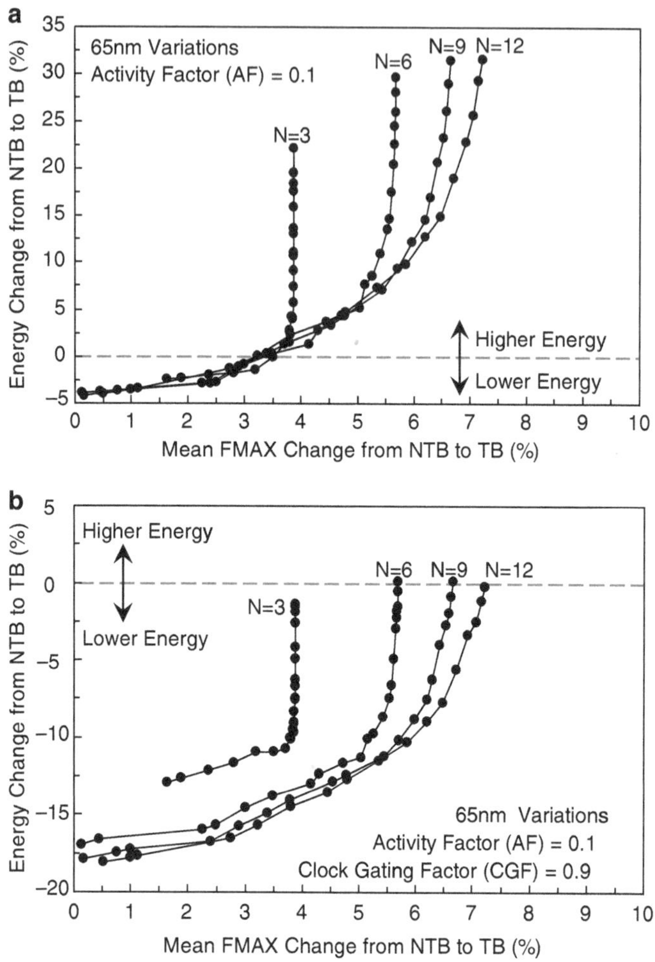

Fig. 7.4. Active (**a**) and average (**b**) energies per cycle vs. mean F_{MAX} change from NTB to TB N-cycle interconnect. Reproduced with permission from [5], ©2006 IEEE

equals 250ps. Monte-Carlo simulations are performed to determine the F_{MAX} distribution for these interconnect designs, including WID transistor, interconnect, and voltage variations, and the effects of transparency windows, clock skew, and jitter. In Fig.7.4, mean F_{MAX}, active energy, and average energy of multi-cycle NTB and TB interconnects are compared for a range of cycles, representing present (N=3&6) and future (N=9&12) interconnect designs. The mean F_{MAX} of TB interconnects is 3.5% higher than NTB interconnects at equal active energy. Additional mean F_{MAX} gains can be obtained by expanding the transparency window at the expense of higher clocking energy until averaging of delay variations approaches the ideal case. Since the impact of additional clocking energy in TB flip-flops is mitigated by clock gating, TB interconnects enable 4–6% mean F_{MAX} gain with a corresponding 10% average

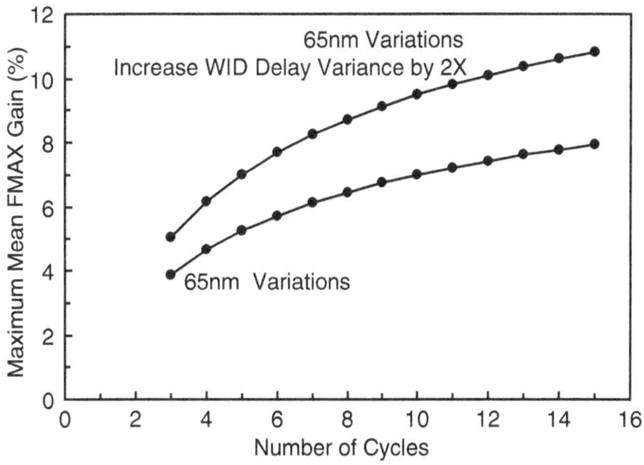

Fig. 7.5. Maximum mean F_{MAX} gain from NTB to TB N-cycle interconnect vs. N. Reproduced with permission from [5], ©2006 IEEE

energy savings as shown in Fig. 7.4b. As process technology scales, WID variations become worse, especially for larger dice and longer interconnects. Figure 7.5 shows that if WID delay variance increases by 2X, the maximum mean F_{MAX} gain rises to 5–10%. Thus, the usage of time-borrowing flip-flops can absorb this increased WID variation in the datapath, and at the same time reduce the impact of clock skew and jitter.

7.3 Variation Reduction Through Tuning

The previous section described techniques that can be employed in the design phase for reducing the impacts of clock uncertainty and process variations on the performance of a microprocessor. While these techniques ideally can hide some of the effects of clock uncertainty, they do not eliminate clock skew and jitter, and cannot adjust to changing dynamic conditions such as supply voltage and temperature fluctuations. To further improve processor performance and energy-efficiency, tuning can be applied post-fabrication, on a per-die or per-lot basis, to compensate variations. This tuning may be performed only on the clock network, to reduce clock skew and jitter, or may be applied to the entire processor to dynamically adapt to voltage and temperature variations.

In this chapter, we describe three types of post-silicon tuning. The first – manufacturing techniques – encompasses all static techniques which can be applied after fabrication to compensate the effects of process variations. These techniques are applied once, and settings do not change over the lifetime of the part. In contrast to this, the other two techniques described in this chapter are dynamic in nature. Active clock deskew is used to dynamically adjust the timing of the clock network to reduce clock skew as the environmental conditions change, and dynamic frequency adjusts the frequency of the entire clock network as a method of responding to dynamic voltage and temperature changes. Both of these techniques allow the processor to respond to operational conditions rather than assuming a worst-case environment and, therefore, improve performance at lower power consumption.

7.3.1 Manufacturing Techniques

Variations that are static in nature can be compensated through a one-time tuning process which is done after fabrication, during the die testing phase. This is typically done for critical analog circuits which are often more sensitive to process variations than the digital circuits in the microprocessor core. Examples of this type of post-silicon tuning include calibration of thermal sensors and I/O receivers, and setting of ideal PLL VCO frequency to minimize PLL random jitter. Once the ideal settings have been determined, they must be permanently programmed into the chip such that they become the default settings each time the microprocessor is powered up. This is typically accomplished through the use of on-chip nonvolatile storage elements such as fuses.

Another way in which post-silicon tuning is used to compensate variations is in the specification of the supply voltage of each processor. Many modern microprocessors include a "voltage identification" (VID) code which allows the processor to specify its supply voltage to the voltage regulator in the system [7]. This VID bus is used, for example, to allow the processor to reduce its supply voltage and frequency during low-power states and then transition back to maximum supply voltage during periods of active computation. However, this feature can also be used to compensate variations by assigning a maximum supply voltage to each processor at test time [8, 9]. Because of process and manufacturing variations, fabricated dice exhibit a range of maximum operating frequencies and standby leakage powers. Those dice that are slow but low-leakage can operate at a higher voltage without violating the maximum power specification, while dice that are fast but high-leakage must operate at a lower voltage. Each die, therefore, has an optimum maximum supply voltage assigned to it that maximizes the processor frequency while ensuring that the power limit is met. This optimum VID setting is fused at test time and becomes the default supply voltage for the part. Modern voltage regulators allow very precise setting of processor supply voltage at a resolution of several millivolts, enabling improvement in microprocessor speed binning and maximizing the number of dice which meet both minimum frequency and maximum power requirements.

7.3.2 Active Clock Deskew

High-performance microprocessors often include clock-skew adjustment circuits that can be tuned at test time to reduce skew or improve processor F_{MAX} by optimizing clock timing. However, adjusting such clock skew buffers on a part-by-part basis is extremely expensive, and such a scheme does not account for dynamic variations such as temperature which can cause clock skew to change. To further improve clock timing, active clock deskewing can be used. Figure 7.6 shows the clocking topology for the IA64 microprocessor in a 0.18 μm CMOS technology, employing

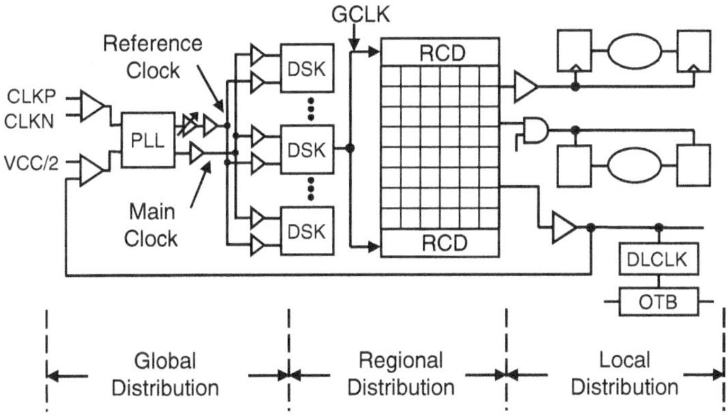

Fig. 7.6. Clock distribution topology for a 0.18 μm, IA-64 microprocessor. Reproduced with permission from [10], ©2000 IEEE

clock deskew buffers (DSKs) to dynamically reduce skew between different regional clock grids [10]. The DSK, shown in Fig. 7.7a, contains a local controller to continuously compare the timing of the regional grid clock to the incoming reference clock, and a variable delay circuit [11] (Fig. 7.7b) to adjust clock delay until skew is minimized. Timing comparison is accomplished through the use of a phase detector in the local controller. This scheme compensates any load variations or WID variations in the core clock distribution. Although the skew of the reference clock, which is distributed to all clock regions on the die, is still included in the final clock skew, the reduced span of the reference clock and the matched load results in significantly less overall skew. Figure 7.8 shows the measured skew. The total skew is 28ps with deskewing and is four times larger with the deskew mechanism disabled. The active deskew technique of the 0.18 μm IA64 processor (in addition to other processors) has also been discussed in Sect. 2.5.2.

The 90nm Itanium® design [12] also includes active deskew circuitry to deliver a low-skew clock without requiring the use of a high-power, dense clock grid. The clocking architecture for the "Montecito" processor (Fig. 7.9) includes regional active deskew (RAD) circuits to actively manage the clock skew between adjacent clock domains [13]. This system adjusts the delay of tunable second-level clock buffers (SLCBs) as shown in Fig. 7.10. Phase comparators are used to compare the clock outputs of two SLCBs, and the delay lines in the tuning buffers are either incremented or decremented until the two clocks are lined up. The complete RAD system is comprised of 30 phase comparators and 26 second-level clock buffers in each core connected to a tree-like structure. In this structure, all of the clock zones in the core are deskewed to a central anchor zone. The hierarchical nature of the system introduces a finite amount of skew but the RAD system manages this skew, to approximately 10ps across each core. Additional details of this hierarchical deskew algorithm are presented in Sect. 2.5.2.

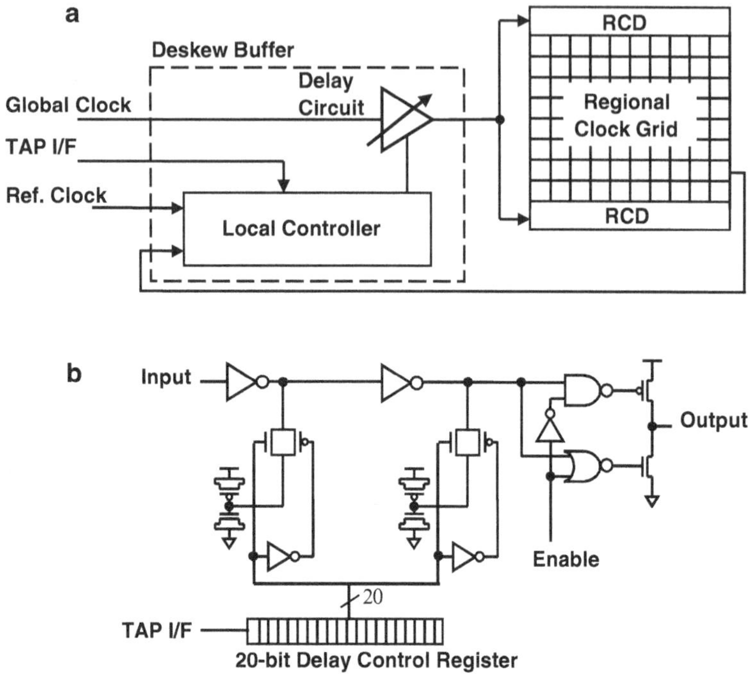

Fig. 7.7. (**a**) Deskew buffer (DSK) architecture. (**b**) DSK variable delay circuit. Reproduced with permission from [10], ©2000 IEEE

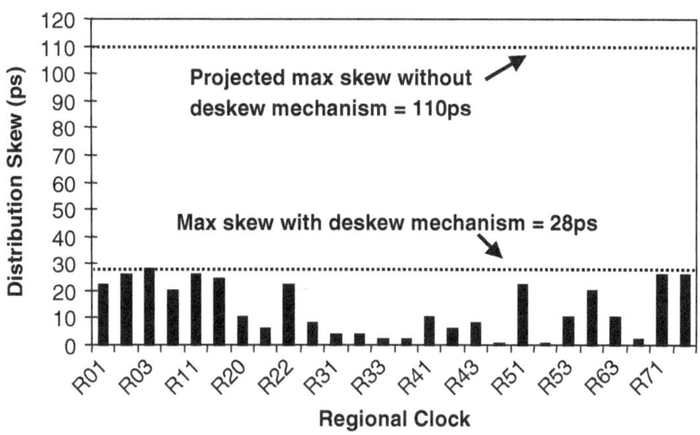

Fig. 7.8. Experimental skew measurements. Reproduced with permission from [10], ©2000 IEEE

7.3.3 Dynamic Frequency

Dynamic variations such as voltage and temperature fluctuations are typically handled by adding a margin to the clock frequency such that when these variations occur, the computation in the datapath is guaranteed to be correct. This "worst-case" design ensures correct results but requires an overhead in terms of power consumption, or a performance reduction. Because most of the time the nominal operating conditions of a microprocessor are not the worst-case conditions that have been designed for, performance or power are left on the table. To reduce this overhead of worst-case design, it is possible to sense and respond to these dynamic variations by changing clock frequency.

The Intel 90nm Itanium processor design [12–14] contains distributed digital frequency dividers (DFDs) for dynamic clock frequency change coupled with clock vernier devices (CVDs) for local delay fine-tuning (Fig. 7.9). The structure of the DFD is shown in Fig. 7.11: the DFD produces an output clock frequency based on the incoming PLL clock frequency and the current divisor value held in the phase compensator state machine (PCSM). For a divisor value of 0, the DFD clock runs at the frequency of the PLL. For all other divisors, the phase mux synthesizes a clock frequency by selecting the appropriate edges from the 64 phases.

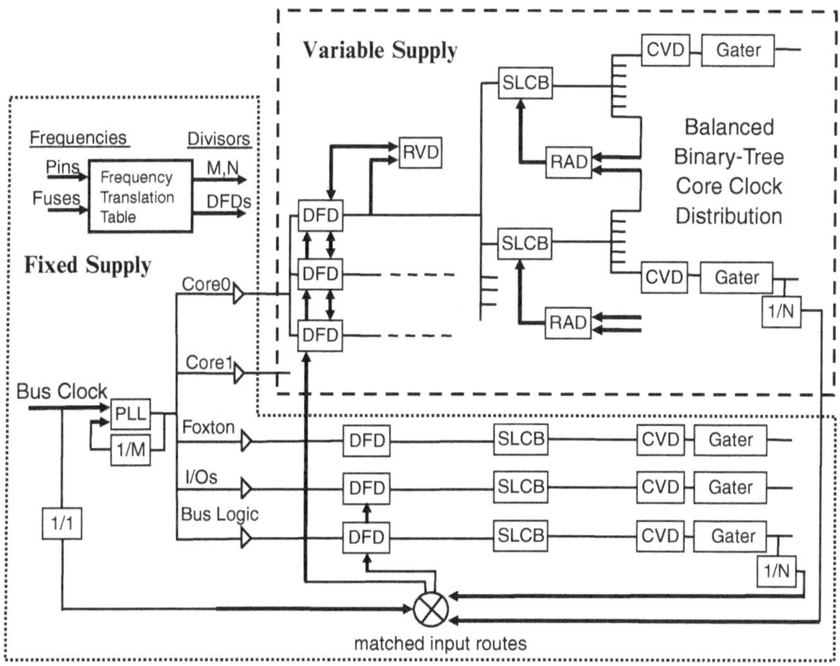

Fig. 7.9. Clock system architecture for the 90 nm Itanium® processor (Montecito). Reproduced with permission from [13], ©2006 IEEE

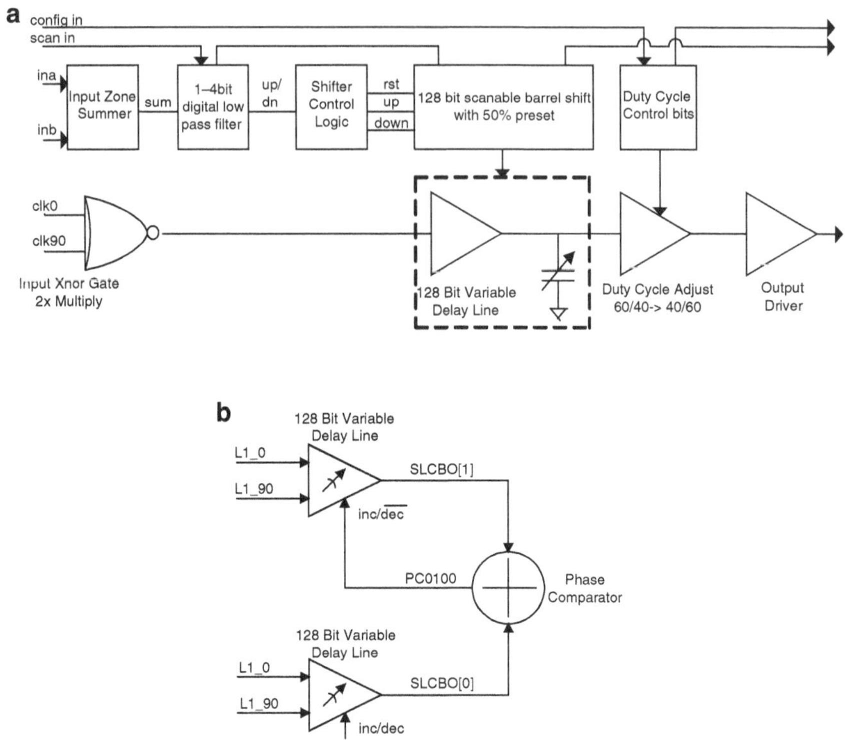

Fig. 7.10. (**a**) Second-level clock buffer (SLCB) architecture. (**b**) RAD SLCB/comparator connections. Reproduced with permission from [13], ©2006 IEEE

Variable frequency mode (VFM) allows the clock of the processor to track local supply voltage changes, reducing the guardband required for supply voltage noise. This is accomplished through the use of voltage-to-frequency converters (VFCs) which allow single-cycle response to droop transients and provide frequency steps of 1.5%. The VFC consists of voltage-locked loops in which a regional voltage detector (RVD) is used to sense voltage and set DFD output frequency. Design of the RVD is shown in Fig. 7.12: four delay lines are configured on a part-by-part basis to match microprocessor critical paths in delay and voltage sensitivity. The RVD outputs are generated by two multiplexing phase detector latches and two configurable delay elements. These outputs denote whether timing margin is available (DOWN), ideal timing margins are present (HOLD), or greater margins are needed (UP). To track regional voltage changes around the die, 12 RVDs are distributed around each core, and three DFD zones are used in each core to reduce frequency adjustment latency in the presence of fast voltage droops.

Measured results of the dynamic frequency system response to a 150 mV first-droop in voltage is shown in Fig. 7.13 [13]. The average pre-droop frequency over 32 cycles is 2,272 MHz, and the frequency averaged over the 16 cycles during the event

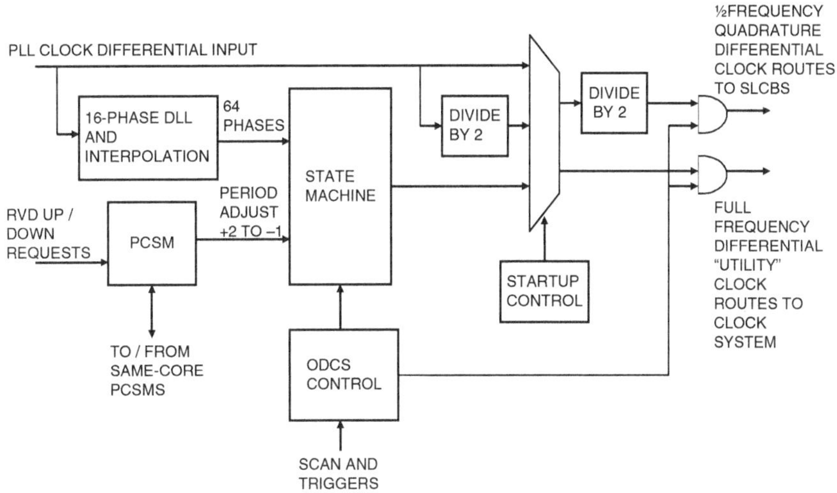

Fig. 7.11. Dynamic frequency divider (DFD) and phase compensator state machine (PCSM) block diagram. Reproduced with permission from [13], ©2006 IEEE

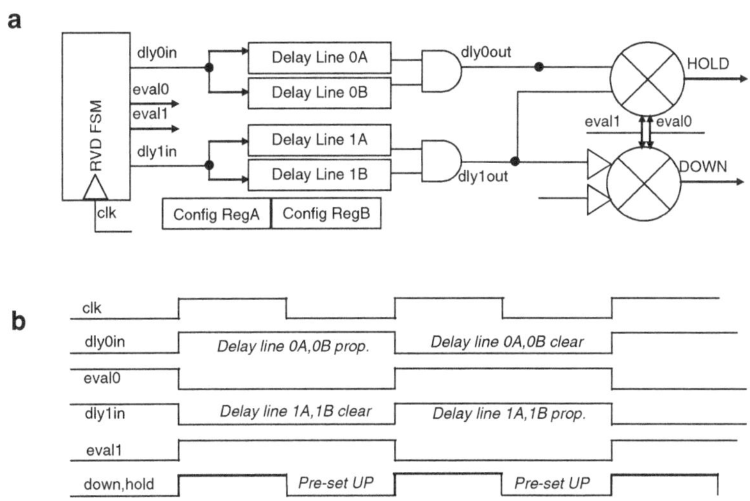

Fig. 7.12. Regional voltage detector (RVD) details. (**a**) RVD architecture block diagram. (**b**) RVD internal timing. Reproduced with permission from [13], ©2006 IEEE

is measured at 2,087 MHz. The result is that code can execute at a higher average frequency using VFM than could be achieved at fixed frequency mode where this voltage droop needs to be margined. Comparison of VFM and FFM for different test cases is shown in Fig. 7.14: the DTB and FPU cases are focused on critical path tests,

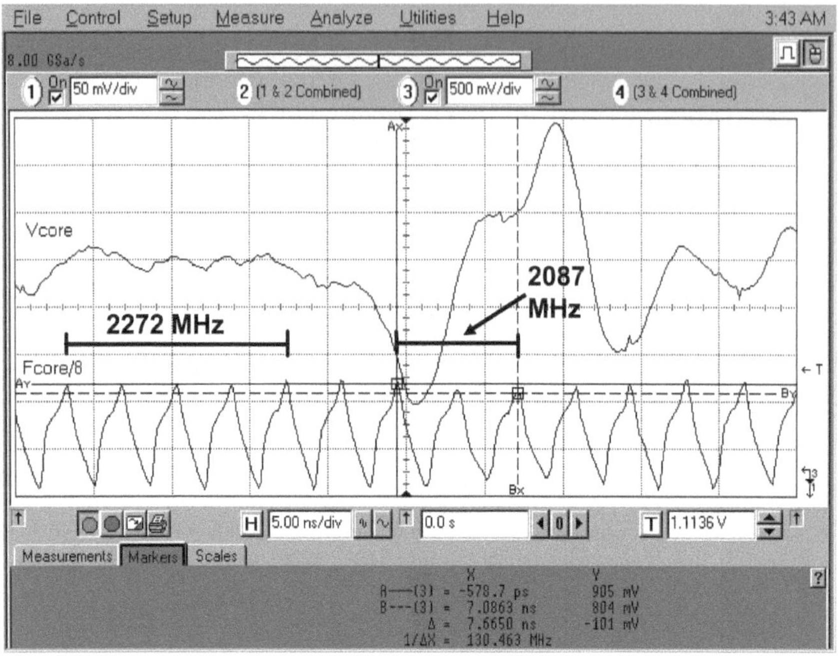

Fig. 7.13. Oscilloscope trace of clock system VFM response to 150 mV droop at 2.27 GHz, 1.2 V. Reproduced with permission from [13], ©2006 IEEE

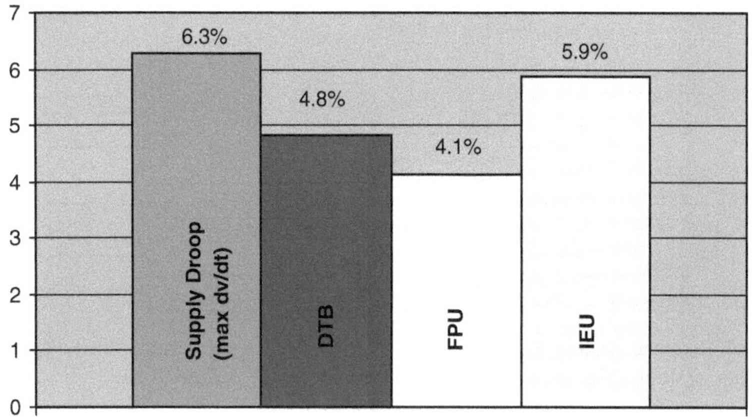

Fig. 7.14. Measured variable-frequency mode (VFM) (%) Frequency increase over fixed-frequency mode for multiple test cases. Reproduced with permission from [13], ©2006 IEEE

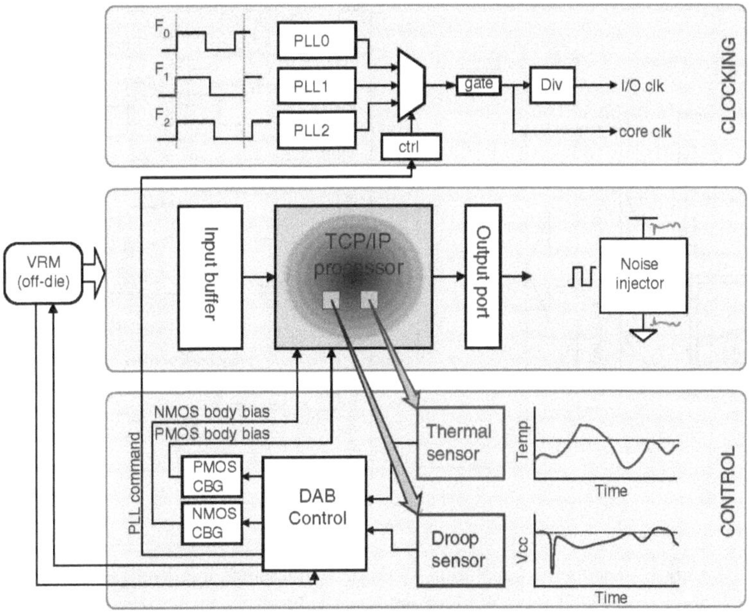

Fig. 7.15. Overview of 90 nm TCP/IP processor test chip with dynamic voltage, frequency, and body bias. Reproduced with permission from [15], ©2007 IEEE

while the supply droop case is the maximum dv/dt test. Here, maximum frequency improvement of 4.1% to 6.3% is shown by enabling the variable frequency mode.

The 90nm TCP/IP processor test chip in [15] uses dynamic frequency in a similar fashion in order to reduce the guardbands applied for voltage droop, temperature, and aging. In this case, however, rather than using dynamic voltage-to-frequency converters, analog sensors are used to continuously sense voltage and frequency, and this information is sent to a dynamic adaptive bias (DAB) controller which adjusts frequency, voltage, and body bias (Fig. 7.15). Details of the DAB control algorithm are shown in Fig. 7.16: multiple temperature and voltage sensors are used to accurately measure the range of voltage and temperature as code is executed on the TCP/IP processor core. These voltage and temperature values are used as an index into a lookup table of die characterization information which is pre-loaded at test time. This lookup table specifies the optimum voltage, body bias, and frequency to be applied for each combination of temperature and voltage measurements. If the DAB control determines that a frequency change is needed, the dynamic clocking block (Fig. 7.17) changes the core frequency within a single core-clock cycle, allowing fast response to first-droop changes in voltage. This fast frequency change is accomplished through the use of multiple PLLs, each locked to a different frequency. When the frequency needs to be lowered due to a voltage droop event or temperature change, the slower PLL is selected by the clock select mux, ensuring a transition to the slower clock frequency without any short cycles or glitches. Once the slow PLL is

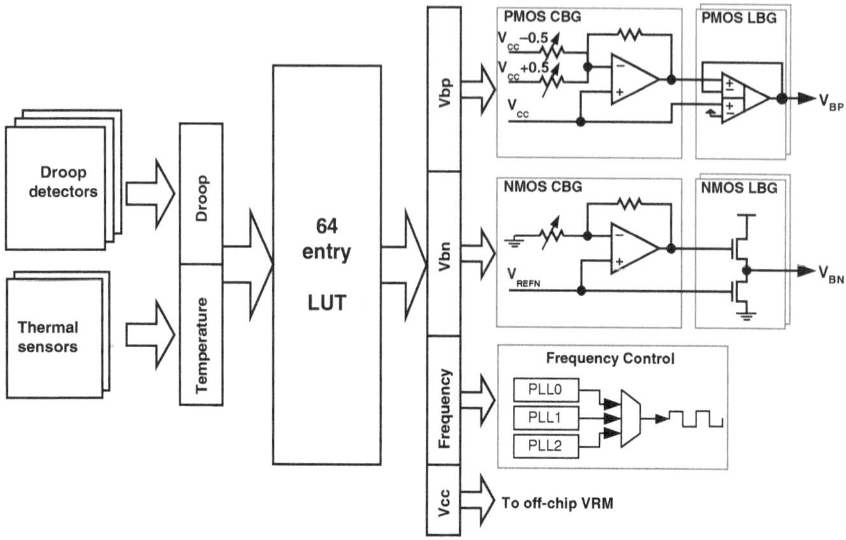

Fig. 7.16. Details of dynamic adaptation method. Temperature and droop sensors are used to index a lookup table (LUT) which gives optimum voltage, frequency, and body bias for the given conditions. Reproduced with permission from [15], ©2007 IEEE

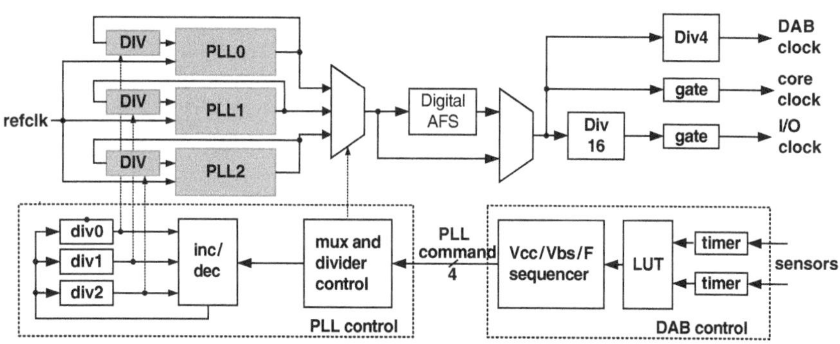

Fig. 7.17. Dynamic clocking system details. Multiple PLLs are used to enable fast frequency changes, allowing response to first-droop in voltage. PLL frequency can be changed when not selected, allowing wide frequency range with fine resolution. Reproduced with permission from [15], ©2007 IEEE

selected, the faster PLL can then be re-locked to a new frequency, ensuring frequency change is possible across a wide frequency range with fine (<3%) resolution.

Results from this test chip for supply voltage droop are shown in Fig. 7.18. In this case, a large 20% voltage droop is detected by the voltage sensor, and the DAB control reduces the clock frequency. As the voltage rises at the end of the droop event, this change is again detected and the clock frequency is raised to a new value which

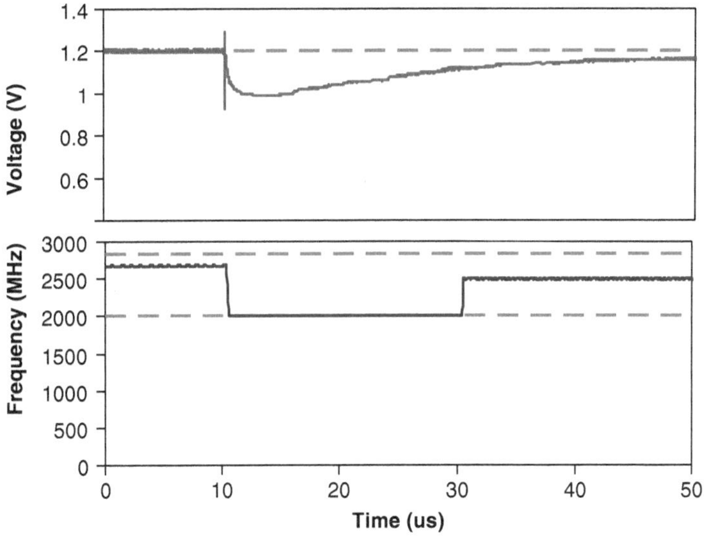

Fig. 7.18. Measured dynamic frequency response to a voltage droop. Maximum frequency is 32% higher for the dynamic processor than would be required for a worst-case design with fixed frequency. Reproduced with permission from [15], ©2007 IEEE

accounts for the IR drop based on the increased current load. Without these dynamic frequency features, the clock frequency would have to be set to the frequency supported at the minimum voltage level, representing a large guardband in frequency for these infrequent voltage droops. By dynamically sensing voltage droop and responding, the average frequency can be increased. Both of these design examples demonstrate the large improvement in performance and energy-efficiency that can be obtained by using the clock as part of dynamic variation response system, rather than simply designing for the worst-case conditions.

7.4 Variation Reduction Through Resiliency

The previous sections have focused on techniques for tolerating or reducing the impact of variations such that the variations do not produce an error in computation. A combination of variation-tolerant design techniques, coupled with dynamic variation detection and response can be used to reduce the guardbands that must be applied to the clock frequency of microprocessors. However, guardbands must still be applied for variations that cannot be predicted, sensed, or which are too fast to respond to. Examples of these types of variations include very fast 1st-order voltage droop, delay variation from crosstalk and noise, and within-die differences in transistor degradation. Reducing these guardbands is the focus of this section.

In contrast to variation tolerance which guarantees that errors in the circuit never occur, resilient designs operate correctly even in the presence of errors. Resiliency implies that any errors which result from variations are sensed and corrected so that the final result is error-free, although there may be errors at intermediate stages in the computation. As variations increase while performance targets and power requirements become even more restrictive, resiliency will become much more important in designs.

Resilient design itself is an extremely broad topic – resiliency can be implemented at the circuit level, in the microarchitecture, in the system, or even in software. In this chapter, we focus on circuit-level resiliency as this is most relevant to clock network design. A complete implementation of a resilient microprocessor will combine these resiliency techniques at the circuit level with error correction at the microarchitectural level and dynamic response at the system level.

7.4.1 Timing Error Detection – Error Detection Sequentials

Dynamic in situ detection and correction of speed path failures were first proposed as part of the Razor project [16, 17]. In the Razor technique, each delay-critical flip-flop in the datapath is replaced by a Razor flip-flop (Fig. 7.19), which includes a shadow latch controlled by a delayed clock. This technique works on the premise that dynamic variations such as voltage, temperature, noise, or data-dependent delay variation will cause the delay of a critical path to increase. If the frequency of the processor is set by the typical condition, this delay increase will cause the data at the receiving flip-flop to arrive after the sampling clock edge, causing an error in the datapath. However, the data are sampled correctly by the delayed clock of the shadow latch, and by comparing these late-sampled data with the data captured by the datapath flip-flop, the error can be detected. Under nominal conditions, data are sampled correctly by both the datapath flip-flop and the shadow latch, and no error is signaled. When a dynamic variation causes the critical path delay to exceed the clock period, the error signal is asserted and the error in the pipeline must be corrected before program execution can continue. In the original Razor design, the correct data always exist in the shadow latch, even if the datapath flip-flop fails. Thus, the correct data can be injected back into the datapath through the multiplexer shown in Fig. 7.19. The Razor technique along with other similar resiliency techniques has also been discussed in Sect. 3.7.2 from the viewpoint of wearout-induced circuit degradation.

One concern for the Razor-style error-detection sequential (EDS) is in the handling of the metastability condition. Because the processor frequency can now be set by the nominal operating conditions and not by the worst-case conditions, no additional setup time margin is necessary in the clock frequency to allow for dynamic variations. Because of this, the data input to the flip-flop may arrive at the same time as the clock edge, leading to a metastable output. The probability of this occurring is very small; however, it is many times larger than the probability of metastability in the original design with guardbands. If the flip-flop datapath output or error output becomes metastable, there is the possibility of an undetected error: the datapath value is sampled incorrectly, but the error signal is not asserted. Obviously an undetected error cannot be allowed, and the Razor flip-flop includes a metastability detector to

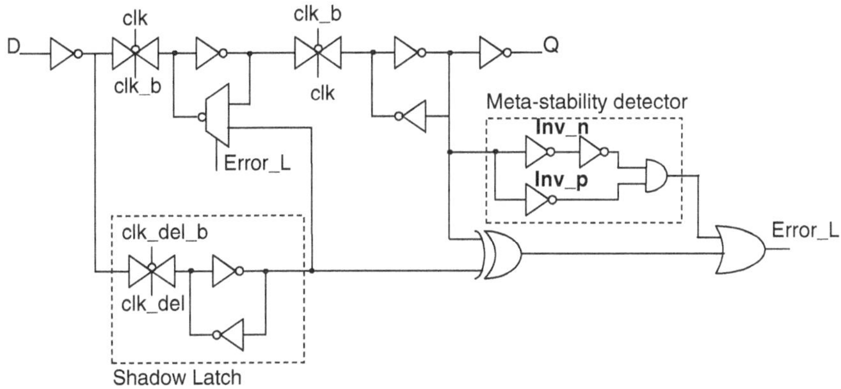

Fig. 7.19. Razor error-detection flip-flop. It consists of a datapath flip-flop, a shadow latch for sampling late-arriving data, and a metastability detector. Reproduced with permission from [16], ©2003 IEEE

signal this condition and initiate error recovery. This metastability detector, while effective, contributes in large part to the area overhead in the error-detection flip-flop, and to the overall power overhead for the error detection system, and thus detracts from the energy-efficiency gains possible through resiliency.

Recently, several techniques were independently developed to address the overhead of the original Razor (Razor-I) approach. The versions shown in [18–20] attempt to reduce the clock power overhead and eliminate the metastability concerns by replacing the pipeline flip-flop with a latch. This approach is motivated by the observation that because the critical paths must be designed such that the data never arrive in the nominal condition in the high phase of the clock, the master stage of the datapath flip-flop is unnecessary. Thus this flip-flop can be replaced by a latch which operates essentially as a pulsed latch, where the length of the high phase of the clock determines the transparency window of the latch as well as specifying the min-delay requirement which must be met by each logic path. Because late-arriving data are not gated by the closing of the master stage of the flip-flop, there is no metastability concern around the rising edge of clock. Late-arriving data are still sampled correctly in the current pipeline stage; however, this may cause a critical path and error in the subsequent pipeline stage and, therefore, this late-arriving data must be signaled as an error.

The transition detection with time-borrowing (TDTB) sequential [18] (Fig. 7.20a) uses a pulse generator to generate a pulse each time data transitions. As shown in the example timing diagram (Fig. 7.21), if this data transition occurs during the high clock phase (which defines the error-checking window), an error signal is generated. Because the datapath flip-flop has been replaced by a latch, performance and power consumption are improved over the Razor design. A similar idea is used in the double-sampling with time-borrowing (DSTB) technique [18] (Fig. 7.20b), which is similar to TDTB except that a shadow flip-flop replaces the transition

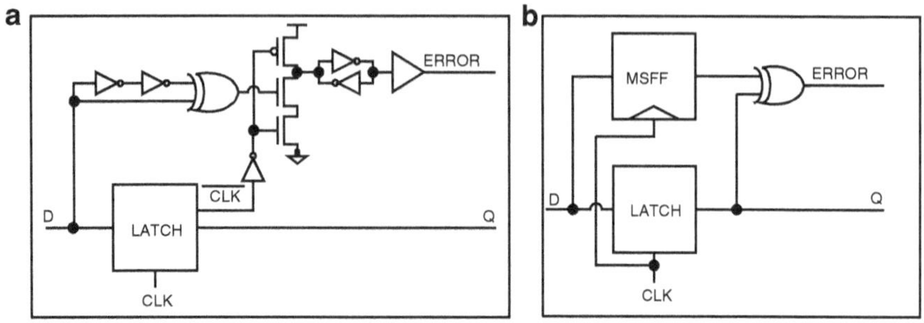

c

	RFF	TDTB	DSTB
Datapath Metastability	Yes	No	No
Clock Energy Relative to MSFF	36% (T) 22% (L)	−9%(T) −13%(L)	34% (T) 14% (L)
Clock Energy Relative to Latch	145% (T) 94% (L)	64% (T) 38% (L)	143% (T) 81% (L)
D-to-Q Delay Relative to MSFF	12%	−14%	−12%
D-to-Q Delay Relative to Latch	62%	23%	27%
Design Complexity	Low	High	Low
SER Tolerant Datapath Sequential	Yes	Partial	Yes
Ability to Turn Off Error Detection & Operate at Low F_{CLK} with 50% Duty Cycle	Yes	No	No

Fig. 7.20. Error-detection sequential circuits. (**a**) Transition-detection with time-borrowing (TDTB). (**b**) Double-sampling with time-borrowing (DSTB). (**c**) Comparison of Razor-I flip-flop (RFF), TDTB, and DSTB in key circuit metrics. Reproduced with permission from [18], ©2008 IEEE

detector. The clock energy overhead for this EDS is slightly lower than for the Razor design because the larger datapath flip-flop has been replaced by a latch, while the minimum-sized shadow latch has been replaced by a flip-flop. The Razor flip-flop (RFF), TDTB, and DSTB error-detection sequentials are compared in terms of key circuit metrics in Fig. 7.20c.

A later version of the Razor design, denoted Razor-II [19], also replaces the datapath flip-flop with a latch to eliminate metastability concerns. This design (Fig. 7.22) also utilizes a datapath latch for improved performance and handling of metastability. A transition detector generates a pulse whenever the output of the latch switches, and this pulse is qualified with a detection clock to determine whether the output has transitioned because of data which arrived after the clock edge. This timing is demonstrated in Fig. 7.23. Data which arrive late cause a transition detector pulse after the detection clock, and an error signal is generated and propagated down the pipeline. Note that in this version of the Razor flip-flop, the correct data are no longer present in the sequential. Therefore, it cannot be injected into the pipeline as in Razor-I, and the offending instruction must be replayed by the microarchitecture.

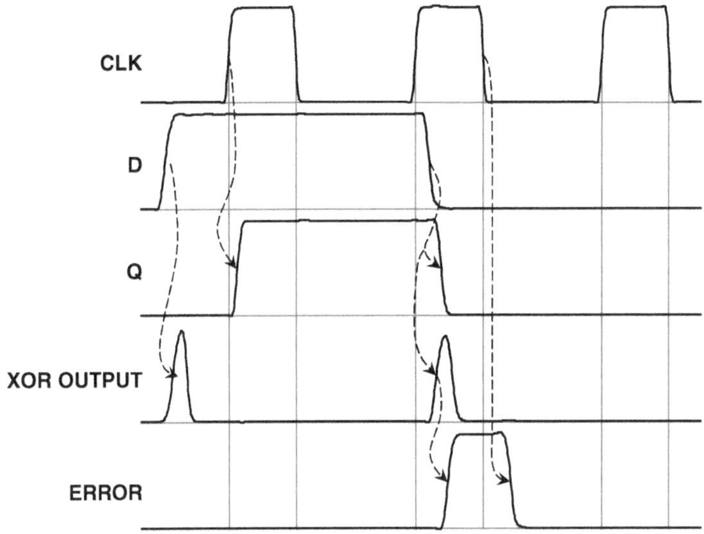

Fig. 7.21. Timing diagram for the transition-detection with time-borrowing (TDTB) technique. Data transitions which occur in the high phase of the clock (which sets the error-checking window for the design) generate an error signal. Reproduced with permission from [20], ©2009 IEEE

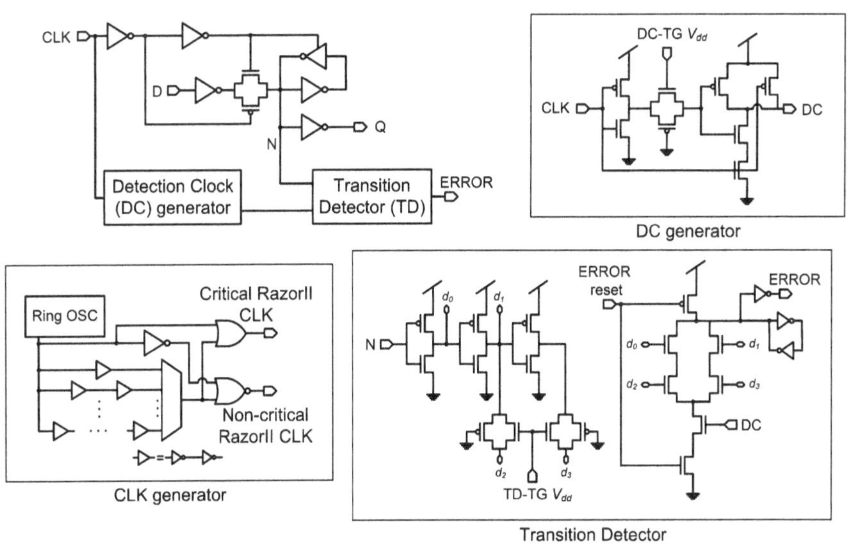

Fig. 7.22. Details of the Razor-II sequential design. It includes the latch, clock generator, transition detector (TD), and detection clock (DC) generator. Reproduced with permission from [19], ©2008 IEEE

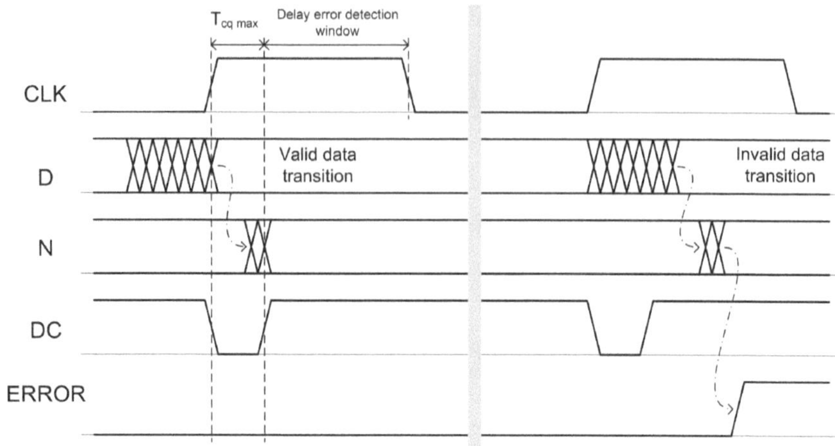

Fig. 7.23. Timing diagram for the Razor-II design. Transitions on latch output node "N" which occur after the allowed region (specified by the detection clock DC) are signaled as errors. Reproduced with permission from [19], ©2008 IEEE

7.4.2 Timing Error Correction and Recovery

Once timing errors have been detected in a circuit, there must be a method for responding to these errors, negating the offending instructions, and re-executing to get the correct result. As in timing error detection itself, this can be done at different points in the design hierarchy – at the circuit level itself, in the microarchitecture, or with software-level checkpointing. In responding to and correcting errors, there is a trade-off between complexity and performance overhead: correcting errors at the circuit level requires complex hardware, but incurs very small performance overhead, while correcting errors at the software level requires minimal hardware support, but may incur a large performance penalty. Thus, the optimal error rate, where energy-efficiency is maximized, is different depending on the error-recovery method which is chosen.

In the original Razor design [16], timing errors are corrected in the pipeline itself, allowing execution to continue with a one-cycle penalty. This technique relies on the fact that the shadow latch has always captured the correct output, even when the datapath flip-flop is in error. When an error occurs, the offending instruction must be nullified as it passes down the pipeline. This ensures that the state of the machine is not corrupted with incorrect data. Next, the pipeline must be stalled for one cycle as the correct value is forwarded from the shadow latch into the main pipeline. Because the execution continues with the stage and following the stage that caused the error, forward progress of the instruction is guaranteed.

In the design in [18], the error signal initiates the error recovery process in the microarchitecture. The error is treated in a similar way to a branch misprediction. As

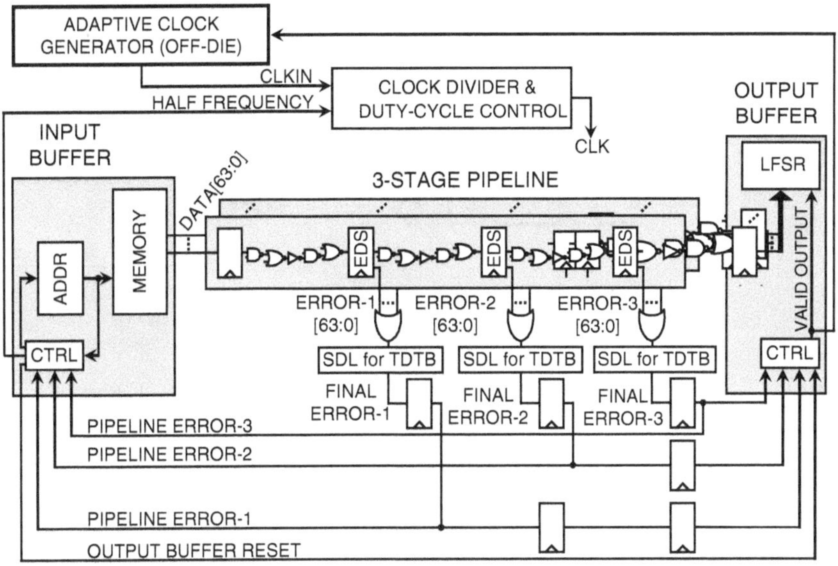

Fig. 7.24. Instruction-replay error-recovery design for one error-detection sequential (EDS). Set-dominant latch (SDL) is only used for TDTB. Reproduced with permission from [18], ©2008 IEEE

shown in Fig. 7.24, all of the error signals in a particular pipeline stage are combined using an OR tree to generate a single error signal for that stage. That error signal is then sent down the pipeline, along with the data, to the output buffer to invalidate the data. This is necessary to prevent the failing instruction from corrupting the state of the machine. At the same time, the error signal propagates to the instruction issue stage (input buffer) to initiate the instruction replay sequence. Similar to a branch mispredict, the program counter is loaded with the address of the failing instruction, and the instruction is repeated. If the error was due to a transient environmental condition, the instruction may execute error-free when it is re-executed, and the processor execution continues as usual until the point of the next error. If, however, the error is due to a condition which is set up by the instruction (or group of instructions) itself, it will fail again on re-execution. For this reason, the chip also contains a clock divider circuit (Fig. 7.25) which reduces the clock frequency by half when replaying the instruction. Re-executing the instruction at half frequency guarantees that the instruction will pass the failing pipeline stage and guarantees forward progress for the workload.

The Razor-II design, which was incorporated in a 64-bit, 7-stage Alpha processor design in 0.13 μm CMOS [19], recovers from errors in a similar way. The last stage of the pipeline is designed to be non-timing critical so that it is known before instruction commit whether an error occurred for that instruction or not (Fig. 7.26). In case

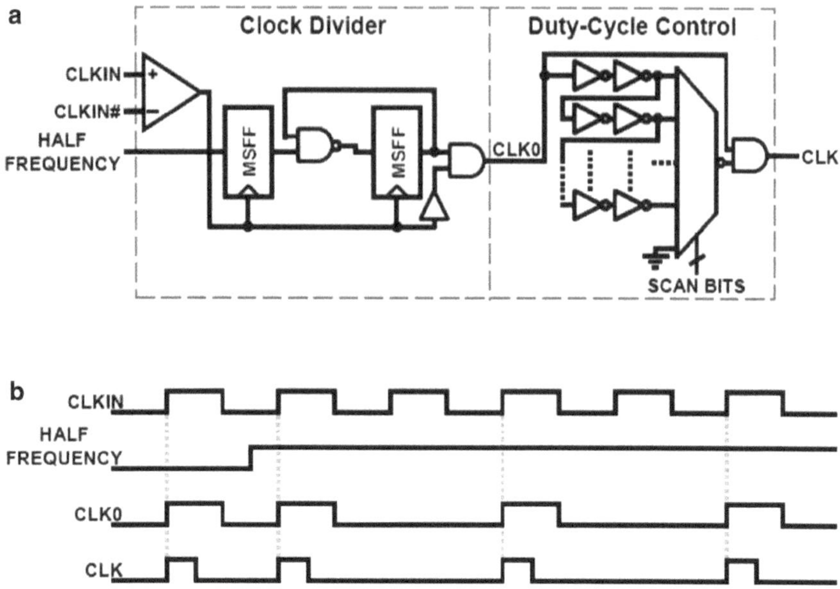

Fig. 7.25. Details of the test chip clock generation. (**a**) Clock generator schematic, including the clock divider which reduces clock frequency by half during error replay, and duty cycle control which is necessary for tuning the error-detection window. (**b**) Timing diagram, showing clock frequency reduction during error replay, while high phase of clock is fixed. Reproduced with permission from [18], ©2008 IEEE

of an error, the pipeline is flushed and the offending instruction is re-executed. The error controller can also reduce the clock frequency by half for 8 cycles to ensure that repeatedly-failing instructions are able to execute correctly.

7.4.3 Results: Guardband Reduction Through Resiliency

To get the maximum benefit from resiliency, error detection and correction techniques must be combined with adaptive voltage and/or frequency at the system level to respond to slow-changing variations. For example, a temperature increase may persist for milliseconds or longer; hence, the performance of the system would suffer if this results in a large increase in errors that must be corrected. Instead, the error rate is continuously monitored by an error control unit which dynamically changes the frequency or voltage of the processor such that the error rate is regulated at a sufficiently low value. As described before, the ideal error rate for energy-efficiency depends on the type of circuit that is implemented and the energy and performance overhead in doing recovery.

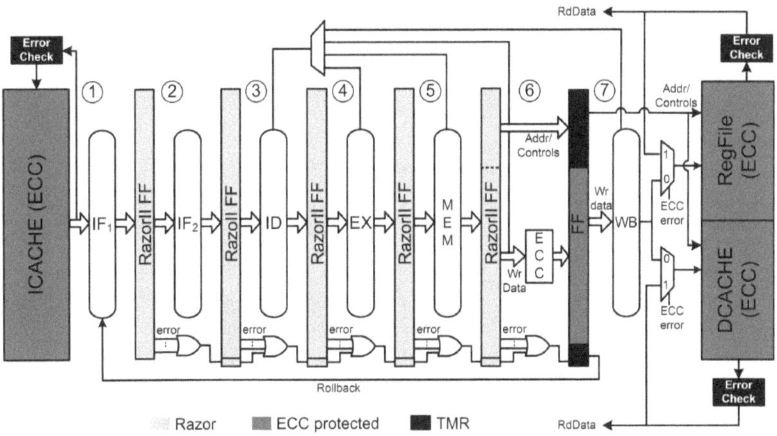

Fig. 7.26. Overview of 0.13 μm, 64-bit, 7-stage alpha processor design incorporating Razor-II error-detection sequentials and instruction replay. Reproduced with permission from [19], ©2008 IEEE

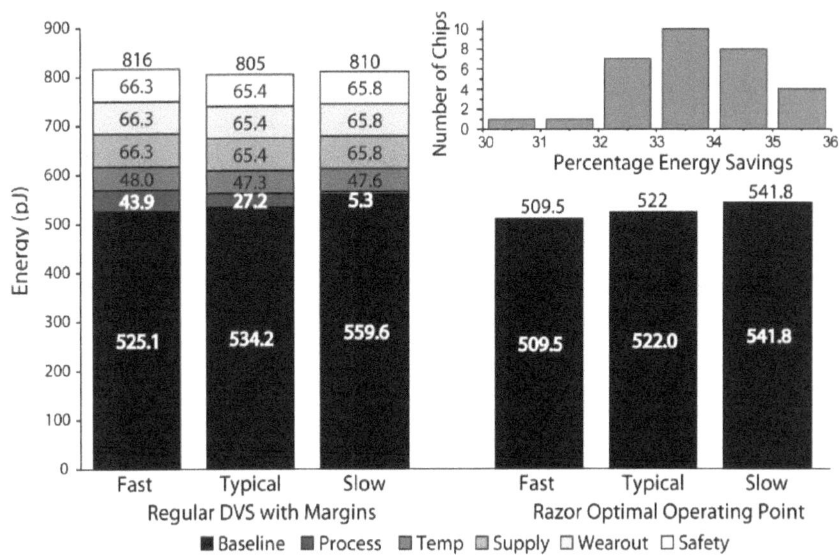

Fig. 7.27. Measured Razor-II energy consumption and distribution of energy savings. Reproduced with permission from [19], ©2008 IEEE

Measured energy dissipation results for three Razor-II chips, operating at 0.04% error rate, are shown in Fig. 7.27 [19]. Compared to the energy consumption when voltage is set such that all 31 fabricated dice operate correctly at 85C with 10% margin for wearout, supply fluctuation, and safety, energy dissipation gains are 33.1% to

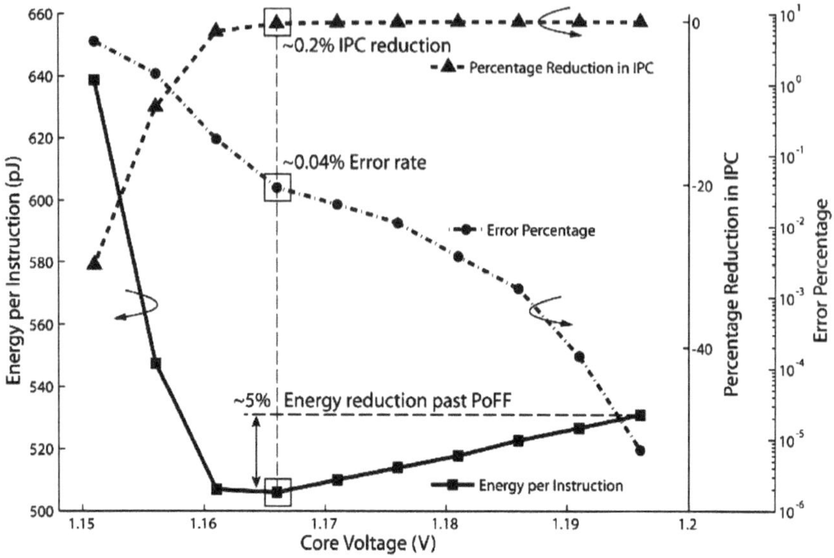

Fig. 7.28. Measured energy per instruction and error rate for the Razor-II processor. PoFF (point of first failure) represents the voltage at which the first error in the circuit occurs. Reproduced with permission from [19], ©2008 IEEE

37.5%. The process of setting the optimum core voltage and error rate is shown in Fig. 7.28. As the core voltage is reduced, the error rate, which is initially zero at high voltage, begins to increase. At the same time, the energy per instruction decreases as well as a result of the lower voltage. The impact on instructions per cycle (IPC) is initially negligible because the error rate is small. However, as the voltage continues to decrease and the error rate increases, the performance penalty for replaying instructions begins to have an impact on the IPC. Further reduction in the voltage actually causes the energy per instruction to reach a minimum and then increase – this occurs because the energy overhead of replaying failed instructions becomes larger than the energy saved by reducing voltage. This minimum in the energy per instruction represents the most energy-efficient point of operation for this system – in this case, this occurs at a 0.04% error rate.

Measurements on the resilient circuits test chip [18] also show the performance and energy benefits from reducing guardbands and operating at the optimal error rate. In Fig. 7.29, throughput (TP) and error rate are measured for the TDTB EDS circuit versus F_{CLK}, and a range of V_{CC} droop magnitudes and durations are inserted based on data from a recent microprocessor along with assumptions on V_{CC} droop-inducing events. The worst-case V_{CC} droop magnitude is 10% and the worst-case temperature is 110C. Nominal V_{CC} is 1.2V and the nominal operating temperature is assumed to be 60C. In Fig. 7.29a, throughput increases linearly as F_{CLK} increases with no errors. Once errors occur, instructions per cycle (IPC) are reduced as a function of error rate and recovery time. Since V_{CC} droop events are assumed to be infrequent, throughput

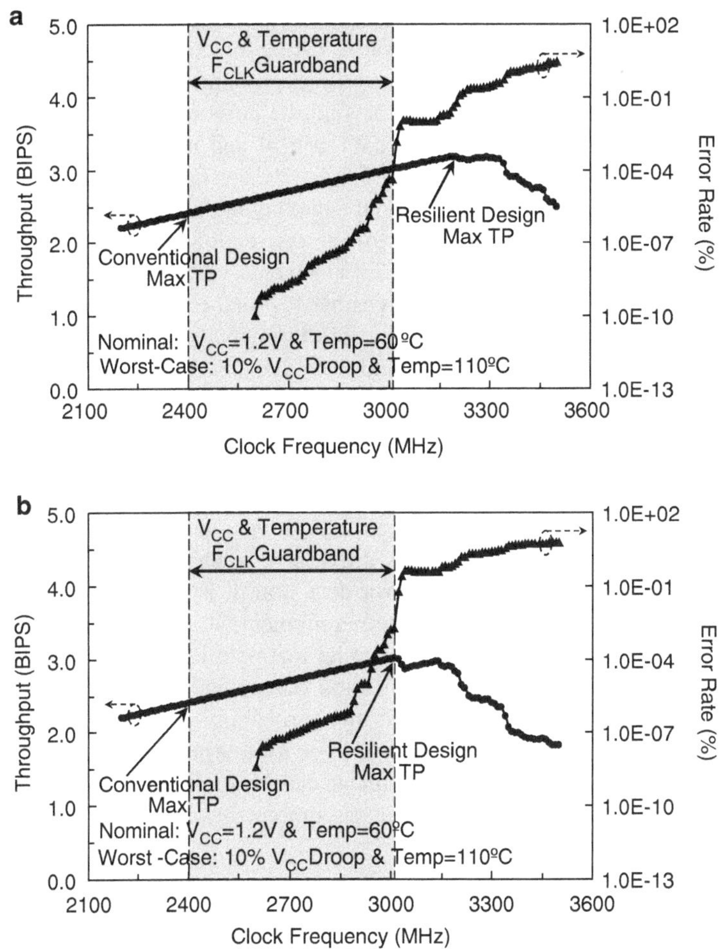

Fig. 7.29. Measured throughput and error rate versus clock frequency for resilient design with TDTB EDS circuits for two different path activation examples. (**a**) Critical paths less-frequently activated than non-critical paths. (**b**) All paths have equal probability of activation. Reproduced with permission from [18], ©2008 IEEE

gains continue as F_{CLK} increases into the V_{CC} and temperature guardband region. When F_{CLK} reaches 3,020 MHz, the first path failure occurs under nominal conditions, resulting in a sharp error rate increase. If the most-critical paths on the die are infrequently activated, further throughput gains are achieved at higher F_{CLK}, as the frequency is increased past the point of the first critical path failure. The maximum throughput of 3.17 billion instructions per second (BIPS) corresponds to a 3,200 MHz F_{CLK}. Increasing F_{CLK} further leads to a larger error rate, where IPC reduction outweighs F_{CLK} gains.

One interesting aspect of built-in error detection sequentials is that the gains achievable depend on the path activation statistics for the critical paths. The previous example in Fig. 7.29a assumed that critical paths are activated less-frequently than non-critical paths, allowing gains beyond the point of first failure. In another scenario, it is assumed that all paths, both critical and non-critical, are activated with equal probability. In this case (Fig. 7.29b), the resilient design cannot exploit the path-activation probabilities, limiting the maximum throughput to 3.01 BIPS. In Fig. 7.29a,b, the maximum throughput for the conventional design with MSFFs is 2.4 BIPS corresponding to an F_{MAX} of 2,400 MHz to guarantee correct functionality within the presence of worst-case dynamic V_{CC} and temperature variations. On the other hand, a resilient design enables 25% throughput gain over a conventional design by eliminating the F_{CLK} guardband from dynamic V_{CC} and temperature variations and an additional 7% throughput increase from exploiting the path-activation probabilities.

These examples demonstrate that by employing error-detection sequentials within the datapath, the processor can reduce or eliminate guardbands that are applied for dynamic variations, critical path activations, and aging. They also point to the necessity of optimizing circuits differently when error detection and correction are used. For example, with error detection, it is no longer necessary that every critical path meets the cycle time requirement. If a critical path is very infrequently activated, it may be desirable to let that path cause an error and execute instead in several clock cycles. Optimization of circuits with error detection and correction needs to consider path activation probabilities to balance the performance and energy-efficiency of the processor as a whole. As demonstrated by these designs, the gains for this investment can be quite large and will become more important as power and performance requirements continue to become more stringent.

7.5 Conclusion

As noted in the introduction to this chapter, the clocking network is both a challenge and an opportunity as part of a variation-tolerant processor design. Variation-tolerant design techniques must first be used on both the clock and the datapath to limit the effects of clock uncertainty such as clock skew and jitter and data uncertainty caused by both static and dynamic variations. On top of these design techniques, dynamic tuning techniques can be used after fabrication to allow the clock to dynamically adjust to the current environmental conditions. By employing detectors or sensors for voltage droop, temperature, and/or transistor degradation, the clock frequency can be continually adjusted, allowing the processor to adapt to the actual operating conditions rather than being limited by the constraints imposed by a worst-case design.

Finally, further guardband reduction and energy-efficiency is made possible by dynamically detecting and responding to timing errors in critical paths of the processor pipeline. A combination of error detection techniques at the circuit level and error-recovery methods at the microarchitectural level ensures error-free computation while allowing errors to be generated and corrected in individual pipeline stages. When combined with dynamic frequency and/or voltage capability, this allows a processor to adapt to changing environmental conditions, data workloads, and device degradation without requiring sensors or safety guardbands. In the future, such error detection capability can also be used to improve reliability of the system and aid in test and debug.

Acknowledgments

The author wishes to acknowledge Keith Bowman (Intel) and David Blaauw (University of Michigan) for the help in the preparation of this manuscript.

References

[1] V. Oklobdzija (ed.), *The Computer Engineering Handbook*. CRC Press, Boca Raton, 2002.

[2] H. Partovi, R. Burd, U. Salim, F. Weber, L. DiGregorio, and D. Draper, Flow-through latch and edge-triggered flip-flop hybrid elements. In: *Digest of Technical Papers IEEE International Solid-State Circuits Conference (ISSCC 1996)*, 1996, pp. 138–139.

[3] V. Oklobdzija, Clocking and clocked storage elements in a multi-gigahertz environment. *IBM J. Res. Dev.*, 47(5/6), 567–583, 2003.

[4] D. Harris and M. Horowitz, Skew-tolerant domino circuits. *IEEE J. Solid-State Circuits*, 32(11), 1702–1711, 1997.

[5] K. Bowman, J. Tschanz, M. Khellah, M. Ghoneima, Y. Ismail, and V. De, Time-borrowing multi-cycle on-chip interconnects for delay variation tolerance. In: *Proc. International Symposium on ISLPED'06 Low Power Electronics and Design*, 4–6 October 2006, pp. 79–84.

[6] P. Bai, C. Auth, S. Balakrishnan, M. Bost, R. Brain, V. Chikarmane, R. Heussner, M. Hussein, J. Hwang, D. Ingerly, R. James, J. Jeong, C. Kenyon, E. Lee, S.-H. Lee, N. Lindert, M. Liu, Z. Ma, T. Marieb, A. Murthy, R. Nagisetty, S. Natarajan, J. Neirynck, A. Ott, C. Parker, J. Sebastian, R. Shaheed, S. Sivakumar, J. Steigerwald, S. Tyagi, C. Weber, B. Woolery, A. Yeoh, K. Zhang, and M. Bohr, A 65nm logic technology featuring 35nm gate lengths, enhanced channel strain, 8 Cu interconnect layers, low-k ILD and 0.57 μm^2 SRAM cell. In *Proc. IEDM Technical Digest Electron Devices Meeting IEEE International*, 13–15 December 2004, pp. 657–660.

[7] A. Sarangi, G. F. Taylor, R. J. Parker, E. P. Osburn, and P. J. Ott, Power level management in an IA32 microprocessor. In: *Proc. Electrical Performance of Electronic Packaging*, 21–23 October 2002, pp. 235–238.

[8] J. W. Tschanz, S. Narendra, R. Nair, and V. De, Effectiveness of adaptive supply voltage and body bias for reducing impact of parameter variations in low power and high performance microprocessors. *IEEE J. Solid-State Circuits*, 38(5), 826–829, 2003.

[9] T. Chen and S. Naffziger, Comparison of adaptive body bias (ABB) and adaptive supply voltage (ASV) for improving delay and leakage under the presence of process variation. *IEEE Trans. VLSI Syst.*, 11(5), 888–899, 2003.

[10] S. Tam, S. Rusu, U. Nagarji Desai, R. Kim, J. Zhang, and I. Young, Clock generation and distribution for the first IA-64 microprocessor. *IEEE J. Solid-State Circuits*, 35(11), 1545–1552, 2000.

[11] G. Geannopoulos and X. Dai, An adaptive digital deskewing circuit for clock distribution networks. In: *Digest of Technical Papers IEEE International Solid-State Circuits Conference (ISSCC 1998)*, 1998, pp. 400–401.

[12] S. Naffziger, B. Stackhouse, and T. Grutkowski, The implementation of a 2-core multi-threaded Itanium®-family processor. In: *Digest of Technical Papers IEEE International Solid-State Circuits Conference (ISSCC 2005)*, 2005, pp. 182–183, 592.

[13] T. Fischer, J. Desai, B. Doyle, S. Naffziger, and B. Patella, A 90-nm variable frequency clock system for a power-managed Itanium architecture processor. *IEEE J. Solid-State Circuits*, 41(1), 218–228, 2006.

[14] P. Mahoney, E. Fetzer, B. Doyle, and S. Naffziger, Clock distribution on a dual-core, multi-threaded Itanium®-family processor. In: *Digest of Technical Papers IEEE International Solid-State Circuits Conference (ISSCC 2005)*, February 2005, pp. 292–293, 599.

[15] J. Tschanz, N. S. Kim, S. Dighe, J. Howard, G. Ruhl, S. Vanga, S. Narendra, Y. Hoskote, H. Wilson, C. Lam, M. Shuman, C. Tokunaga, D. Somasekhar, S. Tang, D. Finan, T. Karnik, N. Borkar, N. Kurd, and V. De, Adaptive frequency and biasing techniques for tolerance to dynamic temperature-voltage variations and aging. In: *Digest of Technical Papers IEEE International Solid-State Circuits Conference (ISSCC 2007)*, 11–15 February 2007, pp. 292–293, 604.

[16] D. Ernst, N. S. Kim, S. Das, S. Pant, R. Rao, T. Pham, C. Ziesler, D. Blaauw, T. Austin, K. Flautner, and T. Mudge, Razor: a low-power pipeline based on circuit-level timing speculation. In: *Proc. 36th Annual IEEE/ACM International Symposium on MICRO-36 Microarchitecture*, 2003, pp. 7–18.

[17] S. Das, D. Roberts, S. Lee, S. Pant, D. Blaauw, T. Austin, K. Flautner, and T. Mudge, A self-tuning dvs processor using delay-error detection and correction. *IEEE J. Solid-State Circuits*, 41(4), 792–804, 2006.

[18] K. Bowman, J. Tschanz, N. S. Kim, J. Lee, C. Wilkerson, S.-L. Lu, T. Karnik, and V. De, Energy-efficient and metastability-immune timing-error detection and instruction-replay-based recovery circuits for dynamic-variation tolerance. In: *Digest of Technical Papers IEEE International Solid-State Circuits Conference (ISSCC 2008)*, 3–7 February 2008, pp. 402–403, 623.

[19] D. Blaauw, S. Kalaiselvan, K. Lai, W.-H. Ma, S. Pant, C. Tokunaga, S. Das, and D. Bull, Razor II: In situ error detection and correction for PVT and SER tolerance. In: *Digest of Technical Papers IEEE International Solid-State Circuits Conference (ISSCC 2008)*, 2008, pp. 400–401, 622.

[20] K. A. Bowman, J. W. Tschanz, N. S. Kim, J. C. Lee, C. B. Wilkerson, S.-L. L. Lu, T. Karnik, and V. K. De, Energy-efficient and metastability-immune resilient circuits for dynamic variation tolerance. *IEEE J. Solid-State Circuits*, 44(1), 49–63, 2009.

8

Physical Design Considerations

Georgios Konstadinidis

Sun Microsystems

8.1 Introduction

Optimization of the structural skew alone, through balanced H or grid clock design and load balancing, is not adequate. Process, voltage, and temperature (PVT) variations dominate in most cases the total clock skew. While active deskew circuits [1–8] (also thoroughly described in Chaps.2 and 7, and the use of asynchronous FIFOs in clock domain crossings [9] can reduce the effect of skew, we still need to reduce the variation in the clock network to minimize the overall complexity and design effort. Higher skew would require larger number and extended range in the deskewing circuits. Increased clock skew would mean larger hold time violations that would require additional delay elements inserted in the critical path. This will increase area and power. This chapter focuses on physical design considerations to help minimize overall skew and to avoid overdesign. At first, we provide an overview of various skew components and explain their dependency on the process, voltage, and temperature variations. We describe the main sources of transistor variation including lithographic, layout, proximity, and strain related effects. Similarly, interconnects suffer from lithographic, process, and pattern density effects that add to the variability. We provide recommendations on how to optimize the layout to minimize both the transistor and interconnect variations, and provide answers to fundamental clock designer questions:

- How should I calculate the total process variation along a chain of clock buffers? Should I just add the variations of the individual stages or should I use a Root Mean Square approach? (It turns out none of the above approaches is correct if used in isolation.)
- What is the best approach in dealing with the voltage variation? Do all clock buffers see the same voltage variation?
- How does the temperature variation affect clock skew, and are there ways to compensate for this?
- What is the impact of inductance?

T. Xanthopoulos (ed.), *Clocking in Modern VLSI Systems*, Integrated Circuits and Systems, DOI 10.1007/978-1-4419-0261-0_8, © Springer Science+Business Media, LLC 2009

Good understanding of the behavior of correlated vs. noncorrelated parameters will help us define the correct methodology for the delay variation estimation using well-established statistical techniques.

8.2 Clock Skew Components

Clock skew is the difference in the clock arrival time at any source-destination pair of clocked elements. The setup time skew refers to the clock arrival difference (modulo the nominal clock period) at two subsequent clock cycles, while hold time skew applies to the same clock edge [10, 11]. Clock frequency reduction can fix setup related failures. However, failures due to hold time skew are independent of the clock frequency and lead to functional failure and yield loss.

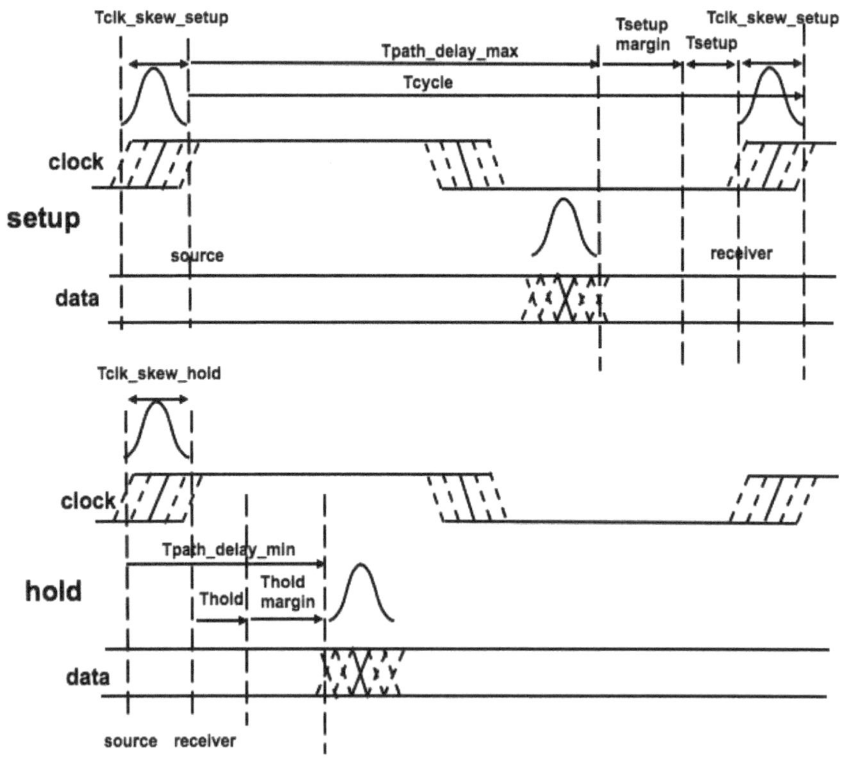

Fig. 8.1. Setup and hold time skew definitions considering variation

Figure 8.1 provides the definition of the setup time and hold time skew. The data need to arrive a certain "setup" time earlier than the clock edge. Process, voltage, and temperature variations may cause the late clock arrival at the source flop and perhaps early arrival at the destination flop. This reduces the usable time for logic operations within a clock cycle. In addition to that, the transistor variation from die to die and within the same die can create extra variation in the datapath. The setup check calculation needs to account for the late arrival of the data and potential early arrival of the clock at the receiver flop as shown in Fig. 8.1. The setup margin is defined as:

$$T_{\text{setup\_margin}} = T_{\text{cycle}} - T_{\text{Path\_delay\_max}} - T_{\text{Clk\_skew\_setup}} - T_{\text{setup}}, \tag{8.1}$$

where T_{cycle} is operation cycle time, $T_{\text{Path\_delay\_max}}$ is the maximum path delay under worst case input conditions in the combinational logic and source flop that result in the largest path delay, $T_{\text{Clk\_skew\_max}}$ is setup time clock skew, and T_{setup} is the setup time of the receiver clocked element. As we will see in the following paragraphs, there is a difference in the clock skews for setup and hold time analysis. Correct functionality is guaranteed for $T_{\text{setup\_margin}} \geq 0$. Silicon setup failures are fixable by increasing T_{cycle}, i.e., by reducing the operational frequency.

Achieving zero skew is not always the target. In many cases, that involve unbalanced timing paths, the designers add intentional skew in the path in order to "steal" time from the previous or the next cycle that has more margin. However, any deviation of the skew from their intended targets in a nonzero skew clock tree will degrade the performance of the design in the same manner as it does for a zero-skew clock tree. This means that in all cases we need to minimize variation. In the subsequent sections, we assume that the load is balanced through appropriate buffer sizing and interconnect clock network optimization and that we need to deal only with the process, voltage, and temperature variations.

In the hold time case, the data need to be stable for a certain time after the clock edge to guarantee correct capture. The clock at the source flop may arrive early while it can be late for the receiving flop. The datapath delay will also have a delay distribution due to PVT variations. The hold time analysis needs to account for the fast datapath delays. The hold time margin defined as:

$$T_{\text{Hold\_margin}} = T_{\text{Path\_delay\_min}} - T_{\text{Clk\_skew\_hold}} - T_{\text{hold}}, \tag{8.2}$$

where $T_{\text{Path\_delay\_min}}$ is the minimum possible path delay considering the worst case inputs for the combinational logic and the source flops, $T_{\text{Clk\_skew\_hold}}$ is the clock skew for the hold time analysis, and T_{hold} the receiving flop hold time. $T_{\text{Hold\_margin}}$ should be greater or equal to zero for correct functionality. Since the hold time check refers to the same clock edge, hold time failures cannot be corrected by reducing the frequency of operation. Due to the larger risk for yield loss, it is typical that the designers add extra margin in the hold time analysis. In order to maximize yield especially in ASIC designs, the timing closure methodology calls for simulations covering the worst case process and environmental variations. While this approach increases yield, it frequently leads to overdesign, and it restricts chip performance due to the use of worst case margins.

Statistical clock skew analysis including both device and interconnect variation is essential in order to calculate more accurately the impact of the numerous variation components and to reduce necessary margins. The chip-mean distribution shown in Fig. 8.2 [12] represents the die-to-die and wafer-to-wafer distribution or so called interdie distribution. For a given point on the chip mean distribution, e.g., corresponding to a fast part, there will be a distribution of transistor and interconnect parameters across the chip as well. These intradie or across-chip variations (ACV) have both systematic and random components.

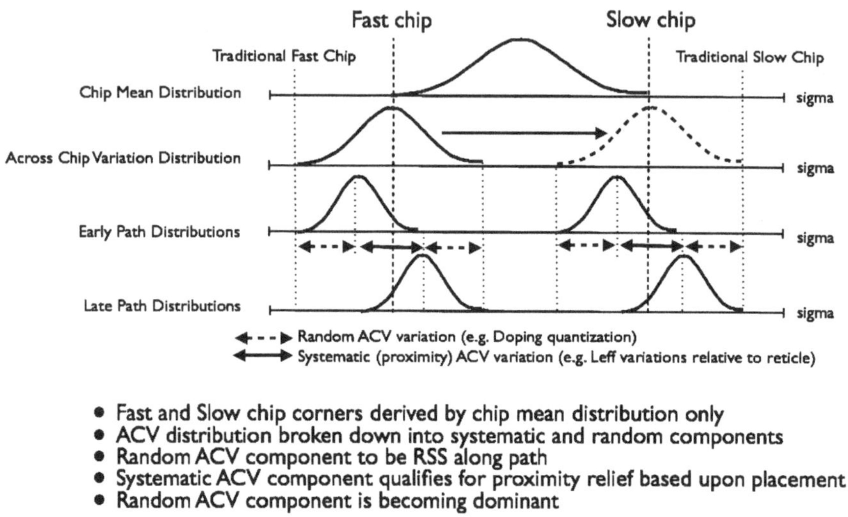

Fig. 8.2. Die-to-die and within-die random and systematic variation distributions (courtesy of N. Rohrer and P. Habitz). Reproduced with permission from [12], ©2006 IEEE

The systematic component is influenced by layout topology [13–15] and global scale process steps like chemical mechanical polishing (CMP). The use of localized SiGe in the PMOS devices and dual stress liners [16–28] to improve mobility add a strong systematic layout dependent component. The transistor performance will vary depending on the gate-to-gate spacing, the gate to diffusion edge distance and proximity to other devices. Scattering of moderate-energy well ion implant species at the edge of the photoresist layer used for well formation has been shown to cause a systematic elevation of threshold voltage [29–32]. Random dopant fluctuations become a substantial part of the threshold voltage variation due to the small gate area as result of aggressive process scaling. Line-edge roughness [33–35], gate oxide variation, interface charge nonuniformities [29, 30, 33–43], pocket implant variations [43] and rapid thermal annealing process steps [42] affect the transistor performance. Some of the effects are spatially correlated. Devices in close proximity and of similar layout style behave similarly to each other, while the behavior becomes less and less correlated as the distance between the devices increases. Figure 8.3 gives an overview, basic classification, and examples of the correlated vs. noncorrelated parameters and their spatial dependencies [44]. The predominant components will be analyzed in more detail in the next sections.

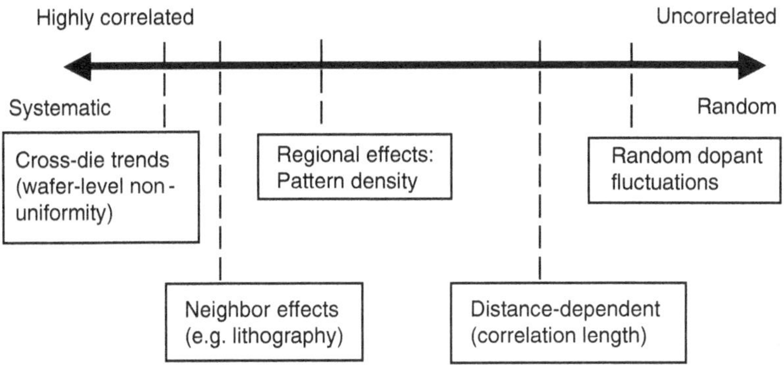

Fig. 8.3. Correlated vs. noncorrelated parameters and their spatial dependencies. Reproduced with permission from [44], ©2008 IEEE

In addition to process induced variations, the transistor drive current variation has a dynamic component that depends on supply voltage variations. The impact on the clock buffer behavior will be described in more detail in Sect. 8.4. The wire delay depends not only on systematic layout related and CMP effects, but it also has an additional component that depends on noise due to the activity of the adjacent nets. This component depends on the workload and is technically deterministic, but due to the computational intractability, it is usually modeled as random. Clock jitter is another substantial component of clock skew. This variation is not correlated to the rest of the skew components and can be treated as random. Chapter 5.1 provides a very detailed PLL jitter analysis.

The best way to cope with the various sources of variations and their effect on clock skew is to use well established statistical methodology techniques and not follow a worst case analysis that would lead to overdesign. A statistical model for clock skew based on Monte Carlo analysis is described in [45]. It describes a skew budgeting methodology incorporating both clock and datapath delay variations. It is shown that considering the variations in the datapath, as well as in the skew calculation, is essential to avoid over budgeting for global skew. The analysis in [45] shows that the jitter induced by power supply noise on the clock buffers is the dominant source of clock skew in H-tree networks. It is also shown that the number of paths between clocked elements has an important influence on the most appropriate clock skew budget used in the design phase.

The approach proposed [46] uses joint probability density functions (JPDFs) that preserve the correlation between minimum and maximum delays. The proposed method computes the skew distribution for the entire clock tree as well as the skew distribution of all subtrees simultaneously and, therefore, allows the designer to identify which portions of the clock tree are most prone to process variations.

An analysis of the impact of die-to-die and within-die parameter fluctuations on the maximum clock frequency is provided in [47]. The maximum operation frequency depends on both the clock skew and the datapath variation. It is shown that the within-die fluctuations primarily affect the mean frequency of operation, while die-to-die fluctuations determine the major component of the maximum frequency variance.

Since the Monte Carlo analysis is very time consuming, especially if applied to the whole clock network, the method proposed in [48] includes a three-phase approach to minimize the overall clock analysis effort. The first phase consists of a pruning of noncritical clock network portions, followed by a standard skew analysis with full extraction. Statistical analysis is applied as a third step only to the most critical portions of the clock network.

Despite the fact that general purpose or clock specific statistical timing simulators [49–51] can be used to analyze the clock network and datapath delay sensitivity under process variations, it is important that the designer acquires very good insight on what to anticipate. This basic understanding of the various interactions and process variation sources will be very useful in making the appropriate tradeoffs to achieve a robust design. In the following, we will describe how the correlated and noncorrelated parameters should be treated in the clock skew and datapath analysis. Furthermore, in subsequent sections we will provide more details on the various sources of variations that can affect a digital network delay and identify ways to minimize their impact.

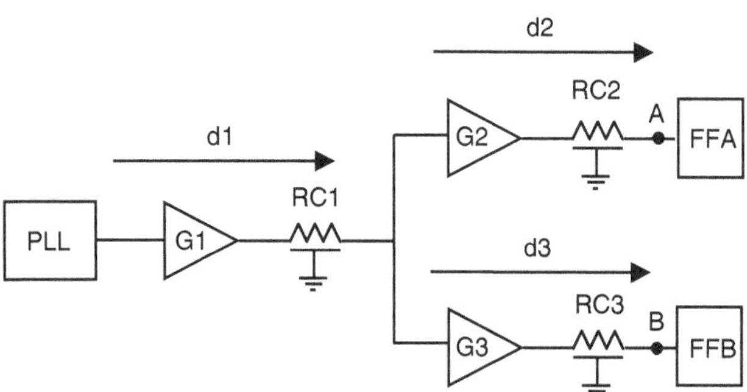

Fig. 8.4. Simplified clock network

Figure 8.4 illustrates a subset of a clock network to help define the various skew components and show the impact of systematic and correlated variations as opposed to the random ones.

8.2.1 Setup Time Skew

Using the sample clock network shown in Fig. 8.4 as an example, the skew between points A and B for the setup time case can be expressed as

$$T_{\mathrm{Clk\_skew\_setup}} = t_{\mathrm{A}} - t_{\mathrm{B}}, \tag{8.3}$$

where t_{A} and t_{B} are the arrival times at points A and B, respectively:

$$t_{\mathrm{A}} = d_1 + d_2, \tag{8.4}$$

$$t_{\mathrm{B}} = d_1 + d_3, \tag{8.5}$$

where d_i represents the combined gate G and RC delay of stage i:

$$d_i = G_i + RC_i, \ i = 1,2,3. \tag{8.6}$$

These stage delays have a systematic and random component and the stage delay can be expressed in the simplified form:

$$d_i = d_{i\_\mathrm{sys}} + d_{i\_\mathrm{random}}, \ i = 1,2,3. \tag{8.7}$$

The systematic component will depend on the layout and load conditions and will remain constant over time for a given chip. The random components may vary over time even for the same chip. Such random variations include voltage, noise, and temperature variations. These effects will be examined in more detail in the next sections. There will also be random variations that are constant over time but still not predictable. One such example is random dopant fluctuations. These variations may not be Gaussian. Proper analysis of the composite variation will require Monte Carlo analysis. Using basic statistical analysis, the skew can be expressed as:

$$\begin{aligned} T_{\mathrm{Clk\_skew\_setup}} &= t_{\mathrm{A}} - t_{\mathrm{B}} = E[t_{\mathrm{A}}] - E[t_{\mathrm{B}}] \\ &\pm M\sqrt{\sigma_{t_{\mathrm{A}}}^2 + \sigma_{t_{\mathrm{B}}}^2 - 2\rho\sigma_{t_{\mathrm{A}}}\sigma_{t_{\mathrm{B}}} + \sigma_{\mathrm{jitter}}^2} \end{aligned}, \tag{8.8}$$

where $E[t_{\mathrm{A}}]$ and $E[t_{\mathrm{B}}]$ are the mean values of the path delays, $\sigma_{t_{\mathrm{A}}}$ and $\sigma_{t_{\mathrm{b}}}$ are the corresponding standard deviations, and ρ is the relevant correlation coefficient. M is the standard deviation multiplication factor selecting the degree of confidence. While (8.8) is valid for even non-Gaussian distributions, the interpretation of the factor M will be different, depending on the actual distribution. For Gaussian distribution, $M = 3$ will correspond to 99.8% confidence level. The exact M number will depend on the number of critical timing paths only (not all paths). Increased number of critical paths will require larger values of M to guarantee a certain final yield. The correlation coefficient ρ is 0 for statistically independent variables and 1 for completely dependent ones. In general, partially dependent variables will have values of ρ between 0 and 1. The delays are partially dependent on each other, since the gate delay will depend on the input slew rate that depends on the delay of the previous stage. On the other hand, the delay of the previous stage depends on the load that varies when the input capacitance for the following stage is varying due to gate oxide or channel length variations. By substituting (8.4)–(8.7) in (8.8), the skew can be expressed as:

$$T_{\text{Clk\_skew\_setup}} = (E[d_1]+E[d_2]) - (E[d_1]+E[d_3]) \pm M\sqrt{\sum_{i=1}^{3}\sigma^2_{d_{i\_\text{random}}} + \sum_{i=2}^{3}\sigma^2_{d_{i\_\text{sys}}} - 2\sum_{i,j=2:i<j}^{3}\rho_{i,j}\sigma_{d_{i\_\text{sys}}}\sigma_{d_{j\_\text{sys}}} + \sigma^2_{\text{jitter}}}. \tag{8.9}$$

As seen in (8.9), the mean value of the common path $E[d_1]$ does not contribute to the clock skew, and this is a very important property we need to take into account. By removing the common $E[d_1]$ delay, (8.9) is simplified to:

$$T_{\text{Clk\_skew\_setup}} = (E[d_2]-E[d_3]) \pm M\sqrt{\sum_{i=1}^{3}\sigma^2_{d_{i\_\text{random}}} + \sum_{i=2}^{3}\sigma^2_{d_{i\_\text{sys}}} - 2\sum_{i,j=2:i<j}^{3}\rho_{i,j}\sigma_{d_{i\_\text{sys}}}\sigma_{d_{j\_\text{sys}}} + \sigma^2_{\text{jitter}}}. \tag{8.10}$$

In the case of gates in close proximity, the systematic variation that mainly depends on the layout style will be in the same direction, indicating a correlation coefficient value closer to 1. In the special case of gates in close proximity with same layout and load conditions ($\rho_{2,3} = 1$ and $\sigma_{d_{2\_\text{sys}}} = \sigma_{d_{3\_\text{sys}}}$), the systematic variation terms in (8.10) will null each other out resulting in reduced local clock skew:

$$T_{\text{Clk\_skew\_setup}}(\text{local}) = (E[d_2]-E[d_3]) \pm M\sqrt{\sum_{i=1}^{3}\sigma^2_{d_{i\_\text{random}}} + \sigma^2_{\text{jitter}}}. \tag{8.11}$$

Equation 8.11 can be applied for skew estimation for gates up to a couple of hundred microns apart. In the other extreme case for gates across the chip, the covariance term will be zero since the gate delays will be completely independent from each other. The skew across chip will be:

$$T_{\text{Clk\_skew\_setup}}(AC) = (E[d_2]-E[d_3]) \pm M\sqrt{\sum_{i=1}^{3}\sigma^2_{d_{i\_\text{random}}} + \sum_{i=2}^{3}\sigma^2_{d_{i\_\text{sys}}} + \sigma^2_{\text{jitter}}}. \tag{8.12}$$

This will be obviously the worst case skew. For intermediate distances where $0 < \rho_{2,3} < 1$, the systematic variation will be partially reduced providing skew values that are within the range defined by (8.11) and (8.12). The root mean square (RMS) term in the (8.10)–(8.12) indicates an averaging effect of the variation components. In the simplified case where all delay components and variations are equal, the skew will be proportional to the stage delay variation times the square root of the number of stages N, rather than N itself. Consequently, simply adding the variation per stage would have resulted in very pessimistic skew estimation.

While the above skew estimation procedure has been applied to a very small network for simplicity, it can easily be extended to large clock tree networks. The total skew will be proportional to the total number of stages in a clock path, and hence proportional to the latency from the PLL output along the clock distribution and grid down to the final clock buffer. This latency is on the order of 700ps to 1ns and the skew is typically on the order of 10% of the cycle time. Typical clock skews values are in the range of 50–70ps when including PVT variations [45].

The final product frequency distribution is affected by skew, setup times, and path delays. Path delays needs to be analyzed in a fashion similar to clock branch delays. The fact that the systematic uncorrelated components are added directly to the cumulative delay causes larger datapath delays overall and will adversely affect frequency distribution. Random variations provide an averaging effect that can reduce total path delay. Statistical timing analysis of both skew and datapath delays is essential to minimize pessimism and prevent overdesign.

8.2.2 Hold Time Skew

The big difference between the setup and hold time case is that the hold time skew refers to the same clock edge. In this case, the PLL jitter and the common path delays will be removed from the skew calculation. This is valid for both random and systematic components since they will have just one value at a given time, and can be removed from the skew as common path delay components. In the case of setup time skew, these components may vary from cycle-to-cycle and, therefore, cannot be removed as common path delays. In the hold time case and for the example of Fig. 8.4, components d_{1_random}, d_{1_sys}, and $\sigma^2_{\mathrm{jitter}}$ are common and, therefore, removed:

$$T_{\mathrm{Clk\_skew\_hold}} = (E[d_2] - E[d_3]) \pm M\sqrt{\sum_{i=2}^{3}\sigma^2_{d_{i\_\mathrm{random}}} + \sum_{i=2}^{3}\sigma^2_{d_{i\_\mathrm{sys}}} - 2\sum_{i,j=2:i<j}^{3}\rho_{i,j}\sigma_{d_{i\_\mathrm{sys}}}\sigma_{d_{j\_\mathrm{sys}}}}. \tag{8.13}$$

8.2.3 Half-Cycle Setup Skew

In this case, the source uses the rising clock edge and the receiver uses the falling edge. The skew in a half-cycle path will be the same as for the single-cycle skew with the addition of the PLL duty cycle variation in the RSS component.

8.2.4 Multiple-Cycle Setup Skew

In the multiple-cycle paths case, the skew will be similar to the single-cycle one. The main difference is the potential extra voltage and temperature variation that may occur during the longer multiple-cycle path.

8.2.5 Grid or H-Tree?

Additional averaging effect will result by using a grid rather than a clock network as long the resistance of the wires connecting the clock buffer stages is smaller than the output impedance of the drivers. In the opposite case, the resistive shielding will keep the buffers in isolation and there will be no skew averaging effect. In such a case, the grid will just increase the total power due to the added capacitive load, without offering any skew reduction benefit. The other issue with the use of a grid is that timing tools have a difficult time analyzing this type of networks. The use of H-tree is also necessary in the case of clock gating, since the clock buffers need to be turned off individually.

8.3 Transistor Variation

This section provides a description of the various sources of variation in the transistor that have an impact on the clock skew. The transistor channel length variation due to lithographic and several layout dependent effects is a major contributor to the clock delay variation and will be described in Sect. 8.3.1. Random dopant fluctuations affect smaller transistors and they will be discussed in Sect. 8.3.2. Layout dependent effects gained significant importance after the 65nm node and have major impact on transistor performance. Advanced technology nodes use strain in the channel region to improve mobility and the layout style has a major impact on the modulation of this effect. Section 8.3.3 provides a brief overview of these effects and potential remedies.

8.3.1 Channel Length Variation

The delay of a clock driver in a first order approximation can be expressed as:

$$\text{gate\_delay} = \frac{C_{\text{load}} \times V_{\text{DD}}}{I_{\text{drain}}}, \tag{8.14}$$

where C_{LOAD} is the capacitive load, and V_{DD} is the supply voltage. The drive current is inversely proportional to the channel length and since the gate capacitance is proportional to the transistor width W and channel length L, the delay is proportional to L^2:

$$C_{\text{load}} \propto W \times L, \tag{8.15}$$

$$I_{\text{drain}} \propto \frac{W}{L}, \tag{8.16}$$

$$\text{gate\_delay} \propto L^2 \times V_{\text{DD}}. \tag{8.17}$$

This means that small variations in the channel length will have a quadratic impact on the gate delay. Short channel effects amplify further this effect by changing the behavior of threshold voltage vs. channel length. Figure 8.5 shows the threshold voltage variation vs. channel length in the linear and saturation regime [26]. At longer channel lengths, the transistor threshold voltage and consequently the drive current variation is less sensitive to channel length variations than in the case of shorter channel

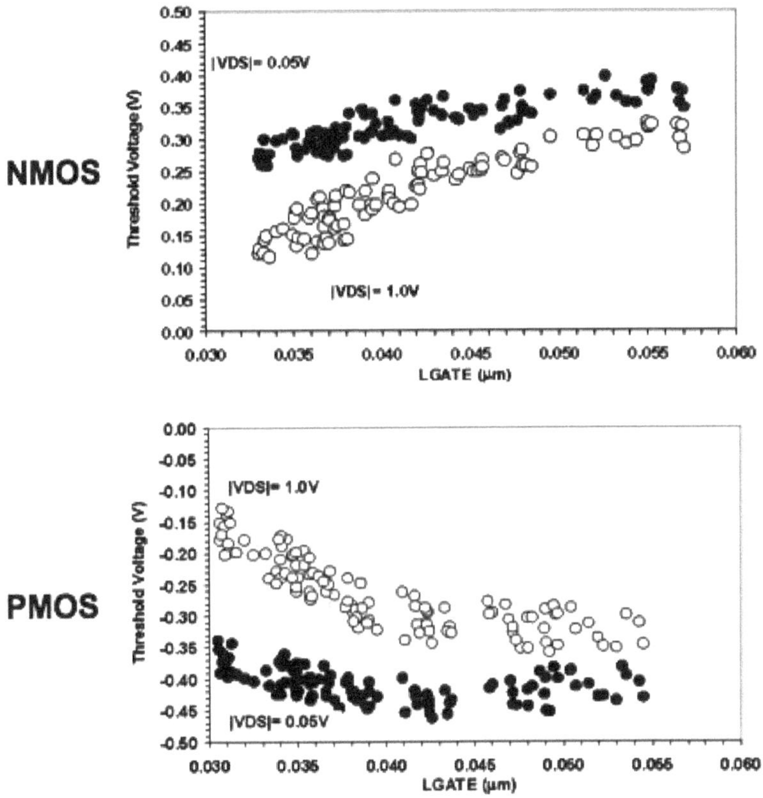

Fig. 8.5. Threshold voltage variation vs. channel length at VDS=0.05V, 1.0V. Reproduced with permission from [26], ©2007 IEEE

lengths. Although it is tempting to use longer channel lengths across the board to reduce variation, this would increase path delay, hold time requirements, and power. Ideally, both the datapath and the clock network should use similar channel length devices so that the minimum delay and the hold time requirement track each other in the presence of variations. This is also important when designing a chip that needs to be portable across process nodes. The datapaths that use long channel devices will not speed up in the new node as much as the paths with short channel devices. While using long channel length in short datapaths will not be a problem, substantial redesign will probably be required if the datapath uses short channel devices while the circuitry that affects hold time uses longer channels. In this case, the hold time will not decrease as much as the short datapath will speed up and hold time violations may be created.

The channel length variation has a systematic and a random component with different values for the die-to-die and within-die cases that depend on the layout style. Local variations are usually small and more systematic and correlated (local photolithographic effects), while the across-chip variations are larger and more random in nature. At the local level, variatons result in limited clock skew, and this is especially important in the case of hold time violations that typically happen locally (back-to-back flops). This type of clock skew relaxation can eliminate thousands of false hold time violations and reduce overdesign. On the other hand, systematic variation across chip may be of different sign, depending on location. Systematic variations at point A may be correlated in one direction, e.g., resulting in increased clock branch delay around point A, while around point B on the other side of chip, the delays may be correlated in the opposite direction, resulting in delay reduction in another clock branch. The consequence will be larger across chip skew due to channel length variations. The following sections provide some background regarding root causes.

Photolithography Challenges

Since the printing features are smaller than the wavelength being used, special optical proximity corrections (OPC) are employed to help print transistor gates [13, 14]. Critical layers in 90nm and 65nm process are exposed with ArF excimer laser light (193nm wavelength). Figure 8.6 [13] shows the minimum device size trend and the exposure wavelength used over time. Starting with the 45nm process node, we need to use immersion technology where the gap between the projection lens and the wafer is filled with water that has a higher refractive index than air [14]. In addition, dual patterning techniques and sacrificial print assist patterns are employed to improve printability at the cost of added process complexity. The 22nm process node will more likely need to use Extreme Ultraviolet Lithography (EUVL). Nanoimprint or Electron Beam Direct Write are the future candidates, but they need further development prior to full deployment in production [13, 14].

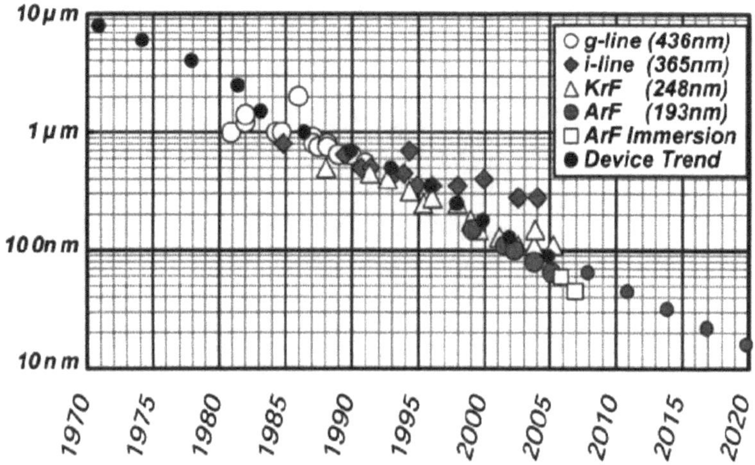

Fig. 8.6. Minimum device size trend and the exposure wavelength used over time. Reproduced with permission from [13], ©2006 IEEE

Apart from the global channel length variation that is more random in nature, there are two very important local components: poly flaring and line edge roughness (LER). Poly flaring is a systematic component that depends on the layout style and can lead to channel nonuniformity and increased leakage. LER is the result of polysilicon grain boundary nonuniformity and is more random in nature. Both of these components are examined in more detail in the following sections.

Poly Flaring and Poly Pullback

The actual structure on silicon is very different from what is drawn in the layout database. Poly flaring as shown in Fig. 8.7 [38] affects the actual channel length at the transistor boundary. The channel width locally is larger causing reduction in drive current. Increase of the distance of the poly jog to active will reduce this effect but it comes at a layout area increase penalty. In addition, the poly extension over active needs to be large enough otherwise the poly pull back during processing will result in a locally very short channel that will cause higher leakage current. This effect will be even worse, if inadequate poly extension is in close proximity to active jogs. The active jog will not be printed as a 90° sharp angle but as a curved edge reducing the poly extension margin and causing higher channel leakage.

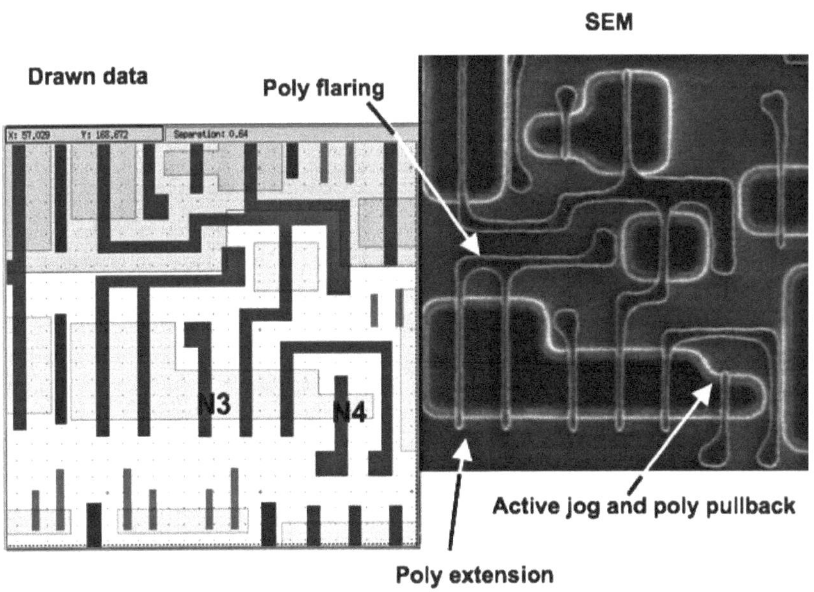

Fig. 8.7. "What you draw is not necessarily what you get on silicon" (courtesy of J. Rosal). Reproduced with permission from [38], ©2006 IEEE

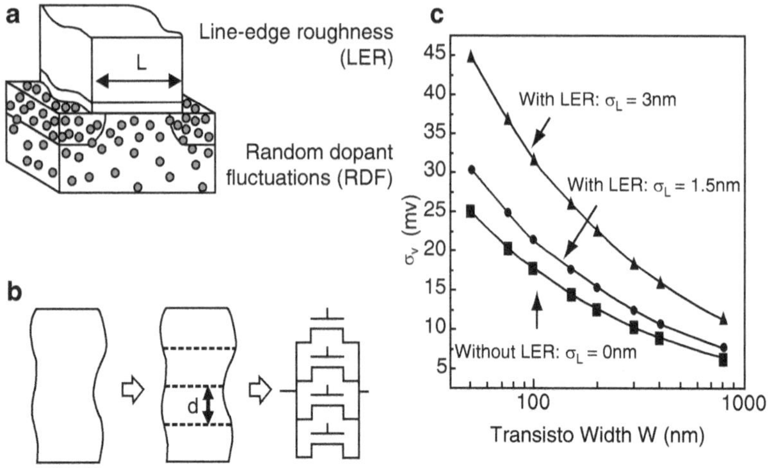

Fig. 8.8. Threshold voltage variation vs. line-edge roughness and transistor width. Reproduced with permission from [35], ©2008 ACM/IEEE

Line Edge Roughness

The poly edge is not a straight line any more exhibiting small curvatures at the edge that are caused by the nonuniformity of the polysilicon grain boundary and photolithographic effects. The end effect is that the channel length is not uniform across the transistor width as shown in Fig. 8.8a, b [35]. This effect is called line edge roughness (LER) and is substantially random in nature. LER impacts the threshold voltage along the transistor width as shown in Fig. 8.8c. The impact is larger for smaller devices and of course it gets worse with larger LER. Figure 8.9 [33] shows actual data from various advanced lithography processes reported by different laboratories indicating that LER does not scale with line-width according to the ITRS roadmap requirements. Circuit simulators that intend to capture the layout dependent effects typically split the transistors in many smaller ones with individual channel lengths per segment. The large number of segments will increase the simulation time. In order to address this issue, instead of multiple segments one can use a single device with the average channel length that is representative of the overall transistor behavior [35].

Channel Length Variation Control

The channel length has a profound impact on speed, leakage, and overall yield. Good control of this variation is essential. This control cannot be achieved by photolithography alone and requires further assistance by design. In the 90 and 65 nm process nodes, some of the design for manufacturability (DFM) rules used to minimize these effects were just guidelines. In 45 nm and beyond, these rules became required design rules. This affects layout density. Some of the rules require a constant poly pitch and

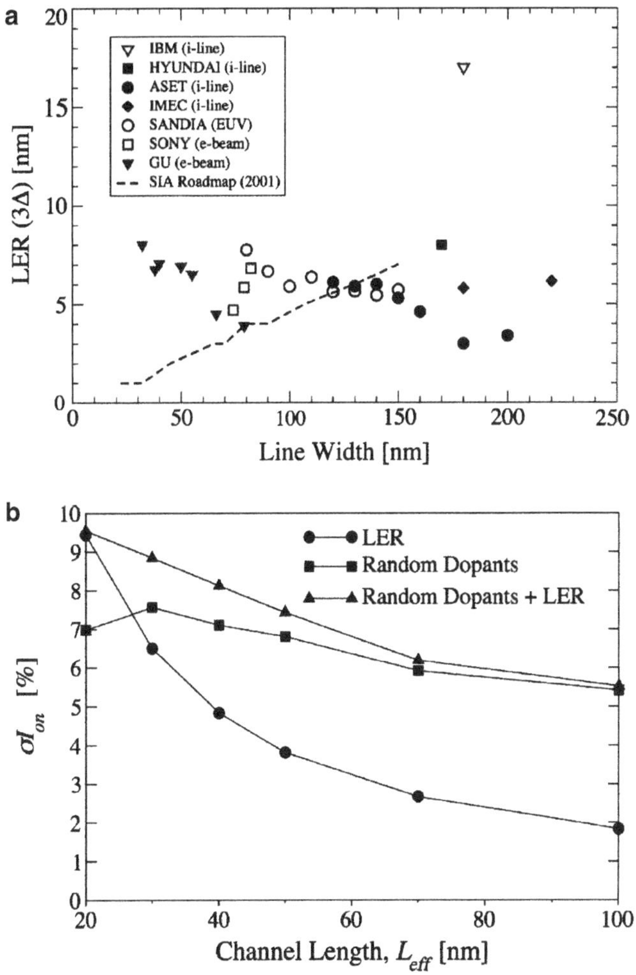

Fig. 8.9. (**a**) Line-edge roughness reported by various labs and ITRS (**b**) Threshold voltage variation vs. channel length, random dopant fluctuation and LER. Reproduced with permission from [33], ©2003 IEEE

same poly orientation (horizontal or vertical) across the chip, adding dummy poly gates [52] next to active gates, and strict poly extension requirements. Jogs of active regions in shared transistors are not recommended since they can cause active flaring and poly pullback. For the same reason, the distance between active edge and poly jog also needs to be increased.

Similar layout style (matching layout) can reduce the systematic variation. Poly shields [52] will help maintain uniformity in the channel length of the external transistor "fingers". Using multi-finger devices for clock drivers is better than single-finger devices, since the intermediate fingers act as a poly shield for the neighbors.

In addition, the random variation within a multi-finger transistor provides an averaging effect on the drive current that minimizes the delay variation in a multi-finger device compared with the delay variation of a single-finger transistor with the same total equivalent width.

8.3.2 Dopant Fluctuation

Narrow width transistors have literally a few atoms in the channel and their fluctuation affects the transistor threshold voltage and consequently the drive strength. This threshold variation is one of the main components of variation in transistors that have the same layout and are close to each other, and otherwise are expected to have similar behavior. The threshold variation is proportional to the inverse of the square root of the gate area $W \times L$ as shown in Fig. 8.10 [39]:

$$V_{\text{threshold\_shift}} \propto \frac{1}{\sqrt{W \times L}}. \tag{8.18}$$

This means that devices with larger gate area will be less impacted by the dopant fluctuations. Therefore, it is preferable to use larger devices in the clock buffers rather than smaller ones. It is a trade-off between power consumption and skew reduction.

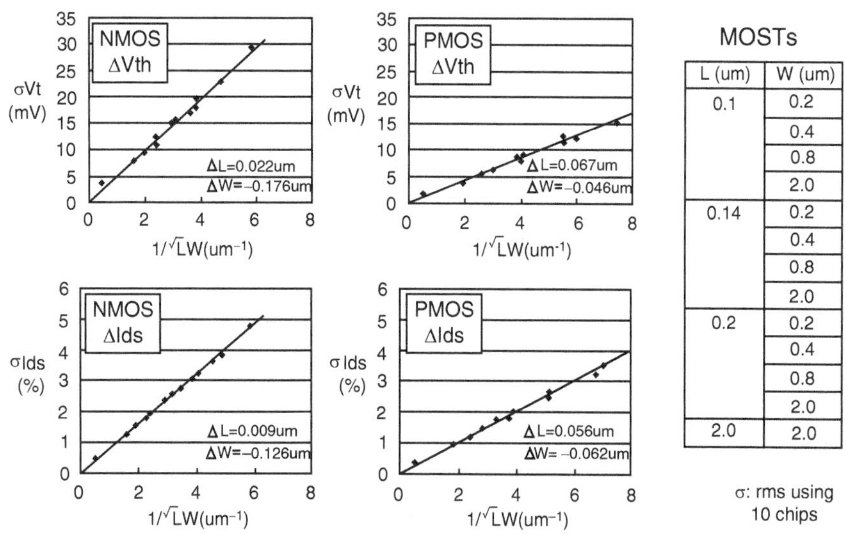

Fig. 8.10. Threshold variation due to dopant fluctuations vs. transistor width and channel length. Reproduced with permission from [39], ©2006 IEEE

The dopant fluctuation issue becomes more significant at thinner gate oxides. The increase in the threshold voltage fluctuations amounts to more than 50% in MOSFET with oxide thickness less than 1.5nm. The poly-Si depletion and the discrete random dopants in the poly-Si gate add an extra 50% in devices with ultrathin oxides [41].

A semianalytical method is proposed in [40] to predict the mean, standard deviation, and probability distribution function of delay in logic circuits considering the random threshold voltage variations, and it was applied to statistical characterization of flip-flops and logic gates.

While this variation is already large and very undesirable, effects like higher local annealing temperature, non uniformities in the gate oxide thickness, and surface states make the problem even worse and require process changes to reduce this variation. Higher local annealing temperature expands junction diffusion both laterally and vertically, resulting in lower V_T due to a combination of short-channel effects and compensation of halo channel doping (for halo/pocket-implanted FETs) [42, 43]. Gate oxide scaling required for better control in the channel area reaches the limits. The gate oxide contains only 4–5 atomic layers. Nonuniformities in the gate oxide thickness and surface states create extra dopant fluctuations [34] and introduce high gate leakage and reliability risks associated with gate oxide integrity. High-K dielectric gates [24, 26] offer a good solution to this problem. This type of gates can achieve the same electrical thickness and control as the traditional gate oxide, but with higher physical gate oxide thickness that reduces gate leakage exponentially [26].

While in the past most of the variability and yield loss came from random defects, in modern technologies yield and performance are substantially impacted by physical layout style and device interaction. Apart from the channel length variation due to lithographic effects as described in previous sections, there are additional effects related to the well proximity effect (WPE), [29–32], and strain effects used to improve the transistor mobility [16–28].

8.3.3 Well Proximity Effect

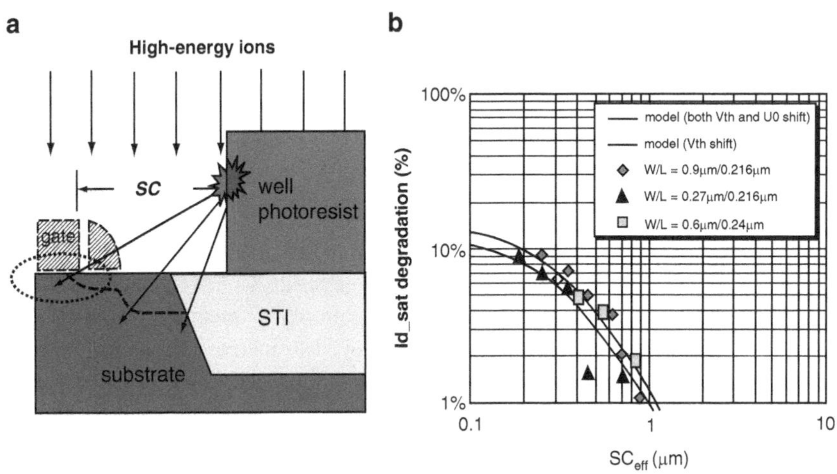

Fig. 8.11. (**a**) Ion scattering at the photoresist edge is the cause of the well proximity effect (WPE) (**b**) Drive current degradation due to WPE vs. gate to well edge distance (SC). Reproduced with permission from [29], ©2006 IEEE

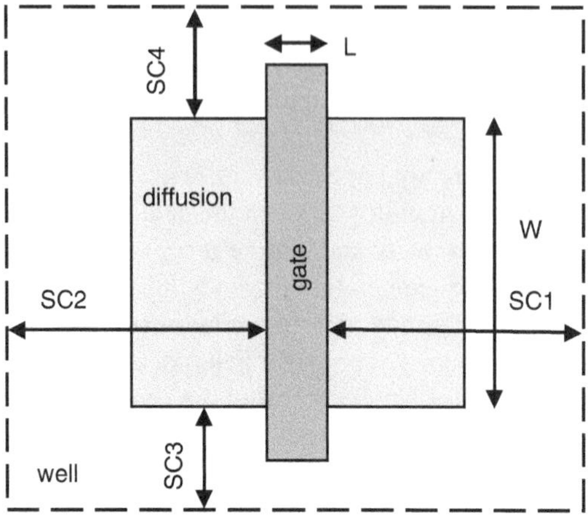

Fig. 8.12. Well to gate distance (SC) definitions [53]. WPE reduces as SC [1–4] increase

Lateral scattering of the ion implants at the photoresist edge when forming the MOSFET wells as shown in Fig. 8.11 can land into the channel area, increasing the threshold voltage of the transistor and, therefore, degrading the transistor performance [29]. The amount of dopants collected in the channel area depends on the distance SC of the poly over active to the well edge as shown in Fig. 8.12. From the physical design perspective, what can be done is to increase the spacing of the well to the active poly. This can usually be done for critical devices but not across the whole chip. Possible process improvement steps are described in [30]. Narrow distances on the order of 100nm can cause a 10% degradation in the transistor drive current as shown in Fig. 8.11 [29]. Similar magnitude of degradation is observed in current mirror devices [31] and in circuit delay [32]. The WPE effects are included in the BSIM 4.6.2 spice model [53]. State-of-the-art layout extractors identify the SC gate distances to the well edge, and these are interpreted as layout input parameters to the spice model.

8.3.4 Strain

Channel length and gate oxide scaling started running out of steam in the 130nm node. Short channel effects cause high source–drain leakage current and gate tunneling effects resulting in exponential gate leakage increase. This prevents further scaling of the channel length beyond $\approx$30nm and gate oxide thickness beyond 1.2nm [16]. New techniques based on strain are employed to increase the mobility while high-K gate dielectrics and new gate materials are now used to reduce gate leakage while maintaining very good drain current control. The methods used to improve the mobility incorporate highly tensile capping layers for NMOS, compressive capping layers for PMOS, stress memorization techniques, SiGe induced strain in the PMOS devices, and shallow trench mechanical stress (STI) [16–28]. All these techniques introduce systematic layout dependent variations affecting both the clock and the datapath delays. The drive strength of a transistor is affected by the poly pitch, the distance of poly gate to STI, and the surrounding devices. Good understanding of these layout and circuit performance interactions is an essential tool to help minimize clock skew and datapath delay variations.

Stress Memorization and Tensile Stress Liner

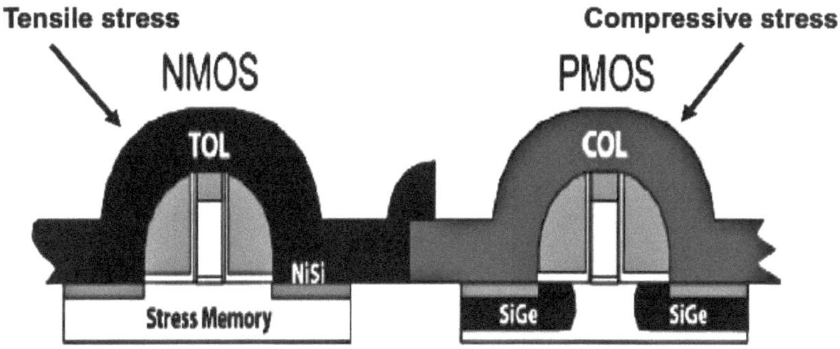

Fig. 8.13. Stress induced mobility enhancement techniques: stress memorization and tensile liner for NMOS, SiGe, and compressive stress liner for PMOS. Reproduced with permission from [22], ©2007 IEEE

The stress memorization technique is applied to the NMOS devices during the source–drain spike anneal [19–22]. This thermal treatment creates stress in the NMOS channel area due to the difference in the thermal coefficients of the Si and the nitride sacrificial capping layer used over the poly gate. This stress is tensile in the direction of the current flow and compressive in the vertical direction [21]. Both of these types of stress enhance the NMOS mobility. Further enhancement is achieved by adding a SiN tensile layer on top of the gate (Fig. 8.13). The stress depends on the intrinsic stress of the material used and its thickness, the poly spacing, the gate height, and the contact to gate spacing. Increased poly spacing will allow for more volume of capping layer to create even more stress [16].

SiGe and Compressive Stress Liner

The PMOS transistor prefers compressive stress parallel to the current flow and tensile stress in the direction perpendicular to the current flow [20]. The compressive cap layer is not as effective as the SiGe [16–19] that is epitaxially grown in the source/drain areas of the PMOS devices as shown in Figs. 8.13 and 8.14. The SiGe has larger lattice constant than Si and introduces compressive stress in the channel. The improvement in whole mobility can be as high as 50% with a 17% Ge composition [17].

The poly pitch optimal distance is not only defined based on the lithographic requirements, but it is also based on the stress inducing capability. Higher gate-to-gate distance will help create more stress due to the higher area of the stress liner

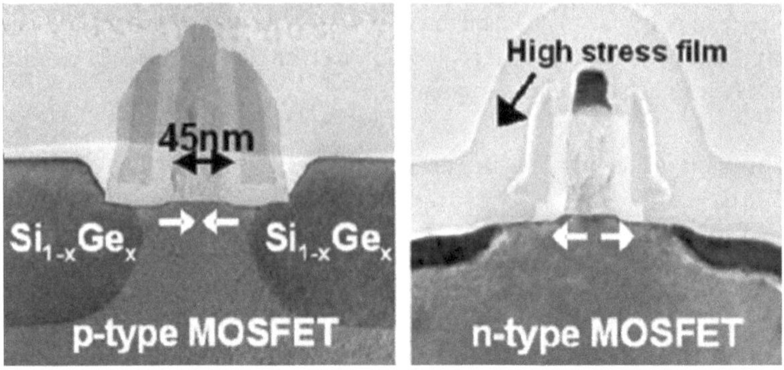

Fig. 8.14. TEM photograph of PMOS with SiGe compressive stress, and NMOS with tensile liner stress in a 45nm Node. Reproduced with permission from [18], ©2004 IEEE

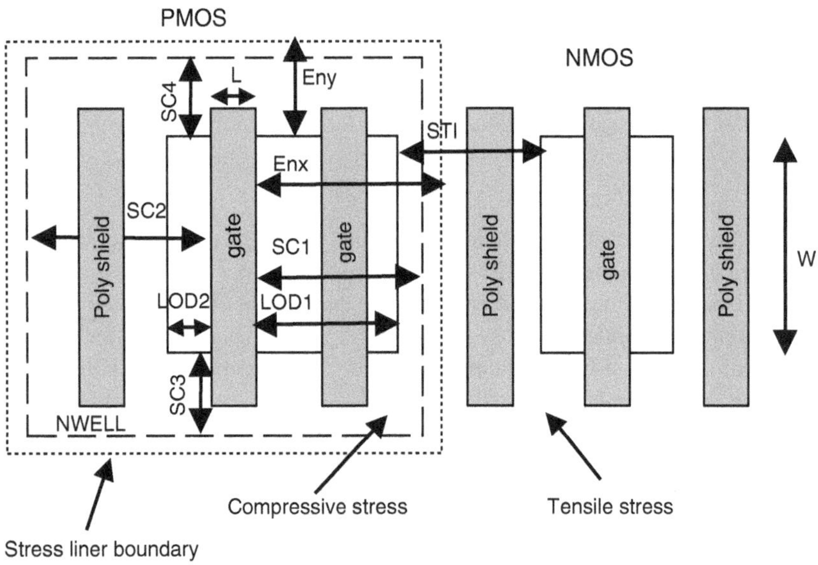

Fig. 8.15. Definition of various gate to diffusion and well distances that impact stress magnitude and consequently transistor performance

and the SiGe volume. Similar arguments apply to the compressive stress liner case. Increased stress liner size in the current flow direction *Enx* as defined in Fig. 8.15 will help increase the compressive stress in the PMOS device. On the other hand the *Eny* size does not have a predominant effect on PMOS performance [20].

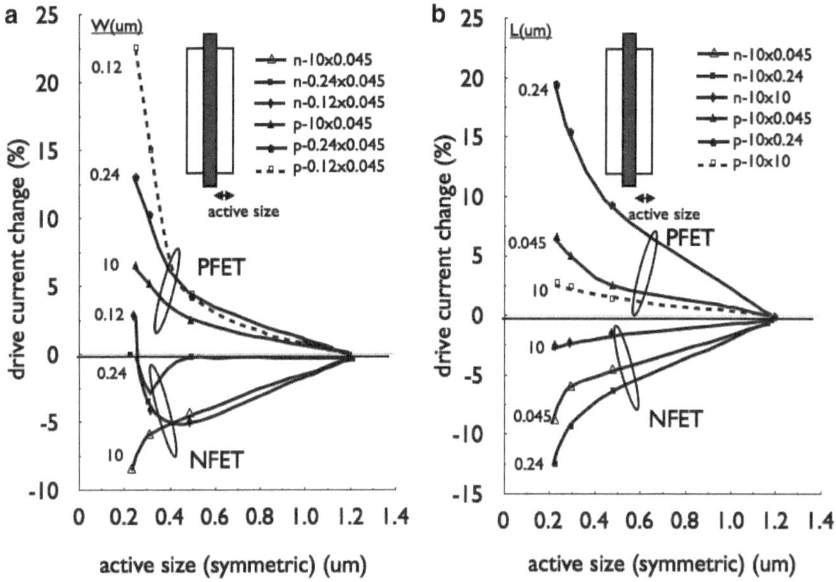

Fig. 8.16. Impact of STI stress effect on transistor drive current vs. active size. Reproduced with permission from [23], ©2003 IEEE

Shallow Trench Isolation

Shallow trench isolation (STI) used to isolate the diffusion areas induces stress between the neighboring devices. This stress is transferred along the device channel and enhances the PMOS mobility, but unfortunately degrades the NMOS mobility. The STI stress depends on the location of the channel within the diffusion region and the STI width defined as STI in Fig.8.15. The STI stress can affect the cell delay by 10–20% depending on the STI-to-poly gate distance [23]. Figure 8.16 shows the NMOS and PMOS current change vs. active area size for various width and channel length combinations. Smaller active size reduces the compressive stress and deteriorates the PMOS device, but it helps the NMOS device that does not fare well under compressive stress. The NMOS is less sensitive to transistor width reduction, but the opposite is true for PMOS. Smaller PMOS width will have more compressive stress and higher drive current. Longer channel devices are relatively insensitive to STI stress. NMOS degrades for intermediate channel lengths and becomes insensitive for small channel lengths. On the contrary, PMOS devices see substantial drive current boost due to STI stress as the channel length becomes shorter. All these layout dependent effects will impact the systematic delay component in the skew calculation as expressed in (8.10).

New Materials

Further enhancement in the transistor performance is achieved by using high dielectric constant gate interface material and metal gates instead of polysilicon. High-K gate dielectrics are used to reduce gate leakage but at the same time offer very good control of the channel field due to the higher gate dielectric material that forms a better capacitor under the gate without requiring further reduction of the gate physical thickness. This improvement combined with metal gates that prevent polysilicon depletion and the strain techniques described above are the predominant performance boosters for the 45, 32, and 22 nm nodes [25, 26]. New, high mobility, narrow bandgap materials like InxGa1-xAs are strong contenders to replace strained-Si channels in technologies beyond the 22 nm node [28].

Guidelines

As explained earlier, the performance improvements based on strain depend strongly on the layout style being used. Matching device layout style will help reduce the mismatch in the clock buffers and minimize skew. Optimized layout to maximize the drive strength will help reduce the buffer size and reduce power. This optimized layout comes typically at the cost of extra cell area and larger parasitics that increase the cell delay and power. Obviously there is an optimal trade-off point beyond which there is substantial area and power penalty without a corresponding speed improvement. State-of-the-art layout extractors take into account all these effects, and spice models like BSIM 4.6.2 [53] incorporate these interactions and model the effect on device performance.

While the layout for the clock headers can be matched and minimize the impact of these systematic layout effects, it is obvious that the generic logic gates can be affected substantially and can introduce systematic delay variations in the datapath delays. Since the next generation technologies are running out of steam in terms of electrical performance improvement, a new layout methodology is needed to achieve the best performance and minimum possible variation. The library cell parasitic extraction and electrical characterization need to take into account the fact that the cell will not be in isolation. The impact of the neighboring devices that can create extra stress and affect the performance of the device being analyzed must be included. Dummy devices need to be placed around the actual cell to be characterized, and the analysis must be performed on a netlist extracted from the complete layout including the dummy devices.

8.3.5 Long Term Effects on Variation

NBTI

The negative bias temperature instability effect (NBTI) [54, 55], impacts the PMOS devices and reduces their current driving capability over time. This is due to threshold voltage increase caused by trapped carriers in the gate oxide especially in the presence of hydrogen. Worst-case bias condition is when the gate of the PMOS transistor is turned on and the supply voltage and temperature are high. There is partial recovery under AC conditions [56–58]. The NBTI shift in the early phases of the stress has an exponential behavior, but after a couple of minutes of stress it follows a power law behavior expressed as

$$V_{\text{Threshold\_NBTI\_shift}} \propto \text{Time}^{\frac{1}{6}}. \tag{8.19}$$

Short testing times may indicate very strong dependency on NBTI due to the initial exponential increase. This early stress recovers over time, but the long term effect will not. NBTI stress will be worse under DC conditions. Lower gate-source voltage will induce less stress and increased switching activity will lead to lower overall stress due to averaging effect of the bias conditions. Lower temperature of operation reduces the NBTI effect in an exponential fashion.

This long-term effect can have major impact on the circuit behavior [59], and proper analysis and margining are required to guarantee correct functionality until the end of the product lifetime [60]. It is typical to add an extra 5% frequency margin to cover for NBTI degradation. While the NBTI impact on the threshold voltage is in the order of 50mV in 65nm for normal SiON oxides as shown in Fig. 8.17, it reduces substantially in the case of high-K dielectric gates [26].

The NBTI degradation will lead to increased latency of the clock network and will increase the rise time at the final clock buffer stage. The amount of skew is proportional to the overall clock network latency and, therefore, NBTI will have an impact on the final skew. However, under symmetrical stress in the clock network the impact on delay will be similar on all clock network branches and, therefore, the impact on skew (difference in clock arrival times) will be minimal. This may not

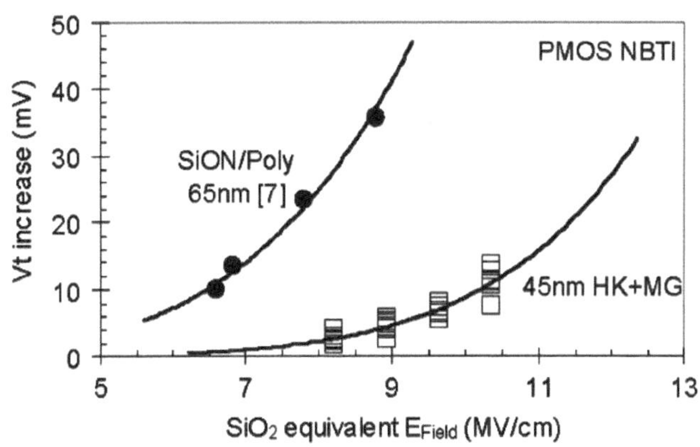

Fig. 8.17. NBTI shift dependency on gate material. High K dielectric reduces NBTI shift substantially at equivalent SiON electric field across the gate. Reproduced with permission from [26], ©2007 IEEE

be the case when extensive clock gating is used. Long term clock gating will affect clock buffers differently, depending on their bias conditions. Every other inverter stage will see a different NBTI stress and this can cause duty cycle shift that reduces margin, and it can lead to timing failures in half-cycle paths.

The NBTI effect represents a systematic delay component and will also have major impact on datapath delay. While the pattern in the clock network is highly predictable if no clock gating is involved, in the case of combinational logic the bias condition patterns will depend on the workload and different workloads will create different degrees of NBTI degradation.

Hot Carrier Injection

At high electric fields in the channel, the carriers may gain sufficient energy and momentum to enter into the gate oxide layer. This effect called hot carrier injection (*HCI*) or channel hot carrier (*CHC*) effect [61, 62] is predominant on NMOS transistors. Electrons are accelerated much more than holes that have higher effective mass. Moreover, the Si–SiO_2 interface barrier is higher for holes. The stored carriers in the oxide increase over time the transistor threshold voltage and reduce the drive current. The degradation will be proportional to the time there is current flow through the channel of the transistor and, therefore, crowbar current, and slow rise times can increase this effect. In the case of clock buffers and repeaters stages in general, one should restrict the output slew rate to minimize HCI. It is also best to use buffer stages built by two inverters rather than single inverter stages since high slew rate at the inverter input will cause higher crowbar current. In the case of two stages that typically use a fanout of 4, the first stage will experience crowbar current but 4 times smaller than in the original single inverter design due to the smaller size of the first stage. The gain from the first stage will help recover the slew rate at the input of the second stage and, therefore, reduce the crowbar current of the output stage. This will also help minimize the electromigration effects at the output of the large driver.

8.4 Voltage Variation

The transistor current drive strength depends strongly on voltage. Typically, we see a 10% speed change for 100mV of voltage variation. In 45nm and beyond, this sensitivity is even larger approaching the 15–30% range at lower supply voltages.

The clock and gate switching during normal operation create voltage transients that affect the actual voltage at the transistor level and consequently affect speed. A dynamic IR analysis will be required to identify the exact voltage variation at the various points of the power grid. The voltage variation in the buffers along the main clock tree depends on the chip switching activity. The largest voltage droop will occur when the final clock headers switch, since the power consumed in this case is a substantial part of the total chip power (typically on the order of 30% as

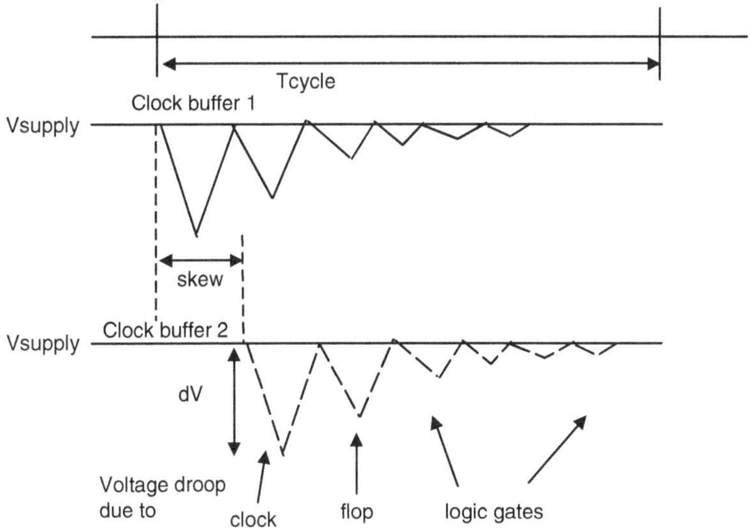

Fig. 8.18. Dynamic voltage drop vs. time, clock, flop, and combinational gate activity

mentioned in Sect. 2.2.3). There will be an instantaneous current and voltage surge at the time the final clock buffers switch as shown in Fig. 8.18. After a certain time, the flops will switch creating a second current surge. Subsequent activity will depend on the logic propagation and will gradually reduce as we reach the next clock edge. The voltage spikes at the main clock buffers can be minimized by appropriate use of decoupling capacitors in close proximity. These decoupling capacitors need to be placed close by so that the RC time constant including the power grid connection resistance is at least 5 times smaller than the rise or falling transition time of the voltage spike. Otherwise, the decoupling capacitors will not be able to reduce these fast voltage transients. Differential structures may also be used for the main clock trunk to minimize voltage variation, but this comes at the expense of increased total power since the bias current source used will consume constant power [63].

In the ideal case where the clock skew is zero, the voltage difference at the final buffer stages would be zero as well. However if the skew is larger than the clock fall/rise time, then the two buffers may see a different voltage variation as shown in Fig. 8.18. The voltage variation to be considered for the final stages is much larger than the variation observed at the main clock tree buffers, since the power surge at the time of final buffer switching is very large. In order to minimize the impact of the voltage variation on the PLL jitter, typically the PLL uses a separate supply than the main supply of the chip. Detailed analysis of the PLL jitter is provided in Chap. 5.1 and an analytical model of supply-induced period jitter for long buffer chains is described in Sect. 6.8.

Apart from the high-speed voltage variation, there is a longer-term voltage variation due to package inductance. Typically, the resonance frequency of the package inductance and chip capacitance (large CPU) is on the order of 50–100MHz. The noise period spans several CPU cycles for a GHz clock design. Package decoupling capacitors are used to reduce the noise amplitude, but they cannot filter out the high frequency spikes on the chip. Even in the case of long term Ldi/dt variations, one should calculate the corresponding voltage variation within one cycle and include it in the setup skew calculation. This is not necessary for the hold time skew since it refers to the same clock edge and the Ldi/dt variation for this short time will be negligible. Special attention should also be paid to multi-core designs with power switches used to power down parts of the design. While this separation reduces overall power, it also reduces the total amount of available decoupling connected to the power supply grid leading to larger Ldi/dt noise [64].

8.5 Temperature Variation

Both transistor and interconnect performance are affected by temperature. Chip temperature depends on workload conditions. The thermal time constant is on the order of a minute, so it does not track rapid workload variations. However, there is typically a spatial temperature gradient across the chip. Floating point units for example will be much hotter than the L2 cache. Figure 8.19 shows a typical thermal map of a multi-core chip under worst case load conditions [65]. The thermal gradient is 40°C across the chip.

In worst case hold time analysis, one should assume that the source flops are at the lower temperature (i.e., fast clock-to-q delay) and the receiving flops are at the higher temperature and lower voltage (i.e., higher hold time). Low temperature will also reduce the wire resistance which helps achieve higher frequency of operation but may cause hold time violations if this effect is not accounted properly in the hold time analysis.

The transistor drive current is proportional to the carrier mobility and (to first order) to the difference of $V_{DD} - V_{threshold}$. The threshold voltage on the other hand is reduced with temperature, so there should be a voltage point for each process where the rate of mobility decrease vs. temperature is compensated by the drive current due to the threshold voltage reduction. At this particular supply voltage, the device delay becomes independent of temperature. This particular aspect of transistor drive current vs. temperature was used in [66] to develop a self-adjusting clock architecture that minimizes the clock skew dependency on temperature variations.

It is important to note that apart from the impact on delay variation there is a positive aspect of the temperature spatial variation. Various failure mechanisms such as gate oxide integrity, electromigration, and NBTI have a strong dependency on temperature. Taking advantage of a spatial thermal map under worst case load conditions as shown in Fig.8.19 can help reduce overdesign since the reliability risk for the areas with lower temperature will be smaller.

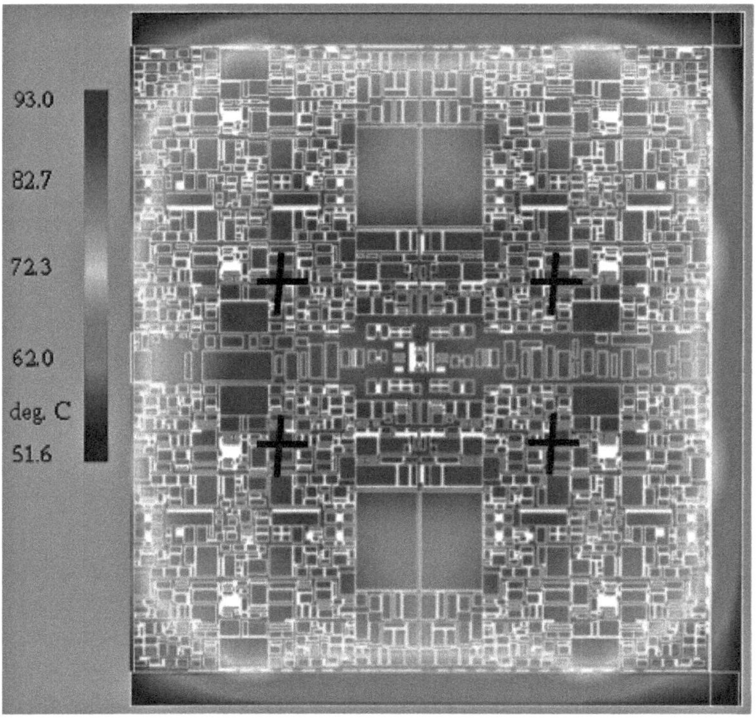

Fig. 8.19. Thermal map of a multi-core microprocessor under worst case workload conditions. the crosses represent the thermal sensor locations. Reproduced with permission from [65], ©2008 IEEE

8.6 Interconnect Variation

Aggressive scaling and continuously higher demand for more dense routing require substantial improvements in the interconnect stack material and physical construction. However, this comes also with the cost of more challenging reliability requirements, and substantial variations in the wire characteristics affecting the wire delay, the clock skew, and consequently the product performance. In this section, we will focus on the nature of these variations and interconnect design requirements for a robust and predictable clock network.

Similarly to the transistor variation, the interconnect variations have systematic and random components. However, in the interconnect case the systematic variations dominate at least for wider interconnects. Appropriate extraction and modeling are required to allow for a predictable clock skew analysis. Critical systematic variations include nonlinear resistance (NLR) effect, selective process bias (SPB) effect, and thickness variations due to etch and chemical mechanical polishing (CMP). As shown in Fig. 8.20 [15], the sheet resistance varies as a nonlinear function of line

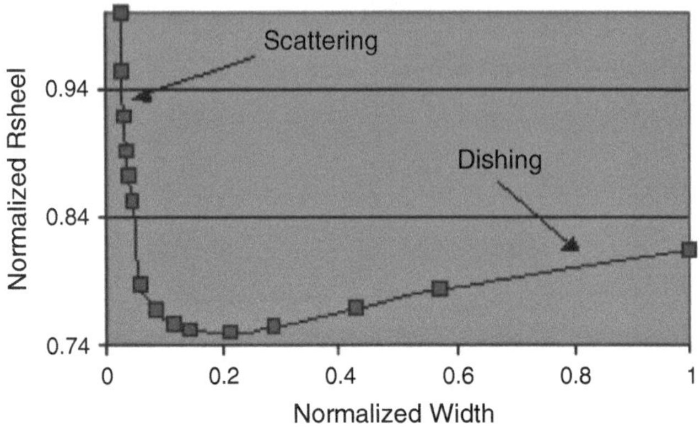

Fig. 8.20. Sheet resistance nonlinear behavior vs. interconnect width. Reproduced with permission from [15], ©2004 IEEE

width. Two phenomena cause significant and systematic changes in sheet resistance as a function of line width for copper technologies. The first one is due to scattering of electrons along the sidewall of the conductor and along the grain boundaries. The second includes systematic changes in copper cross-sectional area as a function of line width that are induced by the damascene process flow. An example of such an effect is CMP-induced copper dishing that would increase resistance. For a given pitch, the width and spacing on silicon will be different from the drawn features. Figure 8.21 shows the difference in the width on silicon vs. the normalized drawn width and spacing [15]. Depending on the width and spacing, the CMP will impact the thickness of the wire. In the end, we have a dependency of R_{sheet} on wire width and spacing as shown in Fig. 8.22 and not only on width as would have been the expectation in the past. Similarly, the coupling capacitance will depend on both the width and spacing and not spacing alone, since the sidewall capacitance depends on the wire thickness that now depends on both wire width and spacing. In addition, R_{sheet} will have a nonlinear dependency on width and thickness, due to the high resistance sidewalls that reduce the effective wire width and thickness. These sidewalls are required to contain the copper and prevent its diffusion to the surrounding materials. The use of ultra low dielectric constant materials ($K = 2.4$) to reduce the overall capacitance has a strong side effect on the mechanical stability [67]. These materials are "porous", not extremely stable and a challenge for packaging. Special thin Nitride layers used as intermediate layers between the metals increase the mechanical stability but also increase the effective dielectric constant and, therefore, increase the final wire capacitance value. Figure 8.23 shows the combined effect of SPB on wire RC vs. width and spacing [68].

Increase of the aspect ratio (metal thickness over width) to reduce the sheet resistance has the side effect of increased sidewall capacitance and sidewall contours as shown in Fig. 8.24 [67] and creates additional variation in the interconnect per-

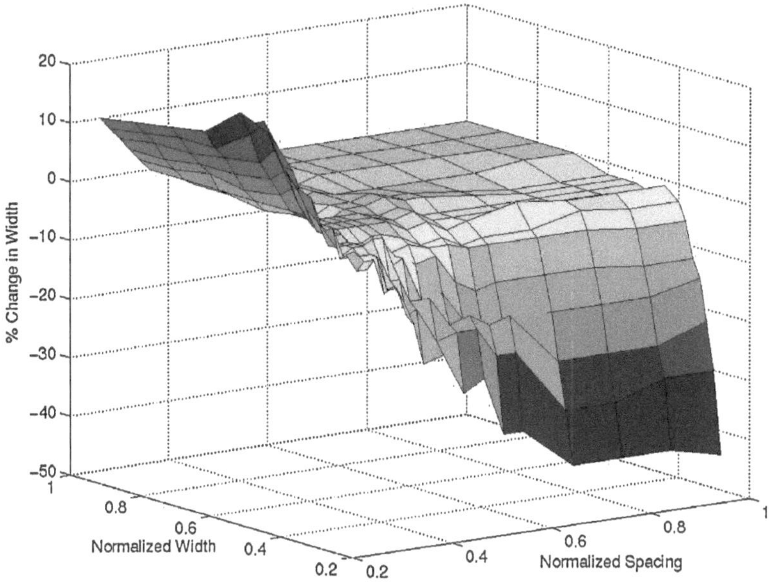

Fig. 8.21. Impact of SPB on interconnect width on silicon vs. normalized drawn width and spacing. Reproduced with permission from [15], ©2004 IEEE

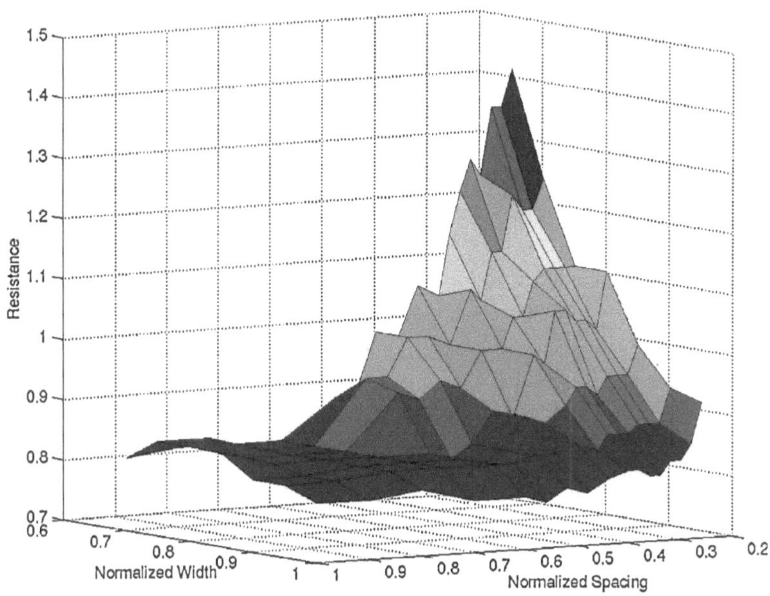

Fig. 8.22. Interconnect sheet resistance dependency on width and spacing – not only width – due to SPB, CMP effects. Reproduced with permission from [15], ©2004 IEEE

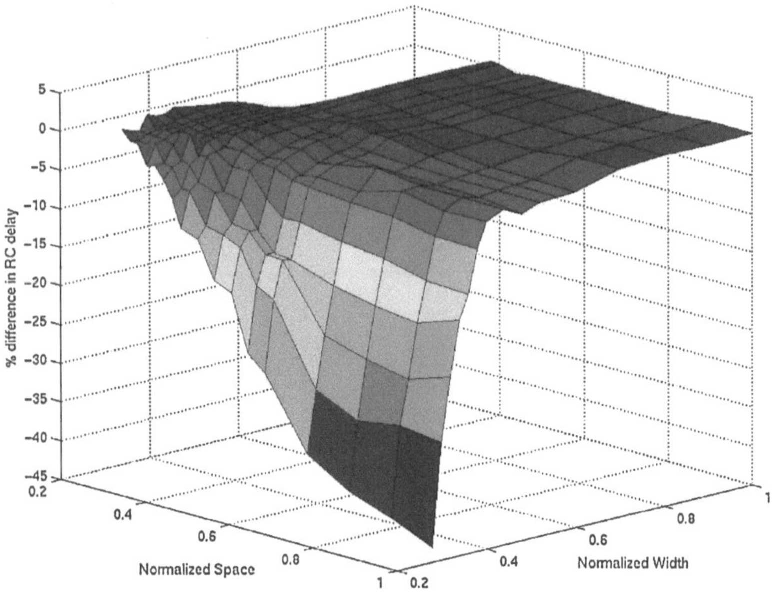

Fig. 8.23. SPB impact on RC delay vs. interconnect width and spacing (courtesy of C. Bittlestone and U. Narasimha). Reproduced with permission from [68], ©2005 IEEE

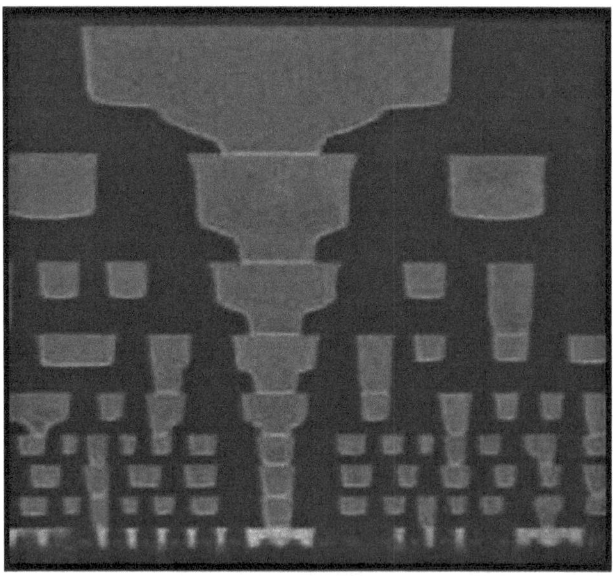

Cross-section of 8 of the 9 Cu interconnect layers

Fig. 8.24. Cross section of interconnect stack of a 45 nm process indicating the contours that affect in a nonuniform manner the sidewall capacitance and sheet resistance. Reproduced with permission from [67], ©2006 IEEE formance. The aspect ratio is typically kept between 1.5 and 2: Lower for the lower level metals where capacitance is more important, and higher for longer upper level interconnects that need to have both low resistance and low capacitance. Due to the higher aspect ratio, the thickness and the sidewall capacitance will be larger, and therefore upper level interconnects will require higher spacing and pitch. This hierarchical stack provides high routing density and low capacitance for local interconnects, and lower routing density but higher performance for upper level interconnects. Typically, the pitch increase in the transition from one routing layer to the next should be accompanied by a similar speed improvement to make the transition worthwhile.

Some examples of the impact of selective process biasing (SPB) on the circuit performance are provided in [69, 70]. Ring oscillator measurements for different interconnect layout patterns show that the stage delay can vary up to 10% [69]. The analysis in [71] shows that the interconnect variations can affect the clock skew by 25%.

The impact of size effects and copper interconnect process variations on the maximum critical path delay of single and multiple core microprocessors is described in [70]. It includes the CMP and edge line roughness effects that will become more important for very narrow interconnects in the future. It looks like that the trend of using multicores instead of monolithic single cores helps contain the issue of variation and speed restrictions posed by long interconnects, since multiple core implementations require typically shorter interconnects. The problem of creating a robust clock automatic routing in the presence of interconnect variations is addressed in [72]. The algorithm calculates the delays and the variations at the various sink points, but instead of using worst case corner approach, it uses a principal component analysis and statistical centering approach to define the optimal clock network implementation considering all sinks simultaneously.

State-of-the-art parasitic extractors account for all these effects, but it is highly desirable for the designer to know in advance what to expect from each interconnect configuration and how to address these issues ahead of time in order to minimize variation. Since these interconnect variations are more systematic, use of similar routing patterns will help minimize variation. Otherwise, one needs to wait for the full extraction and full post layout timing and clock analysis to see the exact result. Isolated lines are impacted more by the CMP and dishing effects and several design rules impose a certain density range per metal layer to minimize this variability and improve yield. In order to achieve this specific metal density, we may need to add metal fill that can be floating or grounded to reduce noise. In order to reduce the coupling capacitance and help speed and power, it is preferable to maximize the spacing of the metal fill to the signal lines. The floating capacitance is typically extracted as grounded and this can affect the final capacitance by a large amount as shown in [15] if the spacing is not adequate.

Additional variation is introduced by capacitive coupling, and this depends on the workload and the resulting signal activity. In setup time analysis in most cases, designers use a worst-case Miller factor of 2 while in hold time analysis, a Miller factor of 0. This can be quite pessimistic and lead to overdesign. Special timing analysis based on timing windows [73] can help reduce this pessimism by calculating the effective Miller factor based on the actual switching activity of the gates in the critical path. Shielding of the clock lines is an effective way to provide a controlled environment for the clock network for both the inductance and noise immunity at minimal expense in routing resources and power for the main clock trunk.

The workload will introduce temperature variations as shown in Fig. 8.19. The wire resistance may vary by 30% going from room temperature to the typical operating condition of 105°C. The temperature gradient will cause additional delay differences that affect both the clock and the datapaths. However, there is a positive aspect in all this. Reduced temperature will help reduce the electromigration risk. A worst case thermal map can be combined with knowledge of the nodal switching activity to develop a more realistic estimate of electromigration risk and help avoid overdesign. In the clock case, we may only need the temperature gradient and not the activity, since in most cases the clock is active most of the time. Still, if we know how much time the clock is gated or is in a low frequency mode (idle mode), we can help prevent overdesign.

In high-speed designs, one should consider the inductive effects as well. There are clock distribution schemes that take advantage of inductive effects to reduce clock power and minimize skew [8, 74–77]. These are described in more detail in Chap. 4. Shielding of the clock lines with V_{DD} and V_{SS} provides both good return paths to control inductance but also shields the clock network from the noise and delay variation due to capacitive coupling from neighboring switching nets. Reducing the variation will improve overall skew and is a strongly recommended practice for high-speed clock networks. The shielding costs a bit of extra power due to the increased capacitive loading. This additional power consumption is not large for the main clock trunk and the shielding is required to minimize noise induced clock skew. In the final stages, the power consumption is higher due to the large fanout, and one could argue that shielding may not be needed since all signals should be stable before the next clock edge. This may be true when there is no cycle stealing involved and negative edge timing is not used architecturally. Yet, there is another reason to maintain shielding across all stages. As has been demonstrated before, the interconnect resistance and capacitance variation can be reduced when using similar dense layout for all wires, and therefore having a well controlled environment helps minimize skew induced by the interconnect performance variation.

In a nutshell, the clock lines need to be designed wide enough to minimize RC variations have good shielding to minimize variation due to noise and inductance, and use a similar routing patterns to minimize the SPB variations. In all cases, full extraction considering all these effects and post layout timing analysis is absolutely required to sign-off the clock design.

8.7 Conclusion: Clock Design and Analysis Guidelines: Putting All Together

As a conclusion, in this section, we collect a summary of all analysis and design guidelines presented earlier on.

8.7.1 Clock Analysis

- Need to use statistical analysis and not rely on worst case process corners that are very pessimistic and lead to overdesign.
- Need to separate the systematic and correlated variation components.
- Systematic/ correlated components include strain and WPE-induced layout dependent effects, portion of the channel length variation, and CMP, SPB effects for the interconnects. All systematic effects should be modeled with accurate post layout extraction.
- Random components include random dopant fluctuation, line edge roughness (LER) effects, voltage and temperature variation, and PLL jitter.
- Use thermal maps to identify worst case temperature variations.
- Use dynamic IR simulations to identify the local and global IR drop for the clock buffers.

8.7.2 Minimizing Variation

- Maximize $W \times L$ to minimize random dopant fluctuations. However, short L is need for speed and power reduction. Therefore, larger W devices at minimum L are appropriate for clock buffers to minimize dopant fluctuations and power dissipation.
- Use same poly orientation and poly shields to minimize poly CD variation.
- Use the maximum allowed poly pitch in technologies using strain to maximize the drive strength through strain.
- Use the same layout style to guarantee the same effect on all transistors.
- Enlarge the spacing of well to active to reduce the well proximity effect that increases transistor V_T and reduces drive strength.
- Use multi-finger devices to create an averaging effect and improve overall strength in processes using strain due to the larger strain in the internal devices.
- Avoid active jogs and increase poly to active spacing to minimize poly flaring and leakage.
- Reduce total clock network latency. Skew will be one way or another proportional to the total number of buffers.
- Reduce voltage variation by using large numbers of decoupling capacitors between supply and ground in close proximity to the clock buffers. The RC time constant of the grid resistance and the decoupling capacitance needs to be at least 5 times smaller than the clock rise/fall times.
- Use clock line shielding for improved noise immunity and good inductive return path to reduce overall delay variation due to noise.

- Minimize the total number of clock buffers used and use large devices of similar total equivalent strength. Variation is proportional to the number of stages and the number of critical paths and clock sink pairs.
- Higher supply voltage will help minimize the impact of V_T variation and partially compensate the impact of NBTI.
- Use H-trees for distribution. Make sure to drive back-to-back flops with the same clock branch. The delay of common branches does not affect the hold time skew.
- For interconnects, use similar layout style with shielding to minimize SPB impact. If possible, use width and spacing combinations that minimize the SPB impact. Use extraction tools that account for CMP and SPB effects.
- Use metal fill to meet density requirements and extract it correctly. Extract grounded metal fill as grounded, floating as floating. If possible, use larger spacing to active signals and clocks to just meet density requirements but minimize the impact on delay due to noise and reduce power consumption by reducing the overall capacitive load.
- Shorting clock buffer outputs can provide some averaging effect, but it increases power and complicates timing analysis. Most timing tools have issues with parallel clock drivers. In order to facilitate clock gating, a tree configuration must be used instead of a shorted-output topology to be able to gate individual clock buffers.
- In the case of clock gating, add enough margin to account for the extra asymmetric NBTI shift. This impacts more the half-cycle paths that depend on both the rise and fall clock transitions.
- Use the same minimum channel length in all clock drivers. Different channel lengths have different sensitivity to CD variations and can impact product speed.
- Use wider metals for clock lines to minimize the RC impact, minimize the non-linear sheet resistance effect, and reduce the electromigration risk.
- In physical composition, use the same footprint (abstract) for all clock buffers. It will be easier to do post layout buffer fine-tuning in place, without affecting the floorplan.

Acknowledgments

I would like to acknowledge my wife Anna for her huge patience and support, and Thucydides Xanthopoulos, Antonietta Oliva, Aaron Barker, and David Greenhill for providing thorough reviews of this manuscript.

References

[1] S. Tam, S. Rusu, U. Nagarji Desai, R. Kim, J. Zhang, and I. Young, "Clock generation and distribution for the first IA-64 microprocessor," *IEEE J. Solid-State Circuits*, vol. 35, no. 11, pp. 1545–1552, Nov. 2000.

[2] T. Xanthopoulos, D. W. Bailey, A. K. Gangwar, M. K. Gowan, A. K. Jain, and B. K. Prewitt, "The design and analysis of the clock distribution network for a 1.2GHz Alpha microprocessor," in *Digest of Technical Papers IEEE International Solid-State Circuits Conference (ISSCC 2001)*, 2001, pp. 402–403.

[3] P. J. Restle, T. G. McNamara, D. A. Webber, P. J. Camporese, K. F. Eng, K. A. Jenkins, D. H. Allen, M. J. Rohn, M. P. Quaranta, D. W. Boerstler, C. J. Alpert, C. A. Carter, R. N. Bailey, J. G. Petrovick, B. L. Krauter, and B. D. McCredie, "A clock distribution network for microprocessors," *IEEE J. Solid-State Circuits*, vol. 36, no. 5, pp. 792–799, May 2001.

[4] P. J. Restle, C. A. Carter, J. P. Eckhardt, B. L. Krauter, B. D. McCredie, K. A. Jenkins, A. J. Weger, and A. V. Mule, "The clock distribution of the Power4 microprocessor," in *Digest of Technical Papers IEEE International Solid-State Circuits Conference (ISSCC 2002)*, vol. 1, 2002, pp. 144–145.

[5] N. Bindal, T. Kelly, N. Velastegui, and K. L. Wong, "Scalable sub-10ps skew global clock distribution for a 90nm multi-GHz IA microprocessor," in *Digest of Technical Papers IEEE International Solid-State Circuits Conference (ISSCC 2003)*, 2003, pp. 346–347,498.

[6] S. Tam, R. D. Limaye, and U. N. Desai, "Clock generation and distribution for the 130-nm Itanium® 2 processor with 6-MB on-die L3 cache," *IEEE J. Solid-State Circuits*, vol. 39, no. 4, pp. 636–642, April 2004.

[7] P. Mahoney, E. Fetzer, B. Doyle, and S. Naffziger, "Clock distribution on a dual-core, multi-threaded Itanium®-family processor," in *Digest of Technical Papers IEEE International Solid-State Circuits Conference (ISSCC 2005)*, 2005, pp. 292–293,599.

[8] M. G. R. Thomson, P. J. Restle, and N. K. James, "A 5GHz duty-cycle correcting clock distribution network for the POWER6 microprocessor," in *Digest of Technical Papers IEEE International Solid-State Circuits Conference (ISSCC 2006)*, 2006, pp. 1522–1529.

[9] G. K. Konstadinidis, K. Normoyle, S. Wong, S. Bhutani, H. Stuimer, T. Johnson, A. Smith, D. Y. Cheung, F. Romano, S. Yu, S.-H. Oh, V. Melamed, S. Narayanan, D. Bunsey, C. Khieu, K. J. Wu, R. Schmitt, A. Dumlao, M. Sutera, J. Chau, K. J. Lin, and W. S. Coates, "Implementation of a third-generation 1.1-GHz 64-bit microprocessor," *IEEE J. Solid-State Circuits*, vol. 37, no. 11, pp. 1461–1469, Nov. 2002.

[10] N. H. E. Weste and D. Harris, *CMOS VLSI Design: A Circuits and System Perspective (third edition)*. Addison Wesley, Reading, MA, 2005.

[11] A. Chandrakasan, W. Bowhill, and F. Fox, *Design of High Performance Microprocessor Circuits*. IEEE Press, 2001.

[12] N. J. Rohrer, "Introduction to statistical variation and techniques for design optimization," ISSCC 2006 Tutorial.

[13] K. Ushida, "Future lithography challenges," in *Proceedings of the IEEE International Symposium on Semiconductor Manufacturing ISSM 2006*, 25–27 Sept. 2006, pp. lv–lx.

[14] S. Sivakumar, "Lithography challenges for 32nm technologies and beyond," in *Proceedings of the International Electron Devices Meeting IEDM '06*, 11–13 Dec. 2006, pp. 1–4.
[15] N. S. Nagaraj, T. Bonifield, A. Singh, R. Griesmer, and P. Balsara, "Interconnect modeling for copper/low-k technologies," in *Proceedings of the 17th International Conference on VLSI Design*, 2004, pp. 425–427.
[16] S. Tyagi, C. Auth, P. Bai, G. Curello, H. Deshpande, S. Gannavaram, O. Golonzka, R. Heussner, R. James, C. Kenyon, S. H. Lee, N. Lindert, M. Liu, R. Nagisetty, S. Natarajan, C. Parker, J. Sebastian, B. Sell, S. Sivakumar, A. S. Amour, and K. Tone, "An advanced low power, high performance, strained channel 65nm technology," in *Proceedings of the IEDM Technical Digest Electron Devices Meeting IEEE International*, 5–7 Dec. 2005, pp. 245–247.
[17] T. Ghani, M. Armstrong, C. Auth, M. Bost, P. Charvat, G. Glass, T. Hoffmann, K. Johnson, C. Kenyon, J. Klaus, B. McIntyre, K. Mistry, A. Murthy, J. Sandford, M. Silberstein, S. Sivakumar, P. Smith, K. Zawadzki, S. Thompson, and M. Bohr, "A 90nm high volume manufacturing logic technology featuring novel 45nm gate length strained silicon CMOS transistors," in *Proceedings of the IEDM '03 Technical Digest Electron Devices Meeting IEEE International*, 8–10 Dec. 2003, pp. 11.6.1–11.6.3.
[18] S. E. Thompson, M. Armstrong, C. Auth, S. Cea, R. Chau, G. Glass, T. Hoffman, J. Klaus, Z. Ma, B. Mcintyre, A. Murthy, B. Obradovic, L. Shifren, S. Sivakumar, S. Tyagi, T. Ghani, K. Mistry, M. Bohr, and Y. El-Mansy, "A logic nanotechnology featuring strained-silicon," *IEEE Electron Device Lett.*, vol. 25, no. 4, pp. 191–193, April 2004.
[19] M. Horstmann, A. Wei, T. Kammler, J. Hntschel, H. Bierstedt, T. Feudel, K. Frohberg, M. Gerhardt, A. Hellmich, K. Hempel, J. Hohage, P. Javorka, J. Klais, G. Koerner, M. Lenski, A. Neu, R. Otterbach, P. Press, C. Reichel, M. Trentsch, B. Trui, H. Salz, M. Schaller, H. J. Engelmann, O. Herzog, H. Ruelke, P. Hubler, R. Stephan, D. Greenlaw, M. Raab, and N. Kepler, "Integration and optimization of embedded-sige, compressive and tensile stressed liner films, and stress memorization in advanced SOI CMOS technologies," in *Proceedings of IEDM Technical Digest Electron Devices Meeting IEEE International*, 5–7 Dec. 2005, pp. 233–236.
[20] P. Grudowski, V. Adams, X.-Z. Bo, K. Loiko, S. Filipiak, J. Hackenberg, M. Jahanbani, M. Azrak, S. Goktepeli, M. Shroff, W.-J. Liang, S. J. Lian, V. Kolagunta, N. Cave, C.-H. Wu, M. Foisy, H. C. Tuan, and J. Cheek, "1-D and 2-D geometry effects in uniaxially-strained dual etch stop layer stressor integrations," in *Proceedings of the Digest of Technical Papers VLSI Technology 2006 Symposium*, 2006, pp. 62–63.
[21] C. Ortolland, P. Morin, C. Chaton, E. Mastromatteo, C. Populaire, S. Orain, F. Leverd, P. Stolk, F. Boeuf, and F. Arnaud, "Stress memorization technique (SMT) optimization for 45nm CMOS," in *Proceedings of Digest of Technical Papers VLSI Technology 2006 Symposium*, 2006, pp. 78–79.
[22] M. Wiatr, T. Feudel, A. Wei, A. Mowry, R. Boschke, P. Javorka, A. Gehring, T. Kammler, M. Lenski, K. Frohberg, R. Richter, M. Horstmann, and

D. Greenlaw, "Review on process-induced strain techniques for advanced logic technologies," in *Proceedings of the 15th International Conference on Advanced Thermal Processing of Semiconductors RTP 2007*, 2–5 Oct. 2007, pp. 19–29.

[23] V. Chan, R. Rengarajan, N. Rovedo, W. Jin, T. Hook, P. Nguyen, J. Chen, E. Nowak, X.-D. Chen, D. Lea, A. Chakravarti, V. Ku, S. Yang, A. Steegen, C. Baiocco, P. Shafer, H. Ng, S.-F. Huang, and C. Wann, "High speed 45nm gate length CMOSFETs integrated into a 90nm bulk technology incorporating strain engineering," in *Proceedings of the IEDM '03 Technical Digest Electron Devices Meeting IEEE International*, 8–10 Dec. 2003, pp. 3.8.1–3.8.4.

[24] C. Auth, A. Cappellani, J. S. Chun, A. Dalis, A. Davis, T. Ghani, G. Glass, T. Glassman, M. Harper, M. Hattendorf, P. Hentges, S. Jaloviar, S. Joshi, J. Klaus, K. Kuhn, D. Lavric, M. Lu, H. Mariappan, K. Mistry, B. Norris, N. Rahhal-orabi, P. Ranade, J. Sandford, L. Shifren, V. Souw, K. Tone, F. Tambwe, A. Thompson, D. Towner, T. Troeger, P. Vandervoorn, C. Wallace, J. Wiedemer, and C. Wiegand, "45nm high-k + metal gate strain-enhanced transistors," in *Proceedings of the Symposium on VLSI Technology*, 17–19 June 2008, pp. 128–129.

[25] P. Ranade, T. Ghani, K. Kuhn, K. Mistry, S. Pae, L. Shifren, M. Stettler, K. Tone, S. Tyagi, and M. Bohr, "High performance 35nm LATE CMOS transistors featuring NiSi metal gate (FUSI), uniaxial strained silicon channels and 1.2nm gate oxide," in *Proceedings of the IEDM Technical Digest Electron Devices Meeting IEEE International*, 5–7 Dec. 2005, p. 4.

[26] K. Mistry, C. Allen, C. Auth, B. Beattie, D. Bergstrom, M. Bost, M. Brazier, M. Buehler, A. Cappellani, R. Chau, C. H. Choi, G. Ding, K. Fischer, T. Ghani, R. Grover, W. Han, D. Hanken, M. Hattendorf, J. He, J. Hicks, R. Huessner, D. Ingerly, P. Jain, R. James, L. Jong, S. Joshi, C. Kenyon, K. Kuhn, K. Lee, H. Liu, J. Maiz, B. Mclntyre, P. Moon, J. Neirynck, S. Pae, C. Parker, D. Parsons, C. Prasad, L. Pipes, M. Prince, P. Ranade, T. Reynolds, J. Sandford, L. Shifren, J. Sebastian, J. Seiple, D. Simon, S. Sivakumar, P. Smith, C. Thomas, T. Troeger, P. Vandervoorn, S. Williams, and K. Zawadzki, "A 45nm logic technology with high-k+metal gate transistors, strained silicon, 9 Cu interconnect layers, 193nm dry patterning, and 100% Pb-free packaging," in *Proceedings of the IEEE International Electron Devices Meeting IEDM 2007*, 10–12 Dec. 2007, pp. 247–250.

[27] S. Narasimha, K. Onishi, H. M. Nayfeh, A. Waite, M. Weybright, J. Johnson, C. Fonseca, D. Corliss, C. Robinson, M. Crouse, D. Yang, C. H. J. Wu, A. Gabor, T. Adam, I. Ahsan, M. Belyansky, L. Black, S. Butt, J. Cheng, A. Chou, G. Costrini, C. Dimitrakopoulos, A. Domenicucci, P. Fisher, A. Frye, S. Gates, S. Greco, S. Grunow, M. Hargrove, J. Holt, S. J. Jeng, M. Kelling, B. Kim, W. Landers, G. Larosa, D. Lea, M. H. Lee, X. Liu, N. Lustig, A. McKnight, L. Nicholson, D. Nielsen, K. Nummy, V. Ontalus, C. Ouyang, X. Ouyang, C. Prindle, R. Pal, W. Rausch, D. Restaino, C. Sheraw, J. Sim, A. Simon, T. Standaert, C. Y. Sung, K. Tabakman, C. Tian, R. Van Den Nieuwenhuizen, H. Van Meer, A. Vayshenker, D. Wehella-Gamage,

J. Werking, R. C. Wong, J. Yu, S. Wu, R. Augur, D. Brown, X. Chen, D. Edelstein, A. Grill, M. Khare, Y. Li, S. Luning, J. Norum, S. Sankaran, D. Schepis, R. Wachnik, R. Wise, C. Warm, T. Ivers, and P. Agnello, "High performance 45-nm SOI technology with enhanced strain, porous low-k BEOL, and immersion lithography," in *Proceedings of the International Electron Devices Meeting IEDM '06*, 11–13 Dec. 2006, pp. 1–4.

[28] S. Suthram, Y. Sun, P. Majhi, I. Ok, H. Kim, H. R. Harris, N. Goel, S. Parthasarathy, A. Koehler, T. Acosta, T. Nishida, H.-H. Tseng, W. Tsai, J. Lee, R. Jammy, and S. E. Thompson, "Strain additivity in III-V channels for CMOSFETs beyond 22nm technology node," in *Proceedings of the Symposium on VLSI Technology*, 17–19 June 2008, pp. 182–183.

[29] Y. M. Sheu, K. W. Su, S. Tian, S. J. Yang, C. C. Wang, M. J. Chen, and S. Liu, "Modeling the well-edge proximity effect in highly scaled MOSFETs," *IEEE Trans. Electron Devices*, vol. 53, no. 11, pp. 2792–2798, Nov. 2006.

[30] I. Polishchuk, N. Mathur, C. Sandstrom, P. Manos, and O. Pohland, "CMOS Vt-control improvement through implant lateral scatter elimination," in *Proceedings of the IEEE International Symposium on Semiconductor Manufacturing ISSM 2005*, 13–15 Sept. 2005, pp. 193–196.

[31] P. G. Drennan, M. L. Kniffin, and D. R. Locascio, "Implications of proximity effects for analog design," in *Proceedings of the IEEE Custom Integrated Circuits Conference 2006*, 10–13 Sept. 2006, pp. 169–176.

[32] T. Kanamoto, Y. Ogasahara, K. Natsume, K. Yamaguchi, H. Amishiro, and T. Watanabe, "Impact of well edge proximity effect on timing," in *Proceedings of the ESSCIRC 33rd European Solid State Circuits Conference*, 11–13 Sept. 2007, pp. 115–118.

[33] A. Asenov, S. Kaya, and A. R. Brown, "Intrinsic parameter fluctuations in decananometer MOSFETs introduced by gate line edge roughness," *IEEE Trans. Electron Devices*, vol. 50, no. 5, pp. 1254–1260, May 2003.

[34] A. Asenov, S. Kaya, and J. H. Davies, "Intrinsic threshold voltage fluctuations in decanano MOSFETs due to local oxide thickness variations," *IEEE Trans. Electron Devices*, vol. 49, no. 1, pp. 112–119, Jan. 2002.

[35] Y. Ye, F. Liu, S. Nassif, and Y. Cao, "Statistical modeling and simulation of threshold variation under dopant fluctuations and line-edge roughness," in *Proceedings of the 45th ACM/IEEE Design Automation Conference DAC 2008*, 8–13 June 2008, pp. 900–905.

[36] S. Borkar, "Designing reliable systems from unreliable components: the challenges of transistor variability and degradation," *IEEE Micro*, vol. 25, no. 6, pp. 10–16, Nov.–Dec. 2005.

[37] T. Karnik, V. De, and S. Borkar, "Statistical design for variation tolerance: key to continued Moore's law," in *Proceedings of the International Conference on Integrated Circuit Design and Technology ICICDT '04*, 2004, pp. 175–176.

[38] J. Rosal, "Multi-core SoC test and yield enhancement – Challenges and solutions," ISSCC Microprocessor Forum, February 2006.

[39] H. Masuda, S. Okawa, and M. Aoki, "Approach for physical design in sub-100nm era," in *Proceedings of the IEEE International Symposium on Circuits and Systems ISCAS 2005*, 23–26 May 2005, pp. 5934–5937.

[40] H. Mahmoodi, S. Mukhopadhyay, and K. Roy, "Estimation of delay variations due to random-dopant fluctuations in nanoscale CMOS circuits," *IEEE J. Solid-State Circuits*, vol. 40, no. 9, pp. 1787–1796, Sept. 2005.

[41] A. Asenov, G. Slavcheva, A. R. Brown, J. H. Davies, and S. Saini, "Increase in the random dopant induced threshold fluctuations and lowering in sub-100nm MOSFETs due to quantum effects: a 3-d density-gradient simulation study," *IEEE Trans. Electron Devices*, vol. 48, no. 4, pp. 722–729, April 2001.

[42] I. Ahsan, O. Glushchenkov, R. Logan, E. J. Nowak, H. Kimura, G. Berg, J. Herman, E. Maciejewski, A. Chan, A. Azuma, S. Deshpande, B. Dirahoui, G. Freeman, A. Gabor, M. Gribelyuk, S. Huang, M. Kumar, K. Miyamoto, D. Mocuta, A. Mahorowala, E. Leobandung, and H. Utomo, "RTA-driven intra-die variations in stage delay, and parametric sensitivities for 65nm technology," in *Proceedings of the Digest of Technical Papers VLSI Technology 2006 Symposium*, 2006, pp. 170–171.

[43] T. Tanaka, T. Usuki, T. Futatsugi, Y. Momiyama, and T. Sugii, "v_{th} fluctuation induced by statistical variation of pocket dopant profile," in *Proceedings of the IEDM Technical Digest Electron Devices Meeting International*, 10–13 Dec. 2000, pp. 271–274.

[44] D. S. Boning, K. Balakrishnan, H. Cai, N. Drego, A. Farahanchi, K. M. Gettings, L. Daihyun, A. Somani, H. Taylor, D. Truque, and X. Xiaolin, "Variation," *IEEE Trans. Semicond. Manuf.*, vol. 21, no. 1, pp. 63–71, Feb. 2008.

[45] D. Harris and S. Naffziger, "Statistical clock skew modeling with data delay variations," *IEEE Trans. VLSI Syst.*, vol. 9, no. 6, pp. 888–898, Dec. 2001.

[46] A. Agarwal, V. Zolotov, and D. T. Blaauw, "Statistical clock skew analysis considering intradie-process variations," *IEEE Trans. Comput.-Aided Design Integr. Circuits Syst.*, vol. 23, no. 8, pp. 1231–1242, Aug. 2004.

[47] K. A. Bowman, S. G. Duvall, and J. D. Meindl, "Impact of die-to-die and within-die parameter fluctuations on the maximum clock frequency distribution for gigascale integration," *IEEE J. Solid-State Circuits*, vol. 37, no. 2, pp. 183–190, Feb. 2002.

[48] E. Malavasi, S. Zanella, M. Cao, J. Uschersohn, M. Misheloff, and C. Guardiani, "Impact analysis of process variability on clock skew," in *Proceedings of International Symposium on Quality Electronic Design*, 18–21 March 2002, pp. 129–132.

[49] J. A. G. Jess, K. Kalafala, S. R. Naidu, R. H. J. M. Otten, and C. Visweswariah, "Statistical timing for parametric yield prediction of digital integrated circuits," *IEEE Trans. Comput.-Aided Design Integr. Circuits Syst.*, vol. 25, no. 11, pp. 2376–2392, Nov. 2006.

[50] Y. Zhan, A. J. Strojwas, X. Li, L. T. Pileggi, D. Newmark, and M. Sharma, "Correlation-aware statistical timing analysis with non-Gaussian delay distributions," in *Proceedings of the 42nd Design Automation Conference*, 13–17 June 2005, pp. 77–82.
[51] Z. Feng and P. Li, "A methodology for timing model characterization for statistical static timing analysis," in *Proceedings of the IEEE/ACM International Conference on Computer-Aided Design ICCAD 2007*, 4–8 Nov. 2007, pp. 725–729.
[52] A. B. Kahng, "Key directions and a roadmap for electrical design for manufacturability," in *Proceedings of the ESSCIRC 33rd European Solid State Circuits Conference*, 11–13 Sept. 2007, pp. 83–88.
[53] W. Yang, M. V. Dunga, X. Xi, J. He, W. Liu, M. C. Kanyu, X. Jin, J. J. Ou, M. Chan, A. M. Niknejad, and C. Hu, "BSIM4.6.2 MOSFET model – User's Manual," Department of Electrical Engineering and Computer Sciences, http://www-device.eecs.berkley.edu/ bsim3/BSIM4/BSIM462/doc/BSIM462_Manual.pdf.
[54] S. Tsujikawa and J. Yugami, "Evidence for bulk trap generation during NBTI phenomenon in pMOSFETs with ultrathin SiON gate dielectrics," *IEEE Trans. Electron Devices*, vol. 53, no. 1, pp. 51–55, Jan. 2006.
[55] A. E. Islam, H. Kufluoglu, D. Varghese, S. Mahapatra, and M. A. Alam, "Recent issues in negative-bias temperature instability: Initial degradation, field dependence of interface trap generation, hole trapping effects, and relaxation," *IEEE Trans. Electron Devices*, vol. 54, no. 9, pp. 2143–2154, Sept. 2007.
[56] H. Kufluoglu and M. A. Alam, "A generalized reaction-diffusion model with explicit H {H}_{2} dynamics for negative-bias temperature-instability (NBTI) degradation," *IEEE Trans. Electron Devices*, vol. 54, no. 5, pp. 1101–1107, May 2007.
[57] S. Mahapatra, P. Bharath Kumar, T. R. Dalei, D. Sana, and M. A. Alam, "Mechanism of negative bias temperature instability in CMOS devices: degradation, recovery and impact of nitrogen," in *Proceedings of IEDM Technical Digest Electron Devices Meeting IEEE International*, 13–15 Dec. 2004, pp. 105–108.
[58] C. R. Parthasarathy, M. Denais, V. Huard, G. Ribes, E. Vincent, and A. Bravaix, "New insights into recovery characteristics during PMOS NBTI and CHC degradation," *IEEE Trans. Device Mater. Rel.*, vol. 7, no. 1, pp. 130–137, March 2007.
[59] W. Wang, S. Yang, S. Bhardwaj, R. Vattikonda, S. Vrudhula, F. Liu, and Y. Cao, "The impact of NBTI on the performance of combinational and sequential circuits," in *Proceedings of the 44th ACM/IEEE Design Automation Conference DAC '07*, 4–8 June 2007, pp. 364–369.
[60] A. Haggag, M. Lemanski, G. Anderson, P. Abramowitz, and M. Moosa, "Realistic projections of product fmax shift and statistics due to HCI and NBTI," in *Proceedings of the Reliability Physics Symposium 45th Annual. IEEE International*, 15–19 April 2007, pp. 93–96.
[61] Y. Taur and T. H. Ning, *Fundamentals of Modern VLSI Devices.* Cambridge University Press, Cambridge, 1998.

[62] K. Bernstein, K. Carrig, C. M. Durham, and P. R. Hansen, *High Speed CMOS Design Styles.* Kluwer, Dordecht, 1998.

[63] Q. K. Zhu and M. Zhang, "Low-voltage swing clock distribution schemes," in *Proceedings of the IEEE International Symposium on Circuits and Systems ISCAS 2001*, vol. 4, 6–9 May 2001, pp. 418–421.

[64] N. James, P. Restle, J. Friedrich, B. Huott, and B. McCredie, "Comparison of split-versus connected-core supplies in the POWER6 microprocessor," in *Digest of Technical Papers IEEE International Solid-State Circuits Conference (ISSCC 2007)*, 11–15 Feb. 2007, pp. 298–299,604.

[65] G. Konstadinidis, M. Tremblay, S. Chaudhry, M. Rashid, P. Lai, Y. Otaguro, Y. Orginos, S. Parampalli, M. Steigerwald, S. Gundala, R. Pyapali, L. Rarick, I. Elkin, Y. Ge, and I. Parulkar, "Architecture and physical implementation of a third generation 65 nm, 16 core, 32 thread chip-multithreading SPARC processor," *IEEE J. Solid-State Circuits*, vol. 44, no. 1, pp. 7–17, Jan. 2009.

[66] J. Long, J. C. Ku, S. O. Memik, and Y. Ismail, "A self-adjusting clock tree architecture to cope with temperature variations," in *Proceedings of the IEEE/ACM International Conference on Computer-Aided Design ICCAD 2007*, 4–8 Nov. 2007, pp. 75–82.

[67] S. Sankaran, S. Arai, R. Augur, M. Beck, G. Biery, T. Bolom, G. Bonilla, O. Bravo, K. Chanda, M. Chae, F. Chen, L. Clevenger, S. Cohen, A. Cowley, P. Davis, J. Demarest, J. Doyle, C. Dimitrakopoulos, L. Economikos, D. Edelstein, M. Farooq, R. Filippi, J. Fitzsimmons, N. Fuller, S. M. Gates, S. E. Greco, A. Grill, S. Grunow, R. Hannon, K. Ida, D. Jung, E. Kaltalioglu, M. Kelling, T. Ko, K. Kumar, C. Labelle, H. Landis, M. W. Lane, W. Landers, M. Lee, W. Li, E. Liniger, X. Liu, J. R. Lloyd, W. Liu, N. Lustig, K. Malone, S. Marokkey, G. Matusiewicz, P. S. McLaughlin, P. V. McLaughlin, S. Mehta, I. Melville, K. Miyata, B. Moon, S. Nitta, D. Nguyen, L. Nicholson, D. Nielsen, P. Ong, K. Patel, V. Patel, W. Park, J. Pellerin, S. Ponoth, K. Petrarca, D. Rath, D. Restaino, S. Rhee, E. T. Ryan, H. Shoba, A. Simon, E. Simonyi, T. M. Shaw, T. Spooner, T. Standaert, J. Sucharitaves, C. Tian, H. Wendt, J. Werking, J. Widodo, L. Wiggins, R. Wisnieff, and T. Ivers, "A 45 nm CMOS node Cu/Low-k/ Ultra Low-k PECVD SiCOH (k=2.4) BEOL technology," in *Proceedings of the International Electron Devices Meeting IEDM '06*, 11–13 Dec. 2006, pp. 1–4.

[68] C. Bittlestone, "Nanometer design effects and modeling," ISSCC Microprocessor Design Forum, February 2005.

[69] M. Kulkarni, N. S. Nagaraj, A. Marshall, and V. Le, "Impact of selective process bias (SPB) of interconnects on circuit delay," in *Proceedings of the IEEE Dallas/CAS Workshop (DCAS-04) Implementation of High Performance Circuits*, 27 Sept. 2004, pp. 155–158.

[70] G. Lopez, R. Murali, R. Sarvari, K. Bowman, J. Davis, and J. Meindl, "The impact of size effects and copper interconnect process variations on the maximum critical path delay of single and multi-core microprocessors," in *Proc. International Interconnect Technology Conference IEEE 2007*, 4–6 June 2007, pp. 40–42.

[71] Y. Liu, S. R. Nassif, L. T. Pileggi, and A. J. Strojwas, "Impact of interconnect variations on the clock skew of a gigahertz microprocessor," in *Proceedings of the 37th Design Automation Conference*, June 5–9, 2000, pp. 168–171.

[72] U. Padmanabhan, J. M. Wang, and J. Hu, "Robust clock tree routing in the presence of process variations," *IEEE Trans. Comput.-Aided Design Integr. Circuits Syst.*, vol. 27, no. 8, pp. 1385–1397, Aug. 2008.

[73] R. Y. Chen, P. Yip, G. Konstadinidis, A. Demas, F. Klass, R. Mains, M. Schmitt, and D. Bistry, "Timing window applications in UltraSPARC-IIIi™ microprocessor design," in *Proceedings of the IEEE International Conference on Computer Design: VLSI in Computers and Processors*, 16–18 Sept. 2002, pp. 158–163.

[74] F. O'Mahony, C. P. Yue, M. Horowitz, and S. S. Wong, "10GHz clock distribution using coupled standing-wave oscillators," in *Digest of Technical Papers IEEE International Solid-State Circuits Conference (ISSCC 2003)*, 2003, pp. 428–429,504.

[75] M. Shen, L.-R. Zheng, E. Tjukanoff, J. Isoaho, and H. Tenhunen, "Concurrent chip package design for global clock distribution network using standing wave approach," in *Proceedings of the Sixth International Symposium on Quality of Electronic Design ISQED 2005*, 21–23 March 2005, pp. 573–578.

[76] S. Chan, P. Restle, T. Bucelot, S. Weitzel, J. Keaty, J. Liberty, B. Flachs, R. Volant, P. Kapusta, and J. Zimmerman, "A resonant global clock distribution for the cell broadband-engine™processor," in *Digest of Technical Papers IEEE International Solid-State Circuits Conference (ISSCC 2008)*, 2008, pp. 512–513.

[77] M. Sasaki, "A 9.5GHz 6ps-skew space-filling-curve clock distribution with 1.8V full-swing standing-wave oscillators," in *Digest of Technical Papers IEEE International Solid-State Circuits Conference (ISSCC 2008)*, 2008, pp. 518–519, 633.

Index

Kolibri 360 GmbH
Ferdinand-Jühlke-Straße 7
99095 Erfurt, Deutschland
produktsicherheit@kolibri360.de